THE LASER COOKBOOK

88 Practical Projects

Gordon McComb

TAB BOOKS

Blue Ridge Summit, PA

Ballast Resistor—Buying and Testing He-Ne Tubes—Powering the Tube—Using the Tube

Acknowledgments

Writing books (like this one) provides numerous rewards, including the chance to meet and exchange ideas with many interesting and helpful people. I am indebted to the help and consideration provided by the following people: Dennis Meredith of Meredith Instruments, Roger Sontag of General Science and Engineering, and Jeff Korman and Sam Frisher of Fobtron Components.

Special thanks to Forrest Mims III for his ideas and suggestions for the pocket-dialed laser, laser transmitter and receiver circuits, and the basic workings of the laser tachometer.

The "behind the scenes" people made a great impact in the preparation of this book and I gratefully acknowledge their assistance. Thanks go to Brint Rutherford and Roland Phelps at TAB BOOKS and to my agent Bill Gladstone. Finally, my brother and father, Lee McComb and Wally McComb, helped answer my endless questions about lasers, surveying, mathematics, and earthquakes.

Introduction

You've just bought a surplus helium-neon gas laser at a local swap meet. You get it home, plug it in, and marvel at the incredibly bright, red spot it projects on the wall. Your friends and family seem interested, but they keep asking you "What's it good for?" You mumble something about holography and gun sights, but your interest soon wanes.

You grow tired of playing "laser tag" with the cat or bouncing the beam off reflective objects in your living room. Your toy soon finds its place in the dusty confines of the closet. You've run out of things to do with the laser and you soon move on to other hobbies.

It's time to get that laser out of the closet! *The Laser Cookbook* shows you over 88 inexpensive laser-based projects from experimenting with laser optics to constructing a laser optical bench to using lasers to make stunning holograms. All the projects are geared towards garage/shop tinkering with a special emphasis on minimizing the budget. The book mixes the history of lasers, how lasers work, and practical applications in an easy-to-read and fun text that's suitable for experimenters of all ages.

Among the topics in this book, you'll learn how to use lasers for:

- ★ holography
- ★ optics and optical experiments
- ★ laser guns
- ★ laser light shows
- ★ laser beam intrusion and detection systems
- ★ aerodynamics and airflow study
- ★ coherent-light seismology

★ laser beam communication
★ laser and fiberoptics computer data link
★ precision measurement

Though the laser is a relatively new invention, it has undergone many refinements and improvements since the first successful prototype was tested by Theodore Maiman in 1960. And its cost has been drastically reduced. Today, lasers are inexpensive, almost throw-away devices and are used in numerous consumer products and electronic systems. Lasers are everywhere—from the phone lines that connect home to office to the electronics that play back sound and pictures encoded on a videodisc to the bar-code scanning system used at the supermarket.

Lasers are available to even the most budget-conscious hobbyists, so if you don't already have one (whether in the closet or not), don't despair. Surplus laser kits can be purchased for less than $100, and the latest semiconductor lasers—the kind used in videodisc and audio compact disc players—cost under $15. The tools are here to bring laser technology to the common masses. All that's needed is a book to show how the pieces fit together. Such a book is *The Laser Cookbook*.

WHO THIS BOOK IS FOR

The Laser Cookbook is written for a wide variety of readers. If you're into electronics, you'll enjoy the many circuits you can build, including one that lets you carry your voice over a beam of light, or the project that can detect the presence of intruders around a campsite. A number of the projects are excellent springboards for science fairs. These include measuring—with astonishing accuracy—the speed of light, seismology, and hydrodynamics. Lastly, this book is a gold mine for the gadgeteer. Lasers represent the ultimate in space-age technology, but *The Laser Cookbook* presents numerous laser-based gadgets that you can readily build in your garage.

In all cases, the designs used in *The Laser Cookbook* have been thoroughly tested in prototype form. I encourage you to improve on the basic designs, but you can rest assured that the projects have actually been tried and field tested.

The projects in *The Laser Cookbook* include all the necessary information on how to construct the essential building blocks of high-tech laser projects. Suggested alternative approaches, parts lists, and sources of electronic and mechanical components are also provided, where appropriate.

HOW TO USE THIS BOOK

The Laser Cookbook is divided into 24 chapters. Most chapters present one or more actual hands-on projects that you can duplicate for your own laser creations. Whenever practical, I designed the components as discrete building blocks, so you can combine the blocks in just about any configuration you desire. That way, you are not tied down to one of my designs. You're free to experiment on your own!

If you have some experience in electronics, mechanics, or lasers in general, you can skip around and read only those chapters that provide the information you're looking for. Like the laser designs presented, the chapters are very much stand-alone modules. This allows you to pick and choose, using your time to its best advantage.

However, if you're new to lasers and the varied disciplines that go into them, you should take a more pedestrian approach and read as much of the book as possible. In this way, you'll get a thorough understanding of how lasers tick and the myriad ways you can use them.

CONVENTIONS USED IN THIS BOOK

You need little advance information before you can jump head-first into this book, but you should take note of a few conventions I've used in the description of electronic parts and in the schematic diagrams for the electronic circuits.

TTL integrated circuits are referenced by their standard 74XX number. The "LS" identifier is assumed. I built most of the circuits using LS TTL chips, but the projects should work with the other TTL-family chips—the standard (non-LS) chips, as well as those with the S, ALS, and C identifiers. If you use a type of TTL chip other than LS, you should consider current consumption, fan-out, and other design criteria, because these factors can affect the operation or performance of the circuit.

In some cases, however, a certain TTL-compatible IC is specified in a design. Unless the accompanying text recommends otherwise, you should use only the chip specified.

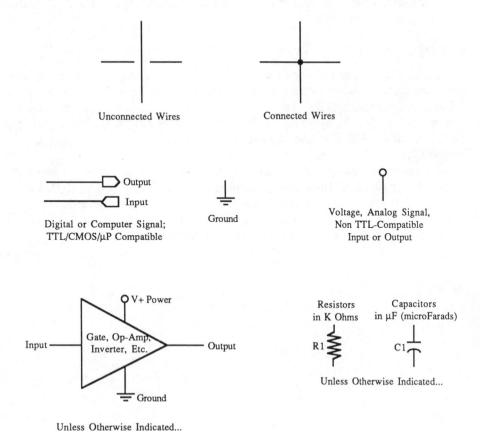

FIG. I-1. *Conventions used in the schematic diagrams in this book.*

Certain CMOS TTL-compatible chips offer the same functions as a sister IC, but the pinouts and operation might differ.

The chart in FIG. I-1 details the conventions used in the schematic diagrams. Note that unconnected wires are shown by a direct cross of lines, a broken line, or a "looped" line. Connected wires are only shown by a connecting dot.

Details on the specific parts used in the circuits are provided in the parts list tables that accompany each schematic. Refer to the parts list for information on resistor and capacitor type, tolerance, and wattage or voltage ratings.

In all schematics, the parts are referenced by component type and number.

* IC# means an integrated circuit (IC).
* R# means a resistor or potentiometer (variable resistor).
* C# means a capacitor.
* D# means a diode, a zener diode, and sometimes a light-sensitive photodiode.
* Q# means a transistor and sometimes a light-sensitive phototransistor.
* LED# means a light-emitting diode (most any visible LED will do unless the parts list specifically calls for an infrared LED).
* XTAL# means a crystal or ceramic resonator.
* S# means a switch, RL# means a relay, SPKR# means a speaker, and MIC# means a microphone.

SAFETY FIRST

It's hard to imagine that something as fun as lasers can be potentially dangerous. While the types of lasers commonly available to experimenters do not pose a great radiation emission hazard, they can do damage if mishandled. This point is reiterated in later chapters, but it's worth mentioning here: **NEVER LOOK DIRECTLY INTO THE LASER BEAM**. The intensity and needle-sharp focus of the beam can do your eyes harm.

Perhaps more importantly, gas lasers, such as the helium-neon type, require high-voltage power supplies. These power supplies generate from 1,000 to 10,000 volts. Though the current provided from these power supplies is low (generally under 7 milliamps), a 5,000- or 10,000-volt jolt is enough to at least knock you down. The power supply of a laser is not a toy and should be considered potentially lethal.

1

Introduction to Lasers

Think how boring science fiction movies would be if lasers did not exist. There would be no gallant laser sword fights between good guy and bad guy; no interstellar spaceships shooting darts of light at one another; no packets of photon energy hurled through the air like matterless handgrenades. Movies and television shows such as *Star Wars*, *Star Trek*, *War of the Worlds*, *Predator*, and *Day the Earth Stood Still*—along with countless others—owe a great deal of their spine-tingling suspense and fast-paced action sequences to the laser.

While lasers are most often regarded by the general public as exotic weaponry, this intriguing and fascinating scientific instrument enjoys a far greater involvement in peacetime applications. Lasers of one type or another are used in:

★ Supermarket scanning systems to instantly read the bar code label on packaged goods.

★ Audio compact disc players to play music with incredible fidelity.

★ Leveling instruments, used at construction sites to assure perfectly flat grading and absolutely straight pipe laying.

★ Light shows, where the beam of the laser dances to the beat of the music and makes beautiful sinuous shapes.

★ Holography, a special type of photography that captures a three-dimensional view of an object.

★ Communications systems, where voice, pictures, or computer data is transmitted through air or optical fibers on a beam of light.

Lasers are also used in gyroscopic inertial guidance systems on board commercial and military aircraft, in rifle and pistol range target practice, for rifle scopes, optical component testing, medicine (such as a high-tech substitute for the scalpel), optical radar, precision rangefinders, and many other diverse products and applications.

In this book, you'll learn how to design and build your own laser systems for holography, light-wave communications, target practice guns, light shows, and more. But first understand some basics. This chapter discusses the fundamentals of lasers: how they work, their history, forms of lasers, and the dynamics of laser light. So much has been written about laser physics that the fine details won't be covered here; only the basics of lasers are presented.

EINSTEIN AS THE SPARK

Albert Einstein was a theoretical physicist; he invented few actual things himself. As with many of the world's great thinkers, Einstein left the practical applications of his pioneering work in physics to other people. In addition to creating the theory of relativity—perhaps his best-known work—Einstein is responsible for first proposing the idea of the laser in about 1916.

Einstein knew that light was a series of particles, called *photons*, traveling in a continuous wave. These photons could be collected (using an apparatus not yet developed) and focused into a narrow beam. To be useful, all the photons would be emitted from the laser apparatus at specific intervals. As important, much of the light energy would be concentrated in a specific wavelength—or color—making the light even more intense and powerful.

Even in Einstein's day, photons could be created in a variety of means, including the ionization of gas within a sealed tube, the burning of some organic materials, or the heating of a filament in a light bulb. In all cases, the atoms that make up the light source change from their usual stable or *ground state* to a higher *excited state* by the introduction of some form of energy, typically heat or electricity. The atom can't stay at the excited state for long, and when it drops back to its comfortable ground state, it gives off a photon of light.

The release of photons by natural methods results in what is known as *spontaneous emission*. The photons leave the source in a random and unpredictable manner, and once a photon is emitted, it marks the end of the energy transfer cycle. The number of excited atoms is relatively low, so the great majority of photons leave the source without meeting another excited atom.

Einstein was most interested in what would happen if a photon hit an atom that happened to be at the excited, high-energy state. He reasoned that the atom would release a photon of light that would be an identical twin to the first. If enough atoms could be excited, the chance of photons hitting them would be increased. That would lead to a chain reaction where photons would hit atoms and make new photons, and the process would continue until the original energy source was terminated. Einstein had a name for this phenomenon, and called it *stimulated emission of radiation*.

RAISING ATOMS

Raising atoms to a high-energy state is often referred to as *pumping*. As already discussed, atoms can be pumped in a number of ways, including charging with electricity or heat. Another form of pumping is optical, with a bright source of light such as an arc lamp or xenon-filled flash tube.

In the common neon light, for example, the neon atoms are pumped to their high-energy state by means of a high-voltage charge applied to a pair of electrodes. The gas within the tube ionizes, emitting photons. If the electrical charge is high enough, a majority of the neon atoms will be pumped to the high-energy state. A so-called *population inversion* occurs when there are more high-energy atoms than low-energy ones. A laser cannot work unless this population inversion is present.

Photons scatter all over the place, and left on their own, they will simply escape the tube. But assume a pair of mirrors are mounted on either end of the tube, so some photons can bounce back and forth between the two mirrors.

At each bounce, the photons collide with more atoms. If many of these atoms are in their excited state, they too release photons. Remember: these new photons are twins of the original and share many of its characteristics, including wavelength, polarity, and phase. The process of photons bouncing from one mirror to the next and striking atoms in their path each time constitutes *light amplification*.

In theory, if both mirrors are completely reflective, the photons would bounce back and forth indefinitely. In reality, the tube would overheat and burn up because the light energy would not be able to escape. Rub a little of the reflective coating off one mirror, however, and it passes some light. Now a beam of photons can pass through the partially reflective mirror after the light has been sufficiently amplified. In addition, because the mirror is partially reflective, it holds back some of the light energy. This reserve continues the chain reaction inside the tube.

LET THERE BE LASER LIGHT

The combination of light amplification and stimulated emission of radiation makes the laser functional. As you probably already know, the word laser is an acronym for *light amplification of stimulated emission of radiation.*

We've used the neon tube to describe the activity of atoms and photons to make laser light, and while neon gas can be made to lase (emit laser light), it is not as efficient as some other materials.

Matter comes in three known states: solid, liquid, and gas. Lasers can be made from all three types of matter. While everyone assumed the first laser would be the gas variety, modeled after the neon sign, the solid laser was the first to be invented.

Theodore Maiman, a research physicist at Hughes Laboratories in Malibu, California, announced the first successful operation of the laser on July 7, 1960. Maiman's contraption was surprisingly simple and compact, consisting of a synthetic ruby rod with mirrored ends, a spiral-shaped photographic strobe lamp, and a high-voltage power supply (refer to FIG. 1-1). The laser head itself—ruby rod and flash lamp—measured only 1.5 inches long and could be held in one hand.

The operation of the ruby laser, still in use today, is straightforward. The power supply sends a short pulse of high-energy electricity to the flash lamp. The lamp flashes on quickly, bathing the rod in white light. Chromium atoms—which give the ruby its red color—absorb just the green and blue spectrum of the light. The absorption of this light raises the energy level of the chromium atoms. Shortly thereafter, the chromium atoms fall back to a transitory level called the *metastable state.*

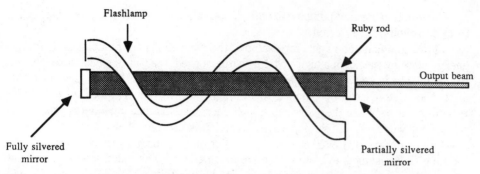

FIG. 1-1. *The main optical components of the ruby laser: a flash tube and a specially-made synthetic ruby rod. The ends of the rods are reflectively coated to provide optical amplification.*

There they stay for a short period of time (a few milliseconds), and then they drop back to the unexcited ground state. In this final drop, as shown in FIG. 1-2, photons are emitted. Some of these photons bounce off of the mirrored ends of the rod. After being amplified by bouncing back and forth between the mirrors, a small stream of photons are emitted out one end. The entire process lasts less than a few hundredths of a second. Because of the way the ruby rod is energized, it is termed an *optically pumped laser*.

Development of Other Types of Lasers

Other forms of lasers were developed after Maiman completed his trials with his ruby device. A more thorough discussion of the history and development of lasers appears below, but it is worthwhile to note that lasers based on the other two forms of matter—gas and liquid—were invented just shortly after Maiman's historical announcement. For instance, the gas laser, using a form of modified neon tube, was tested in 1961. The first successful attempt at the semiconductor laser took place in 1962, using what was essentially a specially-made light emitting diode (LED) suspended in a cold liquid nitrogen bath.

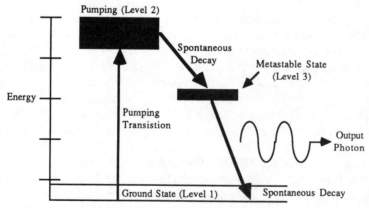

FIG. 1-2. *Energy levels of a ruby laser. After initial pumping by energizing the flash tube, the chromium atoms in the ruby rod spontaneously decay to a metastable state; they then decay once more and output photons.*

Parts of a Laser

Lasers consist of the following components:

★ *Power supply*. All lasers use an electrical power supply delivering a potential of up to 10,000 volts and up to many hundreds of amps.

★ *Pumping device*. Electrical discharge lasers use the high-voltage power supply as the pumping device, but some lasers use a radio-frequency oscillator, high-output photoflash or lamp, or even another laser.

★ *Lasing medium*. The medium is the material that generates the laser light. The lasing medium can be a gas, solid, or liquid. There are thousands of lasing mediums, including specially-treated glass, argon, organic dyes, or even Jell-O.

★ *Optical resonant cavity*. The cavity encloses the lasing medium and consists of mirrors placed at each end. On most lasers, one mirror is completely reflective and the other mirror partially reflective.

TYPES OF LASERS

The three states of matter—gas, solid, or liquid—present a convenient way to generally classify lasers, but it offers no insight into the design and application of various laser mediums. While the list of possible lasing mediums is extensive, most commercial, scientific, and military lasers fall into one of these categories: crystal and glass, gas, excimer, chemical, semiconductor, and liquid. Let's take a closer look at each one.

Crystal and Glass Lasers

As you've already discovered, synthetic ruby makes the basis for a very good laser. *Synthetic ruby* is made with aluminum oxide doped with a small amount of chromium. Why synthetic ruby and not the real thing? Ruby made in the laboratory is far more pure than natural ruby. The purity is necessary or lasing action cannot occur.

Another common crystal laser is the *Nd:YAG* laser. Nd:YAG is composed primarily of the elements aluminum, yttrium, and oxygen, doped with a pinch of neodymium. The name YAG is an acronym for yttrium-aluminum garnet, a synthetic garnet sometimes used in jewelry. The Nd:YAG is similar to the synthetic ruby in that the crystal is the host for the neodymium atoms. These atoms are excited into high-energy states using optical pumping and emit light in the infrared region. Nd:YAG lasers can be operated continuously because the crystal is a good conductor of heat. Synthetic ruby is a poor conductor of heat and will explode if lased continuously.

The element neodymium can be mixed with glass to make a neodymium-glass, or *Nd:glass*, laser. The benefit of the Nd:glass laser is that the glass rod is less expensive than YAG. The largest drawback of this type of laser is that glass is a relatively poor conductor of heat. For this reason, Nd:glass lasers are used in pulsed mode only.

Gas Lasers

Gas lasers represent the largest group of lasers. Their popularity stems from their inexpensive components and ease of manufacture (in comparison to other types of lasers;

all lasers are relatively difficult to manufacture). A number of gases and gas compounds are used, including:

- ✶ Helium-neon.
- ✶ Helium-cadmium.
- ✶ Argon.
- ✶ Carbon dioxide.
- ✶ Krypton.

Over 5,000 types of laser activity in gases are known, and several dozen have been made into working lasers. The *helium-neon* laser, as shown in FIG. 1-3, is the most common, finding wide use in general-purpose bar-code scanning systems in department stores and supermarkets, as well as in holography, surveying, and laboratory experiments. Most all helium-neon lasers emit a characteristic red beam. The power output of He-Ne lasers is limited, and all but unusual laboratory models are air-cooled by convection (some by forced air).

There is nothing exceptionally exotic about the average He-Ne laser, making it affordable and easy to use. Most of the projects in this book are centered around the common helium-neon tube, which you can purchase on the used or surplus market for $35 to $100.

CO_2 lasers are among the most powerful gas lasers. Whether operating in pulsed or continuous modes, a CO_2 laser can produce a beam that's intense enough to cut

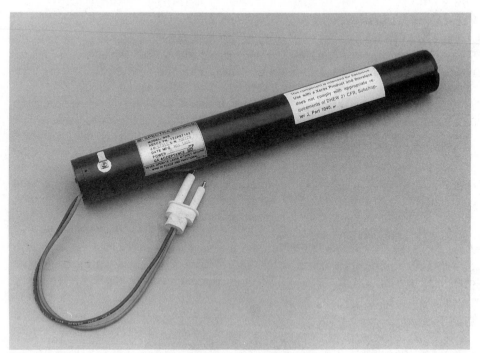

FIG. 1-3. *A commercially made helium-neon laser. The actual laser tube is encased in the aluminum tube; power leads extend from the rear of the laser.*

through almost any metal. CO_2 lasers actually contain a mixture of carbon dioxide, nitrogen, and helium that is continually pumped through the laser tube. To avoid overheating, CO_2 lasers are cooled by running water.

Argon and *krypton lasers* are used for their ability to produce two or more wavelengths of light, particularly in the short blue and green wavelengths. Though both types can produce a great deal of optical energy, depending on the model, they aren't used for cutting materials. They are mainly used for such applications as semiconductor manufacturing, medicine, color holography, and light shows. High-powered argon and krypton lasers are water-cooled; lower powered versions are forced-air cooled.

Excimer Lasers

Excimer lasers are gas lasers with a twist. A rare gas such as argon, krypton, or xenon electrically reacts with a halogen (chlorine, fluorine, iodine, or bromine) to form an excimer. Thinking back to high school chemistry, an excimer is a molecule that exists only in an electrically excited state. When the excimer molecule emits a photon, it doesn't go back to its ground state but breaks up into into constituent atoms. This provides the population inversion necessary for lasing action to occur.

The main benefit of excimer lasers is their ability to emit high-energy ultraviolet light, which is helpful in photochemistry and the manufacture of transistors and integrated circuits. By comparison, most all lasers emit their strongest radiation in the visible spectrum and infrared regions.

Chemical Lasers

Chemical lasers are high-powered beasts favored by the military as weapons against enemy aircraft or missiles. In the typical chemical laser, a flammable mixture of hydrogen and fluorine (or compounds thereof) acts as the lasing medium. The chemicals are pressurized and are sometimes ignited into a flame to initiate the lasing action. The typical chemical laser resembles a jet aircraft engine more than it does a James Bond-type laser. Obviously, chemical lasers are not for hobbyists.

Semiconductor Lasers

Semiconductor lasers are solid-state devices that—while almost as old as ruby and gas lasers—are just now catching on. You are probably aware of the semiconductor (or diode) laser used in compact audio and video discs, such as the one shown in FIG. 1-4. Semiconductor lasers are also used in fiberoptic telephone links, bar-code scanning devices, and military rangefinder equipment.

There are a number of different laser diode designs, with significant variations between each one. For our purposes, however, it is sufficient to say that the laser diode consists of a pn junction, as in a light-emitting diode (LED), but with specially cleaved and mirrored facets. In operation, current applied to the junction causes a glow of light. The mirrors comprise the optical cavity that amplifies the light generated within the diode junction. Laser diodes are made for either pulsed or continuous duty. Both types are widely available in the surplus electronics market.

FIG. 1-4. *A laser diode with heatsink attached. Although not apparent in the photograph, the laser and heatsink measure less than ¾ inch across.*

Liquid (Dye) Lasers

Liquid lasers use molecular organic dyes as the lasing material. The dye is directed through a cavity and pumped by an optical source, such as a CO_2 laser. The unique property of liquid dye lasers is that the output wavelength (see below) is tunable. By varying the mixture of the dyes, it's possible to change the color of the laser beam from a deep blue to a dark red.

LIGHT AND WAVELENGTHS

Light is part *particle* and part *wave*. The particles are called photons and the waves are a part of the *electromagnetic spectrum*. A graphic illustration of the electromagnetic spectrum appears in FIG. 1-5. Note that visible light comprises a relatively small portion of the entire spectrum.

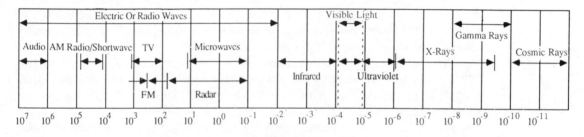

FIG. 1-5. *The electromagnetic spectrum, from long wavelength audio and AM radio signals to ultra-short gamma and cosmic rays. The visible light band is a relatively small section of the spectrum.*

You can best imagine the nature of light by thinking of the photons as bobbing up and down as they travel forward through space. Looking at the path of light sideways shows the photon drawing a sine wave, which is a wave that alternates up and down in a smooth, gradual motion, as shown in FIG. 1-6.

All sine waves—no matter how they are created and what they represent—share similar properties. They are composed of crests and troughs (or valleys). The distance between two consecutive crests determines the wavelength and thus the *frequency* of the wave. If the distance between two crests is small, the wavelength is small and the frequency is high. Increase the distance and the wavelength increases in proportion and the frequency drops. Frequency is most often expressed in *hertz*, or *cycles per second*. That is, frequency is the number of times the wave bobs up and down in one second.

Light travels at a velocity of 300,000 km per second (186,000 miles per hour). The different colors of visible light have different wavelengths and thus frequencies. The frequencies are in the terahertz range (thousand billions of cycles per seconds), as shown in FIG. 1-7. For convenience, light is usually expressed in wavelength, specifically *nanometers* (sometimes Angstroms). One nanometer is one billionth of a meter. TABLE 1-1 lists the different colors of the visible spectrum and their respective wavelengths. TABLE 1-2 compares the units of measurement common in the discussion of light and lasers.

Laser Wavelengths

One of the unique properties of lasers is their ability to emit light at a specific color, or wavelength. This is in contrast to the sun or an incandescent lamp, which both emit all the colors of the rainbow. The output wavelength is determined by the lasing medium. For example, the chromium atoms in synthetic ruby give off light at 694.3 nanometers (or nm), thereby making the wavelength of a ruby laser 694.3 nm.

The laser might emit light at just one specific wavelength or many distinct wavelengths. Light emission at any particular wavelength is called a *line*; if the laser emits light at many wavelengths, each individual wavelength is a *mainline*. The terms line and mainline are derived from the study of optical spectra, where white light passed through a prism is broken down into discrete lines or segments. TABLE 1-3 shows the chief wavelengths emitted by several common types of lasers.

Most helium-neon lasers emit light at one specific wavelength, namely 632.8 nm (632.8 billionths of a meter, equivalent to 6328 angstroms). Argon lasers generate two distinct mainlines, one at 488 nm and the other at 514 nm. Krypton lasers, popular in

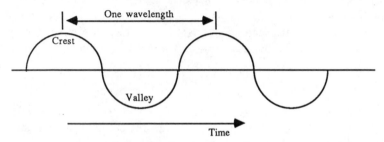

FIG. 1-6. *A sinusoidal wave, showing one wavelength from crest to crest.*

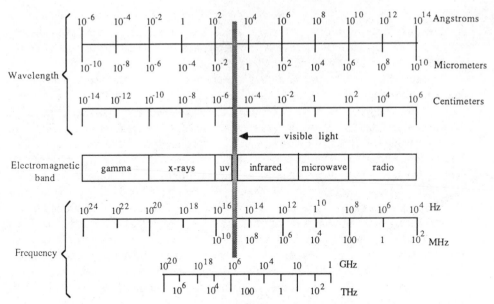

FIG. 1-7. *Comparing wavelength, frequency, and electromagnetic spectrum regions. Wavelengths are shown in three common units of measurement: angstroms, micrometers, and centimeters. A common unit of measurement for light wavelengths is the nanometer—to find the equivalent in nanometers, divide the number in angstroms by 10.*

professional light show systems, produce a mainline at 647 nm, but also produces weaker wavelengths all through the visible spectrum.

Note that mainlines are those wavelengths that greatly contribute to the intensity of the beam. Many lasers emit a whole slew of wavelengths but might be listed as having only one or two lines. These sub-lines are generally too weak to be useful, so they are usually ignored. In many laser systems, light at wavelengths other than at the desired lines are filtered out.

When using a multiple-line laser (such as argon or krypton), the individual colors can be separated by the use of prisms, filters, or diffraction gratings. These and other optical components are discussed in more detail in Chapter 3.

POWER OUTPUT

The intensity of the laser beam is measured in *joules* or *watts*. Both represent the amount of ''work'' that can be done during a particular period of time. In this case, the term *work* is used to denote the amount of energy or power released in some useful form, including heat. Electronic textbooks define one watt as the unit of electrical power equal to one volt multiplied by one ampere. The formula for computing watts is:

$$P = EI$$

where P equals power in watts, E equals EMF in volts, and I equals current in amps. One horsepower is also equal to 746 watts, an archaic but still-used measurement for determining the rate or amount of work performed by some device (usually a motor).

Table 1-1. Color Wavelengths

Color	Wavelength
Red	0.000066 cm; 660 nm
Orange	0.000061 cm; 610 nm
Yellow	0.000058 cm; 580 nm
Green	0.000054 cm; 540 nm
Blue	0.000046 cm; 460 nm
Violet	0.000042 cm; 420 nm

Infrared						Ultraviolet
Nanometers						
1200	700	600	550	500	400	200
Millimeters						
0.0012	0.00070	0.00060	0.00055	0.00050	0.00040	0.00020
Centimeters						
0.00012	0.00007	0.00006	0.000055	0.00005	0.00004	0.00002
	Red	Yellow	Green	Blue	Violet	

Watts, like horsepower, does not take time into consideration; joules does. One *joule* is equal to one ampere passed through a resistance of one ohm for one second (the joule is also a measurement of physical force). When wattage is measured against some factor of time, it is so indicated.

Rather than settle on one standard or another, laser manufacturers use a composite of watts and joules, although they mean different things. Generally, low-power, continuous-duty lasers are measured in watts; high-power, pulsed-duty lasers are measured in joules. And as a rule of thumb, radiant energy is expressed in joules and radiant power is expressed in watts.

Although laser light is most often defined as a pinpoint source, natural divergence and optics can cause the beam to spread. The area of the measured power output of a laser, in watts or joules, is often included in the specifications. A typical specification might be 5.0 joules cm^2. which means 5.0 joules in one square centimeter.

Most laser diodes and helium-neon lasers are rated in milliwatts, or thousandths of a watt. The typical helium-neon laser has a power output of about two milliwatts, or two $1/1000$ of a watt. High-powered gas argon and krypton lasers generate several watts of optical power. Such *multiwatt* lasers are most often used in research, manufacturing, and light shows.

Table 1-2. Metric Prefixes

Prefix	Abbreviation	Power of Ten	Value
tera	T	10^{12}	thousand billion
giga	G	10^{9}	billion
mega	M	10^{6}	million
kilo	k	10^{3}	thousand
deci	d	10^{-1}	tenths
centi	c	10^{-2}	hundredths
milli	m	10^{-3}	thousanths
micro	μ	10^{-6}	millionths
nano	n	10^{-9}	billionths
pico	p	10^{-12}	thousand billionths

1 micrometer:

       1/1,000,000 meter; 10^{-6}m
       1000 nm
       10,000 A units

1 nanometer:

       1/1,000,000,000 meter; 10^{-9}m
       1/1000 μ
       10 A units

1 angstrom unit:

       1/10,000,000,000 meter; 10^{-10} m
       1/10,000 μ
       1/10 nm

The power output of a laser is dependent upon many factors, one of which is the efficiency of the conversion of optical or electrical power into photon power. Most lasers are extremely inefficient—about one or two percent of the incoming energy is converted to usable light energy. But even with poor efficiency, the beam from a laser is far more intense (per given area) than sunlight.

By comparison, a standard incandescent light bulb is roughly 2 to 3 percent efficient, yet its light intensity is only a fraction of that of a low-power laser. Even a fluorescent lamp, with an efficiency of 10 to 15 percent, is not nearly as potent as a laser. CO_2 lasers are among the most efficient of the bunch, which is one reason why they emit such a powerful beam. Typical efficiency of a well-built CO_2 laser is about 30 to 35 percent.

COMPONENT PARTS OF THE LASER

A number of components go into making a laser. Because low-power gas and semiconductor lasers are the thrust of this book, we'll concentrate just on those. The

Table 1-3. Wavelengths of Popular Laser Types

Laser	Mainline	Comments
Argon	488.0 nm, 514.5 nm	Multiline: 351-528 nm
Carbon Dioxide	10,600 nm	
Dye laser	300-1,000 nm (typ.)	Tunable
Excimer	193-351 nm	
Helium-Cadmium	442 or 325 nm	
Helium-Neon	632.8 nm	Other lines at 543, 594, 652, 1,152, and 3391 nm
Nitrogen	337 nm	
Krypton	647 nm	Multiline: 350-800 nm
Nd:YAG	1,064 nm	
Ruby	694.3 nm	
Semiconductor (diode)	780, 840, 904 nm (typ.)	Range from 700-1,600 nm

typical helium-neon laser consists of a tube, high-voltage power supply, and power source, as shown in the block diagram in FIG. 1-8. Most He-Ne tubes are self-contained and include the facing mirrors, but some special tubes are available that use separate mirrors. The mirrors are mounted on a precision optical bench and adjusted so the laser beam properly exits the tube.

The high-voltage power supply converts the juice from the power source (usually either 12 volts dc or 117 volts ac) to between 1,200 and 3,000 volts. The high voltage

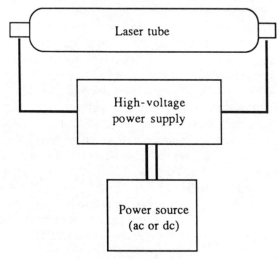

FIG. 1-8. *The major components of a gas laser, including tube, high-voltage power supply, and power source.*

is necessary to ionize the gas within the tube. The current output of the power supply is low—roughly 3 to 7 milliamps. You can calculate the approximate efficiency of the laser by multiplying the voltage (say 2,000 volts) by the current (such as five milliamps). The result is 10 watts of electrical power consumed by the tube. If the laser generates two milliwatts of light energy, the efficiency is only two percent (10 watts in times 0.002 watts out).

Semiconductor lasers consist of the laser chip or element, a power source, and possibly a drive circuit. Because the pn junction of the laser is so small, the chip is attached to a heatsink to dissipate unwanted heat. The chip and heat sink are encased in a metal canister and protected against dust and debris by a clear plastic or glass window. The power source is low-voltage dc. The exact voltage requirements depend on the type of diode used.

THE PROPERTIES OF LASER LIGHT

The light from a laser is special in many respects.

★ *Laser light is monochromatic*. That is, the light coming from the output mirror consists of one wavelength, or in some instances, two or more specific wavelengths. The individual wavelengths can be separated using various optical components.

★ *Laser light is spatially coherent*. The term *spatial coherence* means that all the waves coming from the laser are in tandem. That is, the crests and the troughs of the waves that make up the beam are in lock-step.

★ *Laser light is temporally coherent*. *Temporal coherency* is when the waves from the laser (which can be considered as one large wave, thanks to spatial coherency) are emitted in even, accurately spaced intervals. Temporal coherence is similar to the precise clicks of a metronome that times out the beat of the music.

★ *Laser light is collimated*. Because of monochromaticity and coherence, laser light does not spread (diverge) as much as ordinary light. The design of the laser itself, or simple optics, can collimate (make parallel) the laser light into a parallel beam.

The four main properties of laser light combine to produce a shaft of illumination that is many times more brilliant than the light of equal area from the sun. Because of their coherency, monochromaticity, and low beam divergence, lasers are ideally suited for a number of important applications. For example, the monochromatic and coherent light from a laser is necessary to form the intricate swirling patterns of a hologram. Without the laser, optical holograms would be much more difficult to produce.

Coherence plays a leading role in the minimum size of a focused spot. With the right optics, it's possible to focus a laser beam to an area equal to the wavelength of the light. With the typical infrared-emitting laser diode, for instance, the beam can be focused to a tiny spot measuring just 0.8 micrometers wide. Such intricate focusing is the backbone of compact audio discs and laser discs.

Minimum divergence (owing to the coherent nature of laser light) means that the beam can travel a longer distance before spreading out. The average helium-neon laser, without optics, can form a beam spot measuring only a few inches in diameter from a distance several hundred feet away. With additional optics, beam divergence can be reduced, making it possible to transmit sound, pictures, and computer code many miles on a shaft of light. A receiving station in the path intercepts the signal.

One experiment with the low divergence of laser light was performed during the historic Apollo 11 moon landings. A 1mm beam from a ruby laser was bounced to the moon and back, reflected by a matrix of highly polished mirrors. Although the moon is about 235,000 miles away, the laser beam spread to an area of only 1.5 miles when it reached the lunar surface.

THE BIRTH OF THE LASER

The concept of the laser is said to have occurred in 1951, brainstormed by Charles H. Townes, a physicist at Bell Labs (later a professor at Columbia University). Townes' first idea was to amplify stimulated emissions of microwave radiation. Using ammonia as the lasing medium, this device would emit radiation in the microwave region, below visible light and infrared. Three years later, along with graduate student James P. Gordon and Herbert Zeiger, Townes built and tested the first forerunner of the laser. They dubbed the device the *maser*, an acronym for *m*icrowave *a*mplification of *s*timulated *e*mission of *r*adiation.

In 1957, Townes sketched out an idea for the "optical maser," a device similar to his earlier invention but one that would emit radiation in the visible or at least near-infrared region of the electromagnetic spectrum. Townes joined forces with a researcher at AT&T, Arthur Schawlow, and began work on an optical maser using a large tube filled with gas (as an interesting aside, Schawlow is Townes' brother-in-law). By the time Townes and Schawlow were working on the optical maser in the late 1950s, researchers the world over were quickly drawing up plans of their own.

As mentioned previously, Hughes research scientist Theodore Maiman beat Schawlow and Townes to the punch with his crystalline ruby laser in 1960. The *gas discharge laser*, developed in 1961 by Ali Javan, William R. Bennett, and Dr. R. Herriott of Bell Labs, marked a major discovery, because unlike the ruby laser, it could be powered continuously instead of in short pulses.

The Bell Labs researchers used as the lasing mixture a combination of helium and neon gases. The gas was pumped into a long tube at relatively low pressure and mirrors were mounted on either end. When powered by a charge of high-frequency radio waves, the gas within the tube ionized, and photons began striking against the mirrors. Some photons escaped, and out one end of the tube came an invisible beam of infrared radiation. Later refinements to the helium-neon laser allowed it to emit visible light.

In 1962, a trio of groups including researchers at MIT, IBM, and GE developed the first semiconductor lasers. Shortly thereafter, groups at GT&E, Texas Instruments, and Bell Labs announced similar work. The small size and high energy output of the semiconductor laser required that it be cooled in liquid nitrogen—minus 196 degrees Celsius! Interesting things were done with the semiconductor laser, including transmitting a television picture along the beam of light and changing the characteristics of the diode so that the laser emitted a deep red glow instead of the characteristic invisible infrared radiation.

WILL THE REAL FATHER OF THE LASER PLEASE STAND UP?

Theodore Maiman, the scientists at Bell Labs, and others who built the first lasers are among those people credited with the introduction of the laser. But the actual inventor of the laser — the first person to have thought of it and worked out its principles—has been a debatable question since the late 1950's.

In the summer of 1958, Schawlow and Townes drew up a patent application for the optical maser (the term "laser" didn't come until later). Their patent was granted, and Townes, as well as two Russian scientists Nikolai G. Basov and Aleksander M. Prokhorov, received the Nobel Prize in 1964 for their work in developing the laser.

A graduate student at Townes' university by the name of Gordon Gould later claimed that he developed the basic principles of the laser earlier in November of 1957, more than half a year before the Townes/Schawlow patent application. Although Gould was not a graduate student of Townes, they knew each other and often spoke to one another.

Gould kept his ideas in a notebook which he later had notarized (the first page of this notebook is now at the Smithsonian Institution). Gould claimed he contemplated applying for a patent on his laser ideas, but an attorney gave him the mistaken impression that he needed a working model in order to be granted a patent. Apparently, before Gould could obtain more competent advice, Schawlow and Townes beat him to the punch and filed their application first.

Most textbooks on lasers, particularly those written before 1980, often credit Schawlow and Townes as the sole inventors of the laser. But today, it is Gordon Gould who owns the basic patents on the laser and enjoys a healthy sum from royalties paid by laser manufacturers.

This chapter has merely touched upon the fundamentals and history of lasers. A more thorough discussion of laser principles would have diluted the main purpose of this book—namely, to present a number of affordable and fun projects you can do with a gas or semiconductor laser. If you are serious about lasers and want to learn more, see Appendix B for a list of suggested further reading. A number of the books provide a technical, even scholarly, discourse on laser principles.

2

Working with Lasers

For the uninitiated, the thought of working with lasers means wearing dark tinted goggles and heavy lead-lined gloves while sitting in a concrete, air-conditioned bungalow. Behind a six-inch glass partition are several lab assistants, complete with white coats, clipboards, and solemn faces. Giant computers and monitoring equipment adorn the laser laboratory, soaking up enough electricity to light up Las Vegas. Is that a strain of Hollywood B-movie music in the background? Any moment now a mad scientist will come out and begin the final phase of his quest for world power.

The movies have done a considerable job selling a false and overly dramatic view of lasers (witness the James Bond classic "Goldfinger"). On the contrary, the kinds of lasers available to the electronics hobbyist are so low in power that protective measures are practically unnecessary. The light radiation emitted by a helium-neon laser isn't even strong enough to be felt on skin.

Of course, precautions must still be taken, but for the most part, experimenting with lasers can be done in the comfort of the family living room, under normal temperatures, and with no more electrical power than the current from a set of flashlight batteries.

This doesn't mean hobby lasers are completely harmless. As with all electrical devices, some dangers exist, and it's vitally important that you understand these dangers and know how to avoid them. In this chapter, you'll learn about what you need to know to competently work with lasers, the basics of laser safety, and how to protect yourself and others from accidental injury.

BASIC SKILLS

What skills do you need as a laser experimenter? Certainly, if you are already well-versed in electronics and mechanical design, you are on your way to becoming a laser experimenter *extraordinaire*. But an intimate knowledge of neither electronics nor mechanical design is absolutely necessary.

All you really need to start yourself in the right direction as a laser experimenter is a basic familiarity with electronic theory and mechanics. The rest you can learn as you go. If you feel that you are lacking in either beginning electronics or mechanics, pick up a book or two on these subjects at the bookstore or library. See Appendix B for a selected list of suggested further reading.

Electronics Background

Study analog and digital electronic theory, and learn the function of resistors, capacitors, transistors, and other common electronic components. Your mastery of the subject does not need to be extensive but just enough so that you can build and troubleshoot electronic circuits for your laser systems. You'll start out with simple circuits and a minimum of parts and go from there. As your skills increase, you'll be able to design your own circuits from scratch, or at the very least, customize existing circuits to match your needs.

Schematic diagrams are a kind of recipe for electronic circuits. The designs in this book, as well as most any book that deals with electronics, are in schematic form. If you don't know already, you owe it to yourself to learn how to read a schematic diagram. There are really only a dozen or so common schematic symbols, and memorizing them takes just one evening of concentrated study. A number of books have been written on how to read schematic diagrams (see Appendix B).

Sophisticated laser systems use computers for process control. If you wish to experiment with these control circuits, you need to have at least some awareness of how computers operate. Although an in-depth knowledge of computers and programming is not required, you should have rudimentary knowledge of computers and the way computers manipulate data.

Mechanical Background

The majority of us are far more comfortable with the mechanical side of hobby laser building than the electronic side. It's far easier to see how a motor and lever work than to see how a laser tachometer operates. Whether or not you are comfortable with mechanical design, you do not need to possess a worldly knowledge of mechanical theory. Still, you should be comfortable with mechanical and electro-mechanical components such as motors and solenoids.

The Workshop Aptitude

To be a successful laser hobbyist, you must be comfortable with working with your hands and thinking problems through from start to finish. You should know how to use common shop tools and have some basic familiarity in working with wood, lightweight metals, and plastic.

Ac Electrocution

Some laser supplies and laser projects operate from the 117-volt ac house current. Observe the same precautions when working with ac circuits. A live ac wire can, and does, kill! Exercise caution whenever working with ac circuits and observe all safety precautions. If you are new to electronics and aren't sure what these precautions are, refer to Appendix B for a list of books that can help you broaden your knowledge.

You can greatly minimize the hazards of working with ac circuits by following these basic guidelines:

✶ Always keep ac circuits covered.

✶ Keep ac circuits physically separate from low-voltage dc circuits.

✶ All ac power supplies should have fuse protection on the incoming hot line. The fuse should be adequately rated for the circuit but should allow a fail-safe margin in case of short circuit.

✶ When troubleshooting ac circuitry, keep one hand in your pocket at all times. Use the other hand to manipulate the voltmeter or oscilloscope probe. Avoid the situation where one hand touches ground and the other a live circuit. The ac flows from one hand to the other, through your heart.

✶ When possible, place ac circuits in an insulated Bakelite or plastic chassis. Avoid the use of metal chassis.

✶ Double- and triple-check your work before applying power. If you can, have someone else inspect your handiwork before you switch the circuit on for the first time.

✶ Periodically inspect ac circuits for worn, broken, or loose wires and components, and make any repairs necessary.

PROJECTS SAFETY

A number of the projects in this book require work with potentially hazardous materials, including glass, razor blades, and wood and metal tools. Glass and mirror-cutting should be done only with the proper tools and safety precautions. Wear eye protection whenever you are working with glass. If you can, wear gloves when handling cut glass and mirrors. Finish the edges with a burnisher, flame, or piece of masking tape.

Razor blades provide a well-defined and accurate edge and are helpful in a number of optical experiments. You need new, unused blades; nicks and blemishes on the blade surface can impair your results. Obviously, you must exercise extreme care when handling razor blades.

One good way to handle razor blades is to coat the edges of the blade with a soft wax or masking tape. When you are ready to position and use the blade, peel off the wax or tape. When cutting with razor blades, wear eye protection and gloves. For maximum safety, place the blade behind a large piece of clear plastic. If any pieces break off during cutting, the plastic will protect your body.

An optional experiment in Chapter 22 involves the use of liquid nitrogen, a cryogenic gas that has a temperature of minus 196 degrees Celsius. While liquid nitrogen is non-flammable and non-toxic, its extremely cold temperature can cause frost-bite burns. You should handle liquid nitrogen only in an approved container (a Thermos bottle with a hole drilled in the cap often works), waterproof gloves, and eye protection. Follow the recommendations and handling precautions given in Chapter 22 for more details. When treated properly, liquid nitrogen is safe and actually fun to use.

Construction plans require the use of wood- and metal-working tools. You can use hand or power tools; either way, carefully follow all operating procedures and use the tools with caution. Most power tools provide some type of safety mechanisms so don't defeat them! Thoroughly read the instruction manual that came with the tool. A number of good books have been written on wood and metal working and the tools involved. Check your library for available titles.

Common sense is the best shield against accidents, but common sense can't be taught or written about in a book. It's up to you to develop common sense and use it at all times. Never let down your guard. The laser projects in this book are provided for your education and enjoyment. Don't ruin the fun of a wonderful hobby or vocation because you neglected a few safety measures.

3

Introduction to Optics

Optics are an integral part of laser experiments. They allow you to manipulate the beam in ways similar to how electronic components control the flow of current through a circuit. In this chapter, you'll learn about optics and how they effect laser light. You'll also learn how to care for, clean, and store optics in order to preserve their light-manipulating characteristics. While the following text only scratches the surface of optics, it provides a clear understanding of the fundamentals and helps you start on your way to using optics in your own laser projects.

FUNDAMENTALS OF THE SIMPLE LENS

The speed of light in a vacuum is a constant, which is precisely 299,792.5 kilometers per second or 186,282 miles per second. By definition, a vacuum means the absence of matter; when matter is introduced, light slows down, because the beam of light actually bends. Imagine a ray of light as a line traversing from the sun to the Earth, as in FIG. 3-1. When the ray of light strikes the boundary between vacuum and matter (in this case air), the ray bends.

This is *refraction*, and it occurs whenever light passes through two mediums of different densities (a vacuum has no density; all matter has some density). The angle of the bend depends on three factors:

★ The density of the media.
★ The wavelength of the light.
★ The direction of light travel, whether it is passing from a less dense medium to a thicker one, or vice versa.

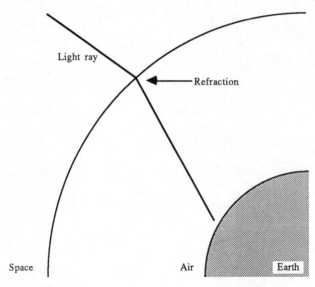

FIG. 3-1. *The effect of refraction on a ray of light entering the Earth's atmosphere.*

Refraction Based on Density

Light rays bend more in heavy density media. The amount of bending is called the *index of refraction*, which is calculated by dividing the density of the medium by that of a vacuum. A vacuum has a refractive index of 1; glass has a refractive index of about 1.6 (depending on the glass); and air has a refractive index of 1.0003. Note that the exact index of refraction for these and other media also depends on temperature, barometric pressure, and impurities, but not on thickness. Forgetting that the atmosphere gets thinner at higher altitudes, one foot of air has the same index of refraction as one mile of air. The slow-down of light rays through a refractive medium is not cumulative, where the light might eventually stop.

Refraction Based on Wavelength

Longer wavelengths of light bend less than shorter wavelengths. As you can see in the chart in FIG. 3-2, red light has a longer wavelength than violet light. If you were to direct a beam of red light through a refracting medium, such as water, its angle of bend would be less than that for a beam of violet light. Of course, this is how a prism, discussed more fully later in this chapter, breaks up white light into its component colors.

Refraction Based on Direction of Light Travel

Draw a line that bisects the border between two different media. Now draw a line at a right angle to the border, as shown in FIG. 3-3. This new line is called *line normal*, and is an important concept in optics. When a ray of light passes from a less dense medium to a more dense one, *the ray bends towards line normal*. The reverse is true when light passes from a dense medium to a less dense one: *The ray bends away from line normal*. So you can refer to this important concept more easily, let's summarize it. We'll use the terms "rare" and "dense" medium to denote less dense and more dense.

28

★ Rare-to-dense transition: light bends towards line normal.

★ Dense-to-rare transition: light bends away from line normal.

LENSES AS REFRACTIVE MEDIA

Lenses are refractive media constructed in a way so that light bends in a particular way. The refractive index of a lens is determined by its chemical makeup. Some lenses refract light more strongly than others. Two common lens glasses are *Schott* and *crown*. At the green light wavelengths of the middle spectrum (approximately 540 nm), Schott glass has a refractive index of 1.79 and crown glass has a refractive index of 1.52. Both crown and Schott glass lenses can be used separately or together.

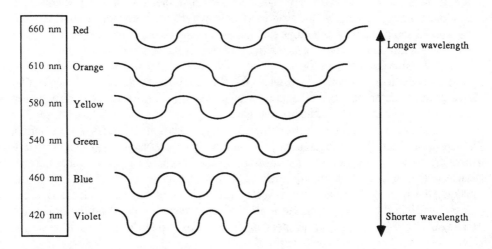

FIG. 3-2. *All colors have a specific wavelength; the wavelength increases as the colors approach the red end of the spectrum.*

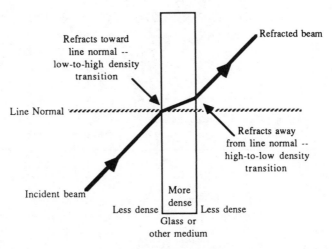

FIG. 3-3. *Light always refracts as it enters or exits mediums of different densities. Line normal shows the effect of refraction in relation to change in density.*

Unless you are a manufacturer of optics equipment like microscopes, telescopes, and binoculars, you won't know the exact makeup of the lenses you buy. Rather, you'll be concerned with three other lens specifications:

★ lens shape
★ lens focal length
★ lens diameter

Lens Shape

More than anything else, the *shape of the lens* determines how light is refracted through it. A flat piece of glass, although still a refractive medium, will bend all the light rays going through it equally. You won't see an apparent change when looking through the glass at some object on the other side (although you can experiment with the effects of refraction through a plane of glass with a laser, as described in Chapter 8).

But if you make the glass thicker in the middle than at the edges, light rays striking the glass are refracted at different angles. Why? Is it not the thickness of the glass that is causing the difference in the bending of light, but the difference in the direction of line normal throughout the radius of the glass.

Take a look at FIG. 3-4 for a better view. The lens curves at the edges, so line normal is bent away from the incoming light rays (assuming parallel light rays). The rays are traveling through a rare-to-dense transition, so the light is bent towards line normal. Because line normal is at a greater angle at the outer radius of the lens than it is at the middle, the rays at the edges are deflected more. Now, when you look through the glass—which in this form is more accurately called a lens—an object on the other side takes on a different appearance. Depending on how close you hold the lens to your eye, the object appears larger than it really is.

There are numerous shapes of lenses, and each shape manipulates light in a slightly different way. The next section discusses these shapes in more detail and how they are used.

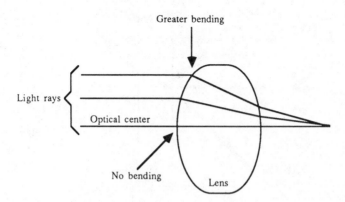

FIG. 3-4. *A lens is a piece of glass whose curvature bends light rays in a certain direction and amount. A common application for a lens is to focus parallel light rays to a point, as shown here.*

Lens Focal Length

The precise definition of *lens focal length* is rather involved and beyond the scope of this book but for our purposes, it is sufficient to say that the focal length of a lens is the distance from the lens where rays are brought to a common point. Rays entering one side of the lens are refracted so that they converge at a common point. Measuring the distance from the optical center of the lens to this point gives you the focal length. Later in this chapter, you'll learn that this definition applies to only certain kinds of lenses. Others behave in an almost opposite manner.

Focal length is an important consideration, because it tells you how much the lens is refracting light. A short focal length means that the light rays are brought to a point very quickly, so the rays must be heavily refracted in the lens. A long focal length means that the rays are gradually brought to a point and that the amount of refraction is mild. You choose a lens of particular focal length depending on the task you want to perform.

Lens Diameter

The *diameter of the lens* determines its light gathering capability. The larger the lens, the more light it collects. For example, you use the largest lens possible at the end of a telescope to bring distant planets and galaxies into view. In laser work, however, bigger isn't always better. The field of view of a laser beam is finite. That means you are interested in dealing with just the pencil-thin beam of the laser and nothing more.

Because the lenses can be small, it's easy to build compact laser optical systems. And perhaps more important to the hobbyist, smaller lenses are much cheaper than larger ones. Most of the lenses you purchase for your laser experiments should be surplus or seconds, but, the preference for small size pays off. Large, unblemished lenses are a rarity in the surplus lens market, and most of the big ones are usually practically useless for anything but paperweights anyway. In addition, most optics experimenters are mainly interested in large surplus lenses for components in telescopes, home-made projectors, and camera attachments. They tend to ignore the smaller ones, so you will probably have more to choose from.

LENS TYPES

There are six major types of lenses, as shown in FIG. 3-5. But before taking a look at each type of lens, let's define the meaning of plano, convex, concave, and meniscus.

- ⋆ *Plano* means flat.
- ⋆ *Convex* means curving outward (with respect to the other side of the lens).
- ⋆ *Concave* means curving inward (with respect to the other side of the lens).
- ⋆ *Meniscus* means curving in on one side and curving out on the other.

The combinations such as plano-convex and double-concave refer to each side of the lens. A plano-convex lens is flat on one side and curves outward on the other. A double-concave lens curves inward on both sides. Negative and positive refer to the focal point of the lens, as determined by its design. In actuality, all lenses are either positive or negative, but only meniscus lenses come in both flavors.

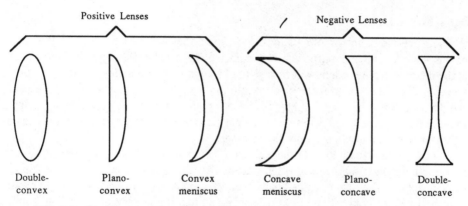

FIG. 3-5. *The six major types of lenses, broken into sub-groups of positive and negative.*

Lenses form two kinds of images: *real* and *virtual*. A real image is one that is focused to a point in front of the lens, such as the image of the sun focused to a small disc on a piece of paper. A virtual image is one that doesn't come to a discrete focus. You see a virtual image behind the lens, as when you are using a lens as a magnifying glass. Positive lenses, which magnify the size of an object, create both real and virtual images. Their focal length is stated as a positive number. Negative images, which reduce the size of an object, create only virtual images. Their focal length is stated as a negative number.

Lenses converge or diverge light. When light is *converged,* the rays are brought together at a common point. When light is *diverged,* it is spread out and the rays don't meet. The capacity to converge or diverge light also depends on the pattern of the light before it enters the lens.

In most cases when dealing with lasers, the light is *collimated*; that is, all of the light rays are traveling parallel (in actuality, the beam is spreading slowly, but the amount is small compared to the light from other sources). In some instances, however, the rays are diverging (spreading), and a given lens can serve to collimate the light (make it parallel).

All lenses have an *optical center*, that is typically in the middle of the lens. Refraction is absent or minimal at the optical center and increases with radius. In most lenses, refraction is the strongest at the outside edge of the lens.

Plano Convex

As noted above, a *plano convex* lens is flat on one side and curved outward on the other. Plano convex lenses have positive focal lengths; that is, they magnify an image when you look through them. Plano-convex lenses converge incident light to a common focal point and they form real or virtual images. They are most often used in telescopes, optical receivers and transmitters, and other applications where it is desirable to focus a beam of light to a common point.

Double-Convex

Double-convex lenses bulge out on both sides. The amount of curvature is typically the same for both sides, but not always. A double-convex lens that is the same on both sides is symmetrical. A lens that has different curvatures is asymmetrical.

Symmetrical lenses are often desired because they suffer from less distortion (covered later in this chapter). Like the plano-convex lens, double-convex lenses have positive focal lengths and converge light to a common point. They create both real and virtual images. Typically, a double-convex lens of a particular diameter has a shorter focal point (higher magnification) than a plano-convex lens of the same size.

Positive Meniscus

A meniscus lens has one surface concave (curved inward) and the other surface convex (curved outward). In a *positive meniscus* lens, the center of the lens is thicker than the edges (the reverse is true in a negative meniscus lens).

A meniscus lens is most often used with another lens to produce a device of a longer or shorter focal length than just the original lens used alone. For example, a positive meniscus lens can be used behind a plano-convex lens to shorten the effective focal length of the optical system. You can also use a positive meniscus lens as a magnifier; like the plano-convex and bi-convex lenses, they converge light rays to a common point, magnify images, and have a positive focal length.

Plano Concave

The *plano-concave* lens is flat on one side and curved inward on the other. Plano-concave lenses have negative focal lengths and diverge (spread) a beam of collimated light. They form virtual images only. Plano concave lenses are often used to expand a laser beam so that it covers a larger area, or they can be used to increase the focal length of optical equipment, such as telescopes.

Double-Concave

Double-concave lenses curve inward on both sides, and are similar in function and application as plano concave lenses. They have negative focal points so they create virtual images only; a double-concave lens can't be used to focus light at a common point.

Negative Meniscus

In a *negative meniscus* lens, the center of the lens is thinner than the edges. See the section above on positive meniscus lenses for a description of possible uses.

Lens Combinations

Lenses are often grouped together to manipulate the light in a way that can't be done by using just one lens alone. As discussed in the section on meniscus lenses, for instance, two or more lenses can be combined to shorten or lengthen the focal length of an optical device. Lenses can also be combined, usually by physical bonding with optically clear cement, to reduce or eliminate aberrations. These aberrations can distort the image or color rendition and are particularly annoying in systems that use large lenses (such as telescopes).

One lens used alone is a *singlet*. The majority of lenses are singlets, but they can be combined with others to create more complicated optical structures. Two lenses used together are called a *doublet* (or duplet); three lenses are called a *triplet*. Often, but not always, the individual lenses that make up a doublet or triplet are made of different substances, each with a different index of refraction.

Because of the monochromatic and pencil-thin nature of laser light, most lenses for laser projects needn't be anything more than singlets. Certain aberrations, particularly those that deal with the problems of faithful color reproduction, aren't of major concern to the laser enthusiast.

LENS COATINGS

The physics of optics says that even under the best circumstances, about 4 percent of incident light on a piece of glass (including a lens) will be reflected. The remaining 96 percent is passed through. This percentage varies depending on the angle of incidence. Reflectance is increased at the outer edges of a convex lens, for example, because the light strikes the lens surface at a greater angle.

The least expensive lenses are composed of bare glass with no special coating. An optical system that uses many uncoated lenses will suffer from an appreciable amount of light loss due to reflection. These reflections must go someplace, and they often strike the inner walls of the optical device or are bounced around in an unpredictable manner. Such reflections decrease contrast and cause flaring, ghosting, and other imperfections.

A thin coating of magnesium fluoride or some other material can decrease reflection to only 1.5 percent or so. That means 98.5 percent or more of the light passes through the lens. The coating applied to better quality lenses helps reduce unwanted reflection, and can also allow only one particular portion of the light spectrum to pass through.

Lens coatings take many forms, and the very expensive lenses have complex, multi-layer coatings applied in specific thicknesses (usually ¼ wavelength of visible light). Unless you order a coated lens directly from a lens manufacturer, you probably won't know the type of antireflection (AR) coating used. You can, however, tell if a lens is coated by tilting it at a 45-degree angle and looking at the reflected light. Coated optics designed to work with visible light have a blue or purplish hue to them. Often, but not always, lenses will be coated on both sides.

THE FUNCTION OF MIRRORS

Mirrors are used in laser experiments to re-direct a beam, to mix a beam with other light sources, and a number of other tasks. Mirrors differ in their reflective material, amount of reflection, flatness, and location of the reflective surface.

The Principles of Reflection

Recall the concept of line normal from the previous discussion of refraction. Line normal is also used in the analysis of reflection. The principle of reflection is simple: the angle of reflectance, in relation to line normal, is equal to the angle of incidence. That is, if you bounce a beam of light off a mirror at a 45-degree angle to line normal, the reflected ray will also be at a 45-degree angle. The reflected ray will be on the opposite side of line normal as the incident ray.

Front or Back Reflective Surface

Most household mirrors consist of a coat of silver applied to the rear side of a sheet of glass. To prevent tarnishing, a lacquer is applied over the silvering. Such a mirror is called *rear-surface* (also back or second surface), because the reflective material is applied to the rear of the glass.

If you look carefully at a back-surface mirror, you'll see two reflections: one from the silver and one from the front of the glass. The amount of reflection from the glass is small—about four to five percent—but it's enough to cause a ghost image when the mirror is used in fine optical equipment. Shining a laser on a mirror produces two beams: the main beam from the silver reflective surface and a ghost beam from the front of the glass.

Front-surface mirrors are coated with a reflective substance on the front of the glass. Looking closely at light reflected from a front-surface mirror, you see only one image — from just the reflective surface. There is no ghost because the glass substrate is behind the reflective layer.

Unless you're after an unusual effect or purposely trying to create image ghosts, you will always use front-surface mirrors in your laser projects. Front-surface mirrors are harder to find than ordinary back-surface mirrors, and they are more difficult to care for. Sources for front-surface mirrors, as well as all optical components discussed in this book, can be found in Appendix A. Look for local sources of front-surface mirrors; check the Yellow Pages under (guess what?!) ''Mirrors.'' Call around until you find what you want. But before you do, be sure to read the section on buying optics below.

Reflective Coatings—Metal

Silver is seldom used as the reflective layer on front-surface mirrors because the exposed metal is liable to tarnish. However, some high-grade mirrors that require excellent reflection at all visible light wavelengths use silver front-surface mirrors that are protected with a thin, optically transparent overcoat.

The most common reflective material of front-surface mirrors is aluminum. Like silver mirrors, many front-surface aluminum mirrors are protected against scratches and marring by a clear overcoating. This overcoating must be optically pure and must be applied in precise layers, usually at a thickness equaling ¼ wavelength of green light (the middle of the spectrum).

Gold-coated mirrors provide the maximum amount of reflection at all visible wavelengths. The gold coating is soft and easily scratched, so a top coating is necessary. Again, the coating is applied in precise layers.

Reflective Coatings—Dielectric

Dielectric (non-electrically conducting) coatings are often used in mirrors designed for use in laser systems. A dielectric coating is extremely thin and semi-transparent and is applied to the glass substrate in a series of layers. The coating reflects light because its index of refraction is higher than that of the substrate (glass) underneath. The amount of reflection varies depending on the angle of incidence, coating type, and thickness of the coating.

Most dielectric coatings are sensitive to wavelength, making the mirrors suitable for different applications. Dielectrically coated mirrors designed for argon lasers have a different coating than those for helium-neon and diode lasers. If given a choice, you should pick the mirror coated for the type of laser you are using.

Note that the mirrors used in a helium-neon laser cavity are most often dielectrically coated. While light passing through these mirrors appears blue, light reflecting off the mirrors appears gold (or blue at some angles). The thickness of the dielectric coating, and hence the degree of reflectivity, is different for the two mirrors. One mirror is designed so that it reflects all or most of the light incident on it. The other mirror is designed so that only some of the light is reflected; the remaining portion passes through as the exiting laser beam.

Amount of Reflection

Even without a coating, glass will reflect about 4 percent of light when the rays are incident at an angle between about 0 and 30 degrees from the surface. Reflectance jumps considerably as the angle of incidence increases. At grazing incidence, reflection approaches 100 percent.

Although plate glass can be used as a kind of mirror, the exact amount of reflection is hard to control, especially without precision mounting equipment. Coatings are used to provide a known reflectance. The reflectance varies depending on the coating, coating thickness, and angle of incidence. Some mirrors are made to be 100 percent reflective at all angles. These generally use silver or aluminum reflective layers. Gold and dielectric coatings tend to be semi-transparent at angles of incidence other than 0 degrees.

OTHER OPTICAL COMPONENTS

There are numerous other optical components you can use in your laser-system building endeavors. Here is short rundown of the more popular ones.

Prisms

Most people learn about *prisms* in grade school as devices to break white light into its component colors. This breakup is more accurately called *dispersion*, as shown in FIG. 3-6, and is caused by refraction. As you learned earlier in this chapter, the longer the wavelength of light, the less it bends due to refraction.

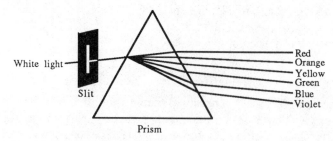

White light — Slit — Prism

Red
Orange
Yellow
Green
Blue
Violet

FIG. 3-6. *A prism refracts (or disperses) white light into its constituent colors. Dispersion allows you to use a prism with white light as a "rainbow maker."*

Prisms are used in laser systems for several important functions:

★ To disperse a multi-line laser beam into its component colors. You'd use a prism to separate the green and blue colors in an argon laser, for example.

★ To redirect a beam at some angle, as with a mirror.

★ To polarize a beam of laser light and direct it in one or more directions.

The most familiar type is the *equilateral prism* that is used mostly to disperse light but also to bend light at some angle. Viewed edge-on, an equilateral prism is an equilateral triangle with 60-degree angles between the equal faces.

Another common type is the *right angle prism* (one angle is 90 degrees; the others are typically 60 and 30 degrees). Right-angle prisms are designed primarily to direct light at a 90-degree angle; other angles are possible by varying the incident angle of the incoming light beam. A prism is often preferred over a mirror when vibration or stress are problems. The prism is a solid piece of glass and withstands mechanical stress better than a mirror. Right-angle prisms can use either an aluminized or coated hypotenuse or total internal reflection to bounce the light between the entrance and exit faces.

A *roof* or *amici roof prism* deflects light through a 90-degree angle. The roof prism is similar to a right-angle prism and almost always uses an uncoated hypotenuse. A *porro prism* (or retroflector) resembles a right-angle prism but is used to turn light around 180 degrees—it's sent back to where it came from. The light enters the hypotenuse of the prism, strikes one side, bounces off the opposite side, and is redirected out the hypotenuse.

Beam Splitters

A *beam splitter* does as its name implies: it takes one beam and divides it into two. Most beam splitters are also beam combiners—when positioned properly, the beam splitter can combine the light from two sources into one shaft of light.

Beam splitters come in two forms: cube and plate glass. Both are shown in FIG. 3-7. *Cube beam splitters* are made by cementing together two right-angle prisms so that their common hypotenuses touch. Usually, some form of reflective or polarizing layer is added at the joint. Anti-reflection coatings are typically applied to the entrance and exit faces to reduce light loss. The basic operation of the cube beam splitter is shown in FIG. 3-8A.

Note that the cube can be made to act as several beam splitters, depending on the coating at the hypotenuse. In many cube beam splitters, an entrance face can also act as an exit face. Most cube beam splitters divide the light equally between the two exit faces. These are called 50/50 beam splitters — 50 percent of the light goes out one face and 50 percent goes out the other (in actuality, less than 50 percent exits the cube at each face due to inherent reflection and transmission losses).

Plate beam splitters use a flat piece of glass to reflect and pass light. Although you can use an uncoated piece of glass, the best plate beam splitters are those designed for the job. An anti-reflection coating is applied to the glass to control the amount of reflection. Plate beam splitters can be made to transmit and reflect light equally (50/50) or unequally. Common ratios are 10/90 (10 percent reflection, 90 percent transmission) and 25/75 (25 percent reflection, 75 percent transmission).

Plate beam splitters often suffer from satellite images—you get two reflected beams instead of one. The first spot of light, as shown in FIG. 3-8B, is the primary beam (the one you want). It is produced when light bounces off the reflective or first surface of the glass. The second spot is the satellite, caused by internal reflection. You can

FIG. 3-7. *Two types of beam splitters: plate and cube (the cube beam splitter is shown attached to a metal mounting bracket).*

sometimes eliminate or reduce the intensity of the satellite by reversing the plate. The reason: some beam splitters are coated on one side only. Turning them over directs the coating toward the laser. You can also eliminate the satellite by carefully placing black tape on the beam splitter. Direct the laser beam so that only one spot is reflected.

Filters

Filters accept light at certain wavelengths and block all others. The color of the filter typically determines the wavelength of light that it accepts, thus rejecting all others. For example, a red filter passes red light and blocks other colors. Depending on the design of the filter, the amount of light blockage can be small or large.

Many filters for laser experiments are designed to pass infrared radiation and block visible light. Such filters are commonly used in front of photo sensors to block out unwanted ambient light. Only infrared light—from a laser diode, for instance—is allowed to pass through and strike the sensor.

Filters for light experiments come in three general forms: colored gel, interference, and dichroic. *Colored gel filters* are made by mixing dyes into a Mylar or plastic base.

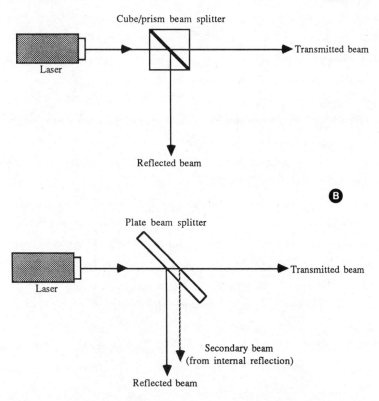

FIG. 3-8. *How cube and plate beam splitters work. The cube beam splitter (A) is composed of two right-angle prisms cemented together. The plate beam splitter (B) is a piece of flat glass that may or may not be coated with a reflective layer.*

Good gel filters use dyes that are precisely controlled during manufacture to make filters that pass only certain colors. Depending on the dye used, the filter is capable of passing only a certain band of wavelengths. A good gel filter might have a bandpass region (the spectrum of light passed) of only 40 to 60 nanometers.

Interference filters consist of several dielectric and sometimes metallic layers that each block a certain range of wavelengths. One layer might block light under 500 nm and another layer might block light above 550 nm. The band between 500 and 550 nm is passed by the filter (see FIG. 3-9). Interference filters are sometimes referred to as bandpass filters, because they are made to pass a certain band of light wavelengths. Interference filters can be made either narrowband, accepting only a very small portion of wavelengths, or broadband, accepting a relatively large chunk of the spectrum.

Dichroic filters use organic dyes or chemicals to absorb light at certain wavelengths. Some filters are made from crystals, such as cordierite, that exhibit two or more different colors when viewed at different axes. Color control is maintained by cutting the crystal at a specific axis. Dichroism is also used to create polarizing materials (colored and uncolored), as discussed more fully in Chapter 8.

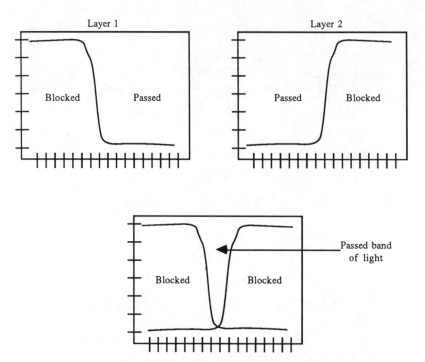

FIG. 3-9. *Interference filters are composed of two (or more) layers that selectively block light beyond a certain wavelength. By sandwiching two complementary layers together, it's possible to pass only light in a very restricted band.*

Both interference and dichroic filters exhibit a rainbow of colors as you tilt them against a white light source. Their coloration gives you no clue to the wavelengths they are designed to pass. Peering through the filter, you might see a bluish or greenish tinge, but tilting the filter or reflecting light off one surface can reveal other striking colors such as gold, purple, or yellow.

Pinholes

A *pinhole* is a small hole drilled or punched into an opaque sheet. The hole can be as small as 1 or 2 micrometers to as large as a millimeter. Pinholes are used to make spatial filters (see below) or to diffract light.

Spatial Filter

Imagine looking at a laser beam head on (don't actually do it— just imagine it!). The spot has a bright central portion and an almost fuzzy outer sheath. Some advanced laser experiments require a near-perfect beam (one that lacks the fuzz). *Spatial filters* are used to "clean up" the beam spot by taking just the center portion and excluding the perimeter noise.

A spatial filter is a pinhole coupled with a microscope objective. As depicted in FIG. 3-10, the lenses in the microscope objective focus the beam of light to a tiny spot, often not more than 25 to 50 micrometers in diameter. That spot is squeezed through a pinhole

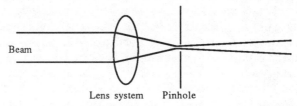

Beam

Lens system   Pinhole

FIG. 3-10. *The basic operation of the spatial filter. A lens system (typically a microscope objective) squeezes the light into a fine point. The light is then passed through a pinhole. The size of the pinhole must complement the magnification power of the lens system.*

of similar diameter. Spatial filters require precision optics and focusing, and though you can make your own, it's often better to buy one (surplus if possible).

The spatial filter won't work without the objective, because the pinhole alone causes the beam to diffract (see Chapter 8 for some experiments with diffraction). Instead of cleaning up the beam, you smear it even more. And without the pinhole, the beam expands too much as it comes out of the objective, producing a large, fuzzy dot.

Slits

A *slit* is a pinhole that has been enlarged to make a long, narrow rectangle. It is designed to produce diffraction in a laser beam. The *width* of the slit is the important consideration, not the length. Slits can be made by precision tooling, with photographic film, and with sharp, unused razor blades (again, see Chapter 8, for ideas on making slits). Slits can either be single or double, depending on the application.

Diffraction Gratings

Diffraction in a spatial filter is an undesirable side-effect, but diffraction in general is a highly useful optical phenomenon. Diffraction is what makes laser holograms work and gives the colorful rainbow to compact discs and some metallic "mood" jewelry.

A *diffraction grating* diffracts light in a controlled manner. The amount of diffraction and the size of the interference fringes that the diffraction produces is determined by the number of lines or scribes made in a transparent or reflective material. Many diffraction gratings are made with up to 15,000 parallel lines scribed into the material.

A diffraction grating can be either transmissive or reflective. *Transmissive gratings* are made by etching a piece of clear film with a precision tool (a diamond or laser). You see the diffraction effects by looking through the material. *Reflective gratings* are made with an opaque metallic sheet. You view the diffraction effects by reflecting light off of it.

Diffraction gratings can be used to split a laser beam into many smaller beams or to experiment with diffraction, light wavelength, metrology (the science of measurement), and much more.

Ronchi Rulings

A *Ronchi ruling* is a coarse diffraction grating but is made with more precision. Instead of 15,000 lines per inch, a Ronchi ruling has from 50 to 400 lines per inch. The lines are etched or scribed in glass and are held to a much tighter tolerance than the lines in a diffraction grating.

You can use a Ronchi ruling for many of the same tasks as a diffraction grating, although the principle use of a Ronchi is to test the flatness of optical components. Holding a Ronchi ruling up to a laser beam causes diffraction, and because the lines in the ruling are precise, you can use the diffraction pattern as an aid in measuring distances, light wavelengths, and more.

Miscellaneous Optics

There are more optical components than the few mentioned above, such as polarizers, quarter-wave plates, and cylindrical lenses. These and others of interest to the laser experimenter are introduced throughout the rest of this book. If you would like to learn more about optics and how to use them, consult Appendix B for a list of books on optical components. Also check your local library for additional titles on the subject.

BUYING OPTICS

Most of the lenses and mirrors you are likely to buy will be surplus, either from equipment taken out of service or no longer in manufacture, or from components that are defective in one way or another. Serious laser endeavors need the best quality optics you can afford, but routine experiments can effectively use second-and third-grade lenses and mirrors. You'll especially want to confine yourself to low-grade optics when just starting out. There's nothing worse than spending $20 on a high-quality lens only to have it ruined because you didn't know how to care for it.

Never-before-used prime surplus optics are the best choice. Why? They are the highest quality, yet they cost less because they are no longer needed by the original manufacturer or purchaser. But such surplus is hard to find, and depending on the source, is not much cheaper than buying components brand new, straight from the factory. Contact a number of optics manufacturers (see Appendix A) and ask for their latest catalog and price list. Use the price list as a "blue book" to know when you are getting a good deal.

Lenses

Optics that are damaged in some way are often referred to as "seconds." The amount of damage can be slight or severe, but in any case, the fault was major enough that the manufacturer or user rejected the component. Lenses suffer from three major faults:

★ *Scratches*. Most scratches are hairline marks that you can't ordinarily see without using a microscope or magnifying glass. The scratch is deep enough that it can't be rubbed out, so it's a permanent scar. If the scratch is large (over about 50 micrometers, visible with a good magnifier), and extends through the optical center, the lens can't be used for laser experiments. Toss the lens or use it for some other less demanding application.

★ *Digs*. Digs are gouges in the lens surfaces. These are usually large enough to see with the unaided eye. Obviously, a dig in the usable portion of the lens makes it worthless. Some digs are small, however, and might not be easily seen. Their presence might only be revealed when shining a laser beam through the lens.

★ *Edge chips*. Most lenses are designed to use the inner 60 to 80 percent of the surface, leaving the outside edge for mounting with retainer rings. Chips at the edges generally don't pose problems. But a large chip can extend into the usable area of the lens and adversely affect the light. Look carefully for edge chips, and reject those lenses where the chip extends beyond 10 percent of the radius.

If you are buying lenses from a surplus dealer, make it a habit to test for the presence of anti-reflection (AR) coating. Although you can use uncoated optics, the coated variety is always better. To test for the coating, hold the lens so that white light glances off the surface. A lens with an AR coating will be tinted.

Mirrors

A scratch on a mirror can smear a laser beam and distort it beyond recognition. Closely inspect the reflective side of the mirror and look for scratches, digs, and chips. Small imperfections around the periphery of the mirror can be tolerated, as long as the center of the mirror is free of blemishes.

Even though an aluminum-or silver-coated mirror might seem 100 percent reflective to you, the metallic coating could be uneven and spotty. Being careful to grasp it by the edges only, hold the mirror up to a light and peer through it. Can you see any small pinholes or other imperfections in the coating? If so, pick another mirror. Even 100 percent reflective mirrors might have a thin enough coating that you can see through it. This is fine as long as the coating appears even.

CARE, CLEANING, AND STORAGE OF OPTICS

The bane of laser optics is dust, dirt, and grime, a trio that can quickly ruin even the best engineered project. Lenses, mirrors, beam splitters, filters, and other optical components must be absolutely free of dust, or the laser beam could be undesirably diffused and diffracted. Obviously, it's impossible to prevent all contamination without a sophisticated "clean-room," but every step you take to control dust goes a long way to assure success in your experiments.

Keep your optics wrapped in lint-free tissue (not facial tissue) until you use them. The tissue paper for wrapping gifts inside boxes is a good choice. Cut the tissue into small, manageable pieces and carefully wrap each lens, mirror, or other component. Use cellophane tape to keep the tissue closed. Don't write directly on the tissue or tape; you'll damage the optics. If you must mark the component, write the description on white paper tape and affix the tape to the outside of the tissue.

Store like items in closeable plastic sandwich bags. Mark the contents on the bag with a label. The seal on the bag prevents contamination from moisture. Leave a little bit of air in the bag as you seal it to provide a soft protective cushion.

Avoid touching optical components unless you have to, and then handle the components by the edges only. Oil from your skin acts as an acid that etches away glass and anti-reflection coatings. Use optics-quality cleaning tissue (or a cotton swab) to remove oil and grime from lenses and mirrors. Never use the tissue dry, as the slightly abrasive surface of the tissue can cause micro-scratches on the surface of the components.

Use an approved lens-cleaning fluid or pure alcohol, but not eyeglass cleaner containing silicone. Apply the cleaner to the tissue, not directly to the optical component. Use only enough cleaner fluid to lightly wet the tissue. Let the component dry on its own; avoid wiping a lens or mirror dry or you might scratch it. Clean plastic lenses with distilled water.

Clean optics with a tissue and cleaning fluid or water sparingly. Use a bulb brush (available at photographic stores) to remove dust from optics each time you use them. A can of compressed air can also be used to blow away light dust particles. Be sure to hold the can upright so that the propellant is not expelled. If the valve has a pressure control, dial it to the lowest setting.

If the optical component is fragile and easily marred, clean it by dipping it in alcohol or optics grade cleaner. If this isn't practical, you may also use an eye-dropper to splash cleaner on the component. Allow to dry or blow away the excess moisture with a can of compressed air.

The best way to prevent dust contamination is to use the optics in the cleanest environment possible. Dusty garages are not the place for laser optics experiments. Choose a place inside that is thoroughly dusted and vacuumed. If possible, aim a small fan *into* the area to create a high-pressure region. Be sure that you don't stir up dust with the fan. Place an air filter in front of the fan to block dust from entering the room.

Laser projects you construct using an optical bench or prototyping board can be protected overnight by taping a large sheet of tissue over the optics. Avoid the use of towels or blankets, as they leave lint on lenses and mirrors. Apply the bulb brush to the optics before continuing the project the next day.

Some laser projects, like the Michelson interferometer described in Chapter 9, are designed to be built with optics left intact. But the optics are delicate and can be ruined if exposed. Find a cardboard box large enough for the project and spray the inside of the box with a clear lacquer. This mats the inside of the box and helps prevent cardboard dust from contaminating the optics. Next, add fiber fill or other bonded interfacing or padding on the bottom and sides of the box and carefully place the project inside. Tape up the top and mark ''Fragile'' on it along with the contents so you know what's inside. Store the box so that nothing heavy will be placed on top of it.

4

Experimenting with Light and Optics

Laser radiation is light, and with few exceptions, all light behaves the same way. One of the best ways to learn about lasers and optics—without having to bother with the cost of laser equipment nor all the sundry safety precautions—is to construct a high-intensity simulated laser. The simulated laser described in this chapter emits a powerful beam of visible light that can be focused, directed, and controlled in much the same way as real laser light.

The parts for the simulated laser, or "simu-laser," are affordable and easy to find, and because you are not dealing with real laser light, there is no worry of accidentally exposing your eyes to potentially harmful radiation. The light from the simu-laser is bright, but it is no more damaging to your eyes than a momentary glint of sunlight from a car mirror.

HIGH-OUTPUT LEDS

The simu-laser is designed around a high-intensity visible light-emitting diode. These exceptionally bright LEDs are enormously efficient, emitting several hundred—and sometimes several thousand—times the light of ordinary LEDs. Visible LEDs are rated by their millicandela output. One millicandela is equal to one thousandth of a standard candle, a common measurement of light intensity. Typical undiffused red LEDs produce from 5 to 20 millicandelas (mcd); diffused LEDs emit even less, often under 2 mcd.

High-output LEDs, usually referred to as "Super Brights" or "Kilo Brights," put out 300 mcd or more. Such LEDs are routinely available at Radio Shack and other hobbyist electronics outlets. You can use a 300 mcd LED for the simulated laser, but you can achieve better results with a component rated at a higher value. A number of companies, including Stanley and Texas Instruments, offer LEDs with light outputs of 1,000, 2,000, even 3,000 mcd. Higher outputs can be obtained—up to 5,000 or 6,000 mcd—when driving the LED with more current.

The simulated laser project detailed in this chapter uses the Stanley Electric Co. H2, which is a 2,000 mcd LED imported by A.C. Interface. This LED, or one like it, is available from a number of industrial electronics outlets. Look in the Yellow Pages under Electronics and call around. Some outlets don't sell on a retail level, but you might have luck striking a deal if you are a part of a group or school.

A number of mail order outfits, such as Allied Electronics and General Science and Engineering, also carry high-output LEDs and are accustomed to dealing with individuals. See Appendix A for a list of mail order companies that offer service to electronics hobbyists. The remaining parts used in the project are commonly available from new and surplus dealers.

The output of LEDs increase linearly as you increase current. The H2 is rated at 2,000 mcd when powered with a current of 20 mA. The output is roughly doubled—4,000 mcd—when the LED is biased with a current of 40 mA.

Although some high-output LEDs can function with currents as high as 50 mA, I found that this much juice heated up and burned out several of the sample H2 LEDs I had. To be on the safe side, don't operate any high-output LED above 45 mA unless the manufacturer's literature says otherwise. As you may expect, high-output LEDs are more expensive than their standard cousins. Typical prices are $2 to $4 for a 2,000 mcd LED. At these prices, you don't want to carelessly burn too many out.

Measurement of light output of LEDs, including the high-output variety, is often made with an *integrating sphere*, a device that measures the total radiant power of the component. The sphere collects light emitted in all directions, including the off-axis radiation that spills out the sides. The simulated laser project detailed here focuses only the light that comes out the top of the LED; no reflector is used to collect the off-axis light. You can increase the brightness of the device by adding a penlight flashlight reflector behind the LED.

BUILDING THE SIMULATED LASER

Follow the schematic in FIG. 4-1 for wiring the high-output LED. The parts list appears in TABLE 4-1. The circuit is simple and runs off a single 9-volt transistor battery. The switch is a miniature single-pole, single-throw (SPST) type. The lens I used for

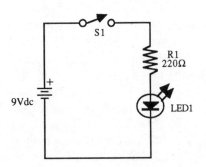

FIG. 4-1. *The basic schematic for the simu-laser. With the supply voltage and resistor shown, forward current through the LED is approximately 34 mA. You can safely increase the forward current to about 45 mA by reducing R1 to 165 ohms (150-ohm and 15-ohm resistors in series).*

Table 4-1. Simu-laser Parts List

R1	220 ohm resistor
LED1	High-output light-emitting diode
S1	SPST miniature switch
1 each	9-volt battery, battery clip, project box (3¼ by 2⅛ by 1 ⅛ inches), 16 mm diameter, 34 mm focal length, positive double-convex lens.

All resistors are 5 to 10 percent tolerance, ¼ watt

the prototype was double-convex, 16 mm in diameter by 34 mm in focal length. You can use a lens with different specifications, but you should avoid a lens with a focal length greater than about 35 mm. A longer focal length means that you must provide more space between the lens and the LED. The project box specified in the parts list is just long enough to accommodate the focal distance between LED and lens.

Mount the components as shown in a small project box. Everything fits in a compact box measuring 3¼ by 2⅛ by 1⅛ inches (Radio Shack catalog number 270-230). Drill a ½-inch hole in one end of the box, and countersink it lightly to conform to the rounded shape of the lens. Use a general-purpose glue (such as Duco cement) to carefully tack the edges of the lens to the outside wall of the box. Use a toothpick to apply small drops of glue to the edge of the lens.

Mount the LED and current-limiting resistor on a small piece of perf board and attach a metal or plastic angle bracket to the back of the board. Suitable brackets are available at hobby stores that specialize in radio-control model airplanes. Cost is under $1 for a set of four or five brackets. You can also salvage the bracket from an Erector Set toy kit.

Measure the diameter of the switch shaft and drill a hole accordingly in the top of the box. Wire the snap-on battery cap, switch, and circuit board and turn the switch on. The LED should glow a bright red (some high-output LEDs emit red-orange light). If all checks out, point the box toward a light-colored wall (no more than a couple feet away), and place the circuit board inside. Aim the LED at the lens and adjust its position and distance until the spot on the wall is bright and well defined. Mark the location on the side of the box with a pencil, then check for proper focal length by measuring the LED-to-lens distance. It should be very close to the rated focal length of the lens.

With the proper mounting position marked off, use a #19 bit to drill a hole in the side of the box and mount the circuit board using ⁸⁄₃₂-by-¼-inch hardware (available from a hardware store or Erector Set). The simu-laser, mounted in its box with all components, is shown in FIG. 4-2.

Place the cover on the box and flick on the power switch. Point the simu-laser at a wall and watch for a bright red spot. Note that the beam spreads out considerably when the wall is more than 4 or 5 feet away. This is due to the wide divergence of the beam, even when optics are used. Later in this chapter you'll see how to add more lenses to control the size of the beam, or even focus it to a bright, pin-point spot.

BEAM COLLIMATION

Depending on the exact distance between the lens and wall, the spot from the simu-laser beam should have a bright center, with concentric light/dark rings around it. This

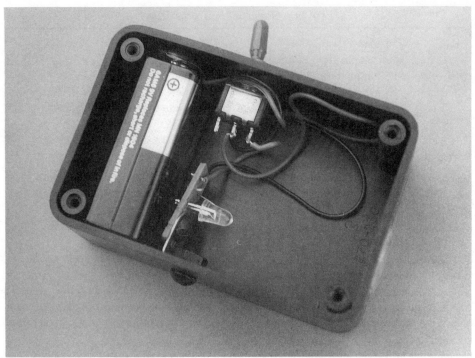

FIG. 4-2. *The simu-laser, with cover removed to show the arrangement of the components.*

"ring" effect is caused by the internal construction of the LED. Place the simu-laser a few inches from the wall and you can see the chip inside the LED, along with a dark spot and line that denotes the contact wire stretching from anode to cathode. You will also see shadows from the glue placed at the edge of the lens.

In a dark room, place the simu-laser on a table and aim the beam at a solar cell. Make sure that all of the spot falls on the surface of the cell. Connect a meter to the cell as shown in FIG. 4-3 and measure the output. Now change the distance between the simu-laser and the cell.

Note that as you increase the distance by a factor of two, the reading on the meter falls off roughly by 50 percent. This is due to the inverse-square law, familiar to photographers. The inverse-square law states that for every doubling in the distance between a light source and subject, the intensity of the light falling on the subject decreases by 50 percent.

What causes this phenomenon? At first glance, it might appear that the light loses its energy the farther it travels. But this can't be the answer, because energy can never be lost, just transformed into something else. The reason behind the apparent weakening in the light intensity is due to the spread—or divergence—of light. Physical law dictates that as light propagates through a vacuum or any medium, the light waves must spread transversely, making it impossible to have a perfectly collimated (parallel) beam.

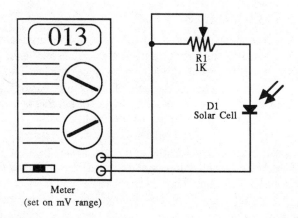

FIG. 4-3. *How to detect and measure light level using a silicon solar cell. Potentiometer R1 allows you to adjust the sensitivity of the cell.*

Light from the simu-laser, though focused by a lens, spreads out as it travels through air or space. The amount of spreading is measured using the metric radians or milliradians (thousandths of a radian). One milliradian is equal to:

★ 0.057296 degrees
★ 3.438 minutes
★ 206.265 seconds of arc

The simu-laser uses a simple lens to focus the light from the LED, so the divergence is wide, on the order of 500 to 750 milliradians. A more elaborate lens system with two or more lenses can decrease the divergence so the light is held in a tighter beam of perhaps 250 milliradians. However, no matter what kind of optics are used, the light beam eventually suffuses to a point where its individual photons are miles apart. Note that the divergence of a real laser is extremely small, on the order of 1 to 5 milliradians, even without optics. This is just one of the properties of lasers that make them so special.

In later experiments in this book, you will find that the inverse-square law does not apply to most lasers when used at close distances. The light beam from a helium-neon gas laser, for example, has so little divergence that its diameter increases an imperceptible amount at close ranges. Only after traversing several hundred meters does the diameter of the beam enlarge to appreciable amounts.

Table 4-2. Parts List for Light Intensity Measurements

R1 1K potentiometer
D1 Silicon solar cell
    Volt ohm meter

OPTICAL EXPERIMENTS

The double-convex lens mounted on the case of the simu-laser does not do an adequate job of collimating the light in a fine beam. A second lens, positioned in front of the simu-laser, can act to make the light rays more parallel. The exact distance between the two lenses is a function of the type of lens you are using and the focal lengths of both lenses. But you can experiment with the proper spacing by adjusting the distance while looking at the spot projected on a nearby wall.

The spot should be well-defined, focused, and smaller than the spot made by just the simu-laser lens. There should be little or no light spilling past the sides of the second lens (light spill is normal in an optical device and is usually countered by the use of apertures or stops). If you can't focus the beam on the wall without a lot of light spilling past it, choose a larger diameter lens. The second lens used in the prototype was a 16 mm diameter, 124 mm focal length positive meniscus placed 5⅜ inches (about 137 mm) in front of the simu-laser. It reduced the beam divergence by about a factor of two.

You can temporarily fix the simu-laser and lens in position with modeller's clay. Use a small piece of wood or plastic (about 12 by 4 inches) and build up the clay to the proper height. Now repeat the test with the solar cell. Note that the simu-laser can be moved farther away without a drastic loss in power. Take another reading without the second lens and compare it to the measurements you just made.

Focusing Light to a Point

Light from the simu-laser can be focused to a sharp point by using two additional supplementary lenses. Remove the meniscus lens used for the previous experiment and mount a 17 mm diameter, 70 mm focal length, double-concave lens approximately two inches from the front of the simu-laser. Use clay to keep it in place. Position a 30 mm diameter, 55 mm focal length, plano-convex lens 6½ inches from the double-concave lens. Adjust the position of the lenses so that their optical centers are aligned and that the lens faces are parallel to one another.

Place a sheet of white paper approximately 2¾ inches in front of the plano-convex lens. What happens? The light is focused to a point a little less than ⅛-inch diameter. Move the paper closer to or farther away from the lens and note that the spot gets larger. A side view of the focused beam is shown in FIG. 4-4. The beam is at its smallest when the paper is located at the focal point.

Table 4-3. Parts List for Simu-laser Optics Experiments

1	Simu-laser
1	16 mm diameter, 124 mm focal length, positive meniscus lens
1	17 mm diameter, 70 mm focal length double-concave lens
1	30 mm diameter, 55 mm focal length plano-convex lens
1	2-by-2-inch front-surface mirror
1	Plate glass beam splitter (approx. 1-by-1 inch); ⅟₁₆- to ¼-inch thick
1	Polarizing film
1	Right-angle prism

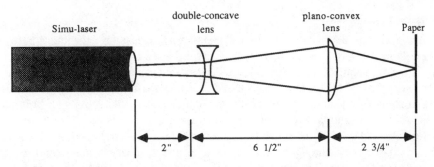

FIG. 4-4. *Arrangement of simu-laser and optics for demonstrating the focusing of light into a point. Experiment with the distances between components if your lenses are different from those specified in the text.*

The double-concave lens acts to diverge the light to approximately the diameter of the plano-convex lens. This beam spreading allows most or all of the surface of the plano-convex lens to be used (and hence the beam is more subject to certain aberrations). You can test the effectiveness of the double-concave lens by removing it from the light path. The beam, still focused in front of the plano-convex lens, is enlarged to about 1/4 inch. Experiment with the effects of different lenses placed at various spots along the length of the test board.

Using Mirrors

Mirrors allow you to direct the beam from the simu-laser in any direction you desire. Note that a flat mirror doesn't alter divergence or convergence of a beam. A converging beam of light still converges after bouncing off a mirror. Concave or convex mirrors, however, act as reflective lenses and spread or focus the light. Concave mirrors are often used in holography to provide a heavily diverged beam in order to cover a large area. The mirror is a more effective diverger than a lens.

Place a 2-by-2-inch, front-surface, aluminized mirror in the path of the light beam, and position it at a 45-degree angle. The light should now bounce off the mirror and be perpendicular to the beam of the simu-laser. Move the mirror closer to the focusing lens and position the paper to catch the light reflected off the mirror, as shown in FIG. 4-5. Note that the focal length between the focusing lens and paper is the same whether or not the mirror is in place. If the focal length is approximately 2¾ inches, the same distance will be covered between lens, mirror, and paper.

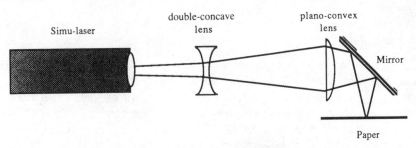

FIG. 4-5. *A flat mirror does not alter the convergence or divergence of light rays.*

Using Beam Splitters

Now exchange the mirror with a small piece of plate glass (any thickness between 1/16- and 1/4-inch will do). When positioned at a 45-degree angle to the simu-laser, the glass acts as a beam splitter. With uncoated glass, approximately 10 to 15 percent of the light is reflected off the first surface and the rest is transmitted through the glass (there is some light loss due to internal reflections). You should see a spot in front of the simu-laser and another, less strong one to the right or left.

Rotate the glass. The transmitted beam remains but the reflected beam will spin in an arc around the beam splitter. Use the solar cell to measure the amount of light reflected from the glass. The amount of reflection should change as the glass is turned. The amount of reflectance is minimal when the glass is positioned between 0 and 30 degrees to the simu-laser, but it begins to increase dramatically at angles greater than 40 degrees. At 90 degrees (glancing angle), reflection is almost 100 percent. Be sure to take your readings in a darkened room.

Light reflecting off a glass medium is partially polarized, and if the angle is just right, one plane of polarization will be almost extinguished. Try this experiment. Place a piece of polarizing film in front of the simu-laser. Position the glass plate at about 55 degrees to the lens of the simu-laser. Locate the solar cell so that it collects the reflected light. Now slowly rotate the polarizing film. See any change? You should. If you don't, change the angle of the glass a small amount and try again. The light should be strongest with the polarizer in one position and weakest when it is rotated 90 degrees.

Future chapters deal with the nature of polarized light in more detail, but it's worth describing here how this phenomenon works. Polarized light has two components, typically identified as p and s. These two components are *orthogonal* to one another, or at 90-degree angles.

When a piece of glass is positioned at a special angle called *Brewster's angle*, the p component vanishes, and only the s component remains. Brewster's angle for most glass is 56 degrees 39 minutes, or roughly 1 radian. By placing a polarizer in the light path, you can control polarization of the light. Rotate it so that the s component passes and you see light reflected off the glass (the p component is "absorbed" by the glass, so only the s component remains). When you rotate the polarizer so that the s component is blocked, the light dims. With both the s and p components reduced or eliminated, the light level drops.

You can visually see the dimming effect as you rotate the polarizer. To assure yourself that you are not just seeing things, use the solar cell to make a graph of the light output. A plastic protractor can be used to help you rotate the polarizer in even 5- or 10-degree intervals.

Using Prisms

Most people think of prisms as rainbow makers—glass objects that break up sunlight into its component colors. While prisms are indeed often used in this role of light dispersion, they also find great application as re-directors of light. Prisms have an advantage over mirrors, in that because they are made of thick glass, they are less susceptible to the effects of stress and vibration. The reflective surface of a prism is typically inside the glass, making it far less susceptible to scratches and blemishes. Prisms can also be designed with unique shapes so that the light passing through it is reflected without flipping it upside down or sideways, as in a regular mirror.

A right-angle prism is used to deflect light at a 90-degree angle. The corner opposite the hypotenuse forms a right angle. Usually, but not always, the sides opposite the hypotenuse are the same length. That leaves a 30- and 60-degree angle for the other two corners. Shine a light into the prism through one face, and it is internally reflected at the hypotenuse and then exits the second face.

You can readily experiment with a right-angle prism by placing it in front of the simu-laser. You'll need a prism that is at least 25 mm wide with faces 25 mm to 30 mm long. Glass and plastic prisms of this size (and larger) are available on the new and surplus markets for under $10. Optical quality for these less expensive prisms might not be high, but prime optics are not required for these experiments.

Position the prism so that the light enters one face and is bounced off the hypotenuse at a 90-degree angle. If the room lights are low, you might be able see the beam in the prism reflecting off of the hypotenuse and exiting the glass.

5

All About
Helium-Neon Lasers

The helium-neon tube is the staple of the laser experimenter. He-Ne tubes are in plentiful supply, including in the surplus market. They emit a bright, deep red glow that can be seen for miles around. Although the power output of He-Ne tubes is relatively small compared to other laser systems—such as CO_2, argon, and ruby—the helium-neon laser is perfectly suited for most any laser experiment. Its moderate power supply requirements coupled with its slim, coherent beam lend the He-Ne laser to inexpensive projects in holography, interferometry, surveying, lightwave communications, and much more.

In this chapter you'll learn all about helium-neon lasers: what they are, how they work, and what you need to put a complete system together. In Chapter 6 you'll learn how to place a bare He-Ne laser tube in an enclosure to make it easier to use, along with plans on building a He-Ne laser experimenter's system. With the experimenter's system, you'll be able to perform numerous optical experiments with your helium-neon laser.

ANATOMY OF A HE-NE LASER TUBE

The helium-neon laser is a glass vessel filled with 10 parts helium with 1 part neon and is pressurized to about 1 mm/Hg (exact gas pressure and ratios vary from one laser manufacturer to another). Electrodes placed at the ends of the tube provide a means to electrify (ionize) the gas, thereby exciting the helium and neon atoms. Mirrors mounted at either end form an *optical resonator*. In most He-Ne tubes, one mirror is totally reflective and the other is partially reflective. The partially reflective mirror is the output of the tube.

The first helium-neon tubes were large and ungainly and required external cooling by water or forced air. The modern He-Ne tubes, such as the one in FIG. 5-1, are about the size of a cucumber and are cooled by the surrounding air. The length and diameter of He-Ne tubes varies with their power output, as detailed later in this chapter.

Most helium-neon laser tubes are composed of few parts, all fused together during manufacture. Only the very old He-Ne tubes, or those used for special laboratory experiments, use external mirrors. The all-in-one design of the typical He-Ne tube means they cost less to manufacture and the mirrors are not as prone to misalignment.

Helium-neon lasers are actually composed of two tubes: an outer *vacuum* (or plasma) tube that contains the gas, and a shorter and smaller inner *bore* or capillary, where the lasing action takes place. The bore is attached to only one end of the tube. The loose end is the output and faces the partially reflective mirror. The bore is held concentric by a metal element called the *spider*. The inner diameter of the bore largely determines the diameter width of the beam, which is usually 0.6 mm to 1 mm.

The ends, where the mirrors are mounted, typically serve as the *anode (positive)* and *cathode (negative)* terminals. On other lasers, the terminals are mounted on the same end of the tube. A strip of metal or wire extends the cathode (sometimes the anode) to the other end.

Metal rings with hex screws are often placed on the mirror mounts as a means to tweak the alignment of the mirrors. Unless you suspect the mirrors are out of alignment, you should *not* attempt to adjust the rings. They have been adjusted at the factory for maximum beam output, and tweaking them unnecessarily can seriously degrade the performance of the laser.

The partially silvered mirror, where the laser beam comes out, can be on either the anode or cathode end. I found that on the many tubes I've tested, the beam extends out the cathode end. Many manufacturers prefer this arrangement, claiming it is safer and provides more flexibility. You can usually tell the output mirror by holding it against a light. You should see the blue tint of the anti-reflective coating. The totally reflective

FIG. 5-1. *A bare helium-neon laser tube. This one measures about 1¼ inches in diameter by 12 inches in length.*

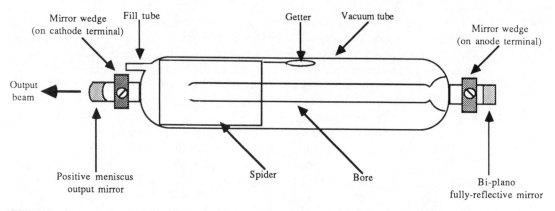

FIG. 5-2. *An x-ray view of a typical He-Ne laser tube, with component parts indicated. The arrangement and style of your laser tube might be slightly different.*

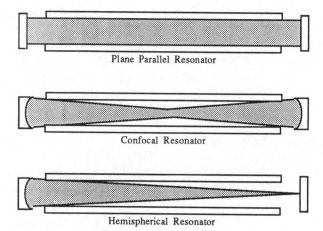

Plane Parallel Resonator

Confocal Resonator

Hemispherical Resonator

FIG. 5-3. *Three ways to implement the optical cavity in a laser—plane parallel, confocal, and hemispherical. Most laser tubes use the latter method or a derivative of it.*

mirror generally is not treated with an AR coating. A cutaway view of a laser tube with all the various components is shown in FIG. 5-2.

The facing mirrors of the He-Ne tube comprise what is commonly referred to as a *Fabry-Perot interferometer* or *resonator*. With the mirrors aligned *plane parallel* to one another, as shown in FIG. 5-3, the light bounces back and forth until the beam achieves sufficient power to pass through the partially reflected output mirror. In practice, the plane parallel resonator is seldom if ever used, because it is unstable and suffers from large losses due to diffraction.

A *confocal resonator* uses two concave spherical mirrors of equal radius, each placed at the center of curvature of the other. The cavity uses a large portion of the gas volume and produces high power, but mirror adjustment is relatively critical. Yet another approach is the *hemispherical resonator*, which is primarily a plane mirror coupled with a spherical mirror. This type of resonator is very stable and easy to align, but its design wastes plasma volume, so power output of the laser is reduced.

The exact configuration of the resonator mirrors in a laser is not a major consideration to the hobbyist experimenter. You'll have little choice of the engineering of the tube when buying parts through surplus outlets. However, you might have a choice if you buy your tubes new or if you purchase a particular type of tube for a special application.

LASER VARIETIES

He-Ne lasers are available in three forms: bare, cylindrical head, or self-contained. *Bare tubes* are just that—the plasma tube is not shielded by any type of housing and should be placed inside a tube or box for protection. *Cylindrical head* lasers (or just "laser heads") are housed inside an aluminum tube. Leads for power come out one end of the laser. The opposite end might have a hole for the exiting beam or be equipped with a safety shutter. The shutter prevents accidental exposure to the beam. Both the bare tube and cylindrical head laser require an external high-voltage power supply (discussed below).

Lastly, *self-contained* (or lab) lasers contain both a laser tube and a high-voltage power supply. To use the laser, you simply plug it into a wall socket and turn on the switch.

Each form of laser has its own advantages and disadvantages:

★ Bare tubes are ideal for making your own self-contained laser projects, such as laser pistols and rifle scopes, and fit in confined spaces. But because the tube and high-voltage terminals are exposed, they are more dangerous to work with. You must exercise considerable caution when working with bare tubes to avoid electrocution and injury from broken glass.

★ Cylindrical head lasers are easy to use because the tube is protected and the high-voltage terminals are not exposed (but care must still be exercised to avoid shock from power applied to the leads). Laser heads are ideal for optical benches and holography (with the right type of mount). On the down side, the tubes tend to be large and are not easily mounted for use in hand-held devices.

★ Self-contained lasers are designed to provide protection against tube breakage and high-voltage electrocution. They are often used in schools and labs where they can be easily set up for optical experiments. The built-in power supply operates from a wall socket, however, so the laser cannot be used where 117 volt ac current is not available. Also, the bulk of the self-contained laser prohibits it from being used in hand-held devices.

THE POWER SUPPLY

Owning a plasma tube doesn't mean you have a laser. The tube is only half the story; just as important is the power supply. You also need electricity for the power supply, either directly from a 117-volt ac wall socket or a 12-volt dc battery. Power supplies for lasers generate a great deal of volts but relatively few milliamps. The typical power supply generates from 1,200 to 3,000 volts at 3.5 to 7 mA. Generally, the larger the tube (and the higher the power output), the more juice it requires.

You have a number of choices for the power supply:

★ *Commercially-made* laser supply, either ac or dc operated. These are the easiest to use and often come completely sealed as a precautionary measure. A common type

FIG. 5-4. *The latest commercially made He-Ne power supplies are available in convenient, sealed packages, such as the one shown in the photograph. The power supply is about the size of a sandwich.*

of ac-operated commercially made He-Ne power supply is shown in FIG. 5-4. Cost on the retail market is about $225; surplus is between $50 and $100. A high-voltage Alden connector (see end of chapter) is common on cylindrical head lasers. The female connector on the power supply matches with the male connector on the laser.

★ *Home-built* power supply kit. Low-current power supplies can be built in your shop but require some specialized, hard-to-find parts. Building your own supply saves money and helps teach you about laser power requirements. Cost: $10 to $25 for parts; pre-packaged kits of parts and circuit board (available from some sources listed in Appendix A) cost $30 to $150.

★ *Salvaged high-voltage TV flyback transformer or module* power supply. Many compact TVs use a self-contained flyback transformer that operates from 12 to 24 Vdc. The transformer, intended as the high-voltage source for the picture tube, generates up to 15 kV at a few milliamps. The modules are cheap ($10 to $20 on the surplus market) but their low current output makes them suitable only for small tubes.

If you are just starting out with lasers, your first power supply should be a commercially made and tested unit, preferably new and not a take-out from existing equipment. Armed with a tube and ready-to-go supply, you'll be able to start experimenting with the laser the moment you get it home. Then as you gain experience,

58

you might want to build one or two supplies as extras or to power the various tubes you are bound to acquire. Chapter 11 presents several power supply circuits you can use to power tubes with outputs from 1 to about 5 mW.

Inside a Power Supply

Although the exact design of He-Ne power supplies varies, the principle is generally the same. He-Ne power supplies are familiar to anyone who has dabbled in high-voltage circuits, such as those required for amateur radio.

Here's how a typical power supply works. An input voltage, say 12 Vdc, is applied to an oscillator circuit, such as a simple transistor, resistor, and capacitor. The values of the resistor and capacitor determine the time constant of the oscillator circuit. The oscillating input is fed to the windings of a step-up transformer. The voltage is stepped up to somewhere between 300 to 1,000 volts by the transformer and then passed through a series of high-voltage diodes and capacitors. These components act to rectify and multiply the voltage presented by the transformer. Depending on the number of capacitors and diodes used, the voltage multiplier doubles, triples, or quadruples the potential.

Power supplies that operate from 117 Vac usually don't need the front-end oscillator circuit since the juice is already in the alternating current format required by the transformer (most transformers can't pass dc current). The transformer steps up the voltage to anywhere between 300 and about 2,000 volts. If the output voltage is high enough, voltage multiplication by means of diodes and capacitors is not required, but the ac component is removed by one or more diodes.

Some of the more advanced power supply circuits use a separate 6 to 10 kV *trigger transformer* to initially start or "ignite" the laser. The trigger transformer, similar to the kind used in photoflash equipment, fires only when the tube first turns on (but might continue firing if the tube doesn't ignite). A circuit in the power supply senses when the laser starts to draw current (indicating that it has started), and shuts the trigger transformer off. A silicon-controlled rectifier (SCR) serves as the switch to turn the trigger transformer on and off.

High-voltage triggering is not required for all tubes, especially those under 2 to 3 mW, but it is usually needed with higher output types. If your tube is hard to start—either refusing to ignite at all or just flickering—you may need to use a supply that has a high voltage trigger or at least one that supplies extra current. Most all commercially-made supplies have a built-in trigger transformer (or something equivalent) but home-brew supply circuits generally do not.

Using a high-powered laser with a power supply that can't deliver the required current may actually damage the supply. TABLE 5-1 lists the average voltage and current requirements for a variety of tubes. The tubes are rated by their output only so the chart should be used only for estimating power requirements. Many other factors, such as polarization of the beam and operating mode, can affect the power output and change the voltage and current requirements. Obtain descriptive literature from the seller of the laser tube if you need more precise information.

Note that most He-Ne tubes can be safely operated over a range of currents. Manufacturer's specifications usually list the recommended operating current for optimum performance. You can often safely increase or decrease the current slightly, for example,

Table 5-1. Voltage/Current Levels for Typical He-Ne Tubes

Power Output	Dimensions	Voltage w/Ballast	Tube Voltage	Typical Current
0.5 mW	5.00/1.00	1250	900	3.5 mA
0.5 mW	6.00/1.12	1390	1050	4.5 mA
1.0 mW	8.90/1.12	1890	1400	5.0 mA
2.0 mW	8.90/1.45	1890	1400	5.6 mA
2.0 mW	10.60/1.45	1990	1500	6.5 mA
5.0 mW	13.80/1.45	2390	1900	6.5 mA
7.0 mW	16.15/1.45	2930	2400	7.0 mA

Notes:
- Dimensions are in inches, length by diameter.
- Tube voltage is without ballast resistor.
- Current rating is recommended maximum; many tubes will fire and lase at currents 20 to 30 percent less.
- Note higher current and voltage requirements for larger tubes in same power output class.

to produce a more powerful beam or to conserve battery power. Unless the specifications state otherwise, you should not exceed 7-8 mA operating current.

A Warning About High Voltage Power Supplies

You've undoubtedly read this before in this book, but the warning can't be stressed enough: *beware of high voltage power supplies*. Never touch the output terminals of the supply when it is on or you may receive a bad shock. Turn the supply off and unplug it before working with the laser.

The capacitors in the voltage multiplier section of the supply can retain current even when the system is turned off. Before touching the tube or power supply, *short the anode and cathode terminals together*. If you can't easily bring them together, keep a heavy-duty alligator-clip test cable handy. String it across the anode and cathode and short the leads.

The laser tube itself can also retain current after power has been removed. *Always short out the anode and cathode terminals of the laser before handling the tube*. Cylindrical head lasers with a male Alden connector can be discharged by touching the prongs of the connector against the metal body of the laser.

POWER OUTPUT

The greatest difference among helium-neon laser tubes is power output. There are some He-Ne tubes designed to put out as little as 0.5 mW of power, while others generate

10 mW or more of light energy. The difference in power output is not always visible to your eye because the spot made by a laser beam is brighter than your eye can register.

By far, the majority of helium-neon tubes are rated at 1-2 mW. This is adequate for most laser experiments and you rarely need more. In fact, lower power lasers are often easier to work with because the power supply requirements are not as stringent. You can get by with a smaller, lighter-weight power supply with a 1-2 mW tube. The higher power comes in handy, however, if you are engaged in holography (the more power the faster the exposure), outdoor surveying, and other applications where a bright beam is necessary.

Cylindrical head lasers, generally designed for use in telecopiers, laser facsimile machines, and supermarket bar code scanners, are generally engineered for high output. Most tubes in this class are rated at 5-8 mW, though a few—such as those made for laser facsimile devices—generate in excess of 10 mW. If you need lots of power check these out. Just be sure that the power supply delivers sufficient current and voltage to the tube.

Bear in mind that power output varies with tube age. A tube that originally produces two milliwatts when new may only generate 1.5 mW after several thousand hours of use. Careless handling also reduces the power output. Every shock or jolt may tweak the mirrors out of alignment, which reduces the power output. If the laser is abused, as it often is in industrial or commercial applications, the mirrors may become so out of whack that the tube no longer generates a beam.

The loss of output power is important to remember when buying used or surplus tubes. Even though the tube may have been rated at 2 mW, there is no guarantee that it's still providing that much power. You can readily measure the power output of a laser if you use a calibrated power meter, such as the Metrologic 45-450.

PHYSICAL SIZE

Helium-neon laser tubes come in a variety of sizes, depending on power output. Most are about 1 to 1¼ inches in diameter by 5-10 inches long. Some very small tubes are designed for use in hand-held bar code readers and generate less than 0.5 mW. Such "pee-wee" tubes are available from a few surplus sources and are fun to play with, but they are extremely fragile. They can be permanently damaged by even a moderate jolt.

Few He-Ne lasers put out more than 10 to 15 milliwatts. These tend to be the largest are usually enclosed in a cylindrical head. The tube may measure about 1.5 inches in diameter and 13 inches long; the entire enclosure is 1.75 inches in diameter by 15 inches in length. A typical aluminum-housed laser head is shown in FIG. 5-5.

Self-contained lab lasers can be most any size depending on the power output. Average size is approximately 24 by 4 by 3 inches (LWH).

BEAM CHARACTERISTICS

The beam emitted from most helium-neon tubes doesn't vary much between laser to laser. Except in special cases, the light has a wavelength of 632.8 nm, and can measure between 0.5 to 2 mm in diameter. The diameter (or *waist*) of the laser beam is measured in a variety of ways and under different operating modes of the laser; hence the wide

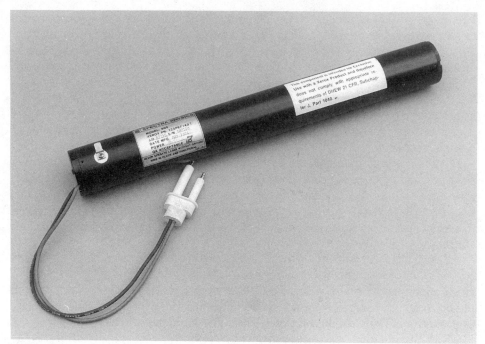

FIG. 5-5. *A typical cylindrical laser head, removed from a Xerox laser printer. Laser heads such as these are common finds in the surplus market.*

disparity in sizes. The diameter is less than the actual side-to-side measurement of the laser beam. Most manufacturers eliminate the outer 13.5 percent of the beam diameter, leaving the bright inside core.

As background, *single-mode* operation (sometimes called TEM_{00}-mode) of the laser produces a solid beam from side to side (looking from head-on). *Multi-mode* operation, which provides higher output power, causes the beam to separate into bands, as shown in FIG. 5-6. Note that most He-Ne tubes work in TEM_{00} mode only and the tube must be specially built to operate in multi-mode.

In TEM_{00} mode, the round beam of a laser has a plane wavefront and a Gaussian transverse irradiance profile. This mode experiences the minimum possible diffraction loss, has minimum divergence, and can be focused to the smallest possible spot.

The plane wavefront, as illustrated in FIG. 5-7, is a natural by-product of spatial coherence. All the waves in the beam are in lock-step, as if it were one big wave. The locking is constant across the diameter of the beam. This is in contrast to a circular wavefront, also shown in the figure, where the waves toward the outside radius of the beam are slightly behind the center waves. Note that you can easily convert the plane wavefront of laser light to circular wavefront simply by placing a lens in the path of the beam.

The Gaussian irradiance profile of a laser operating in TEM_{00} mode is shown in FIG. 5-8. The center 86 percent of the beam is the brightest; the irradiance of the beam falls off as you approach the edges of the beam. You can visually see this effect when the beam is spread using a bi-concave lens. Shine the light at a green or black card and note how the intensity of the beam is greatest at the center and less at the edges. If the

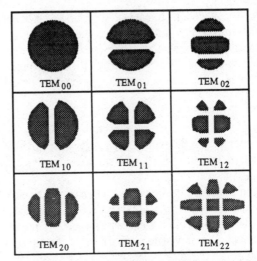

FIG. 5-6. *Only the single mode TEM_{00} provides a solid beam throughout its diameter. Multi-mode operation increases the overall power output of the laser but splits the beam into many segments. The blank areas in the beam are nulls.*

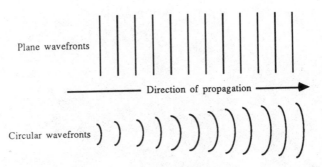

FIG. 5-7. *A comparison of plane and circular wavefronts. The plane wavefront geometry of the typical laser decreases divergence and provides coherency across the entire diameter of the beam.*

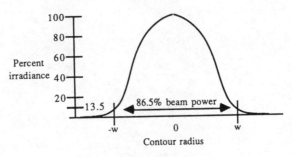

FIG. 5-8. *The Gaussian irradiance profile of an He-Ne laser (in TEM_{00} mode). The peak of the irradiance profile is the center of the beam.*

63

beam is so bright that it causes a halo even on a green or black card, wear dark green glasses or pass the beam through a deep green filter.

Even without external optics, the divergence (spreading) of the laser beam is typically not over 1.2 milliradians (mrads). That means that at 100 meters, the beam will spread 120 millimeters, or about 4.7 inches. Computing beam divergence is more thoroughly covered in Chapter 13.

He-Ne lasers exhibit *random plane polarization*. That is, the beam is plane polarized (as discussed in Chapter 8) in either one direction or 90 degrees in the other. The polarization randomly toggles back and forth, but the switch-over is extremely fast and generally not detectable. In this regard, all laser tubes are polarized, but there is no control over the plane of polarization.

Some helium-neon tubes are designed with internal polarizing optics that block one of the planes. These are called *linearly polarized* tubes. Polarization of the beam is accomplished not with filters but with a clear window placed at *Brewster's angle* at the rear of the tube. The window is visible at the rear of the tube, as shown in FIG. 5-9, and is a clear indicator that the tube is linearly polarized. In some lasers, the rear of the tube is terminated with the Brewster window but lacks a totally reflective mirror. You must then add an external mirror to operate the laser.

Linearly polarized tubes are more expensive than randomly polarized ones and are generally low in power—typically less than 2 mW (higher power polarized tubes do exist,

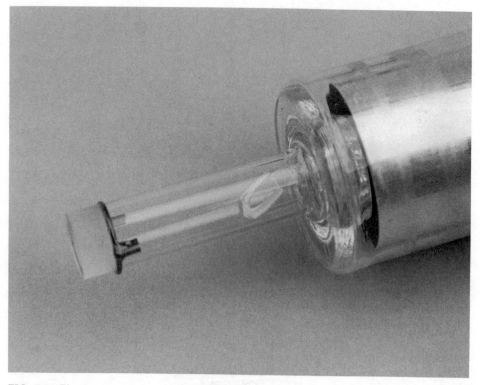

FIG. 5-9. *The tiny piece of glass positioned near the plane parallel mirror in this tube causes polarized beam output.*

but at a cost). However, polarized tubes are particularly handy in advanced holography, interferometry, high-speed modulation, and other applications where you need a stable output and a linearly polarized beam.

Of course, you can always polarize a laser by placing a polarizing filter in the path of the beam, but this significantly reduces the intensity of the beam. In most applications you need as much beam intensity as possible.

Most linearly polarized tubes are rated by their polarization purity. The purity is expressed as a ratio, such as 300:1 or 500:1. That is, under normal operating conditions, the tube is 300 or 500 times per likely to emit one plane of polarization over the other. Few applications, even precisely controlled laboratory experiments, require polarization purity greater than 500:1.

A laser beam might appear grainy or spotty when reflected off a white or lightly colored wall. This effect is called *speckle* and is caused by local interference. Even a smooth, painted surface has many hills and valleys in comparison to the wavelength size of the laser beam. These small imperfections in the surface bounce light in many directions. The uneven reflection causes constructive and destructive interference, where the light waves from the beam either reinforce or cancel one another.

Speckle is also produced in your eye as the beam strikes the surface of the retina. If you move your head, the speckle appears to move, too. Interestingly, beam speckle can be used to detect near- or short-sightedness. The speckle moves in the same direction as the head when viewed by a person with normal vision. But the speckle will appear to move in the opposite direction as the head when viewed by persons suffering from near- or far-sightedness. If you and any of your friends wear glasses, you can try this experiment for yourselves.

HE-NE COLORS

A He-Ne tube generates many different colors. You can see these colors by looking at a bare tube with a diffraction grating (an example is shown on the cover of the book, taken with a criss-cross diffraction grating). The diffraction grating disperses the orange glow of the plasma tube into its component colors. Note that just about every primary and secondary color is present, with dark lines between each one. The dark areas represent in-between colors that are not generated within the tube.

Most all colors (plus infrared wavelengths you can't see) are transmitted out of the tube before they can be amplified. That prevents them from turning into laser light. Here's how it's done: The cavity mirrors of the laser are coated with a highly reflective material that reflects the wavelength of interest (such as 632.8 nm) but transmits the others. As the beam inside the laser grows in strength, it overcomes the reflectance of the output mirror and exits the tube. You can verify the purity of the output beam by examining the red spot of the beam with the diffraction grating. You will see only one, well-defined color.

He-Ne's emit a deep red beam at 632.8 nm because it is the strongest line. The other colors are weak or may not be sufficiently coherent or monochromatic. Yet there are some special helium-neon lasers that are made to operate at different wavelengths, namely 1.523 micrometers (infrared) and 543.5 nm (green). Green and infrared He-Ne lasers are exceptionally expensive (rare in the surplus market) and are designed for special applications.

You can see the 543.5 nm green line by examining a regular red He-Ne tube with a diffraction grating. You will also notice a second, darker red line at 652 nanometers. These and other spectra are created as the helium and neon atoms drop from their raised, excited state to various transition levels. FIGURE 5-10 shows a diagram of the energy levels within a typical He-Ne tube. As an aside, the invisible 1.523 micrometer line is the wavelength of the first helium-neon tubes that were developed by Javan, Bennett, and Herriott at Bell Telephone Labs.

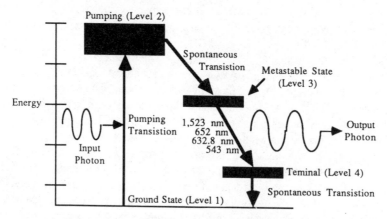

FIG. 5-10. *This simplified view of the energy levels of the He-Ne laser shows how photons are created as the atoms decay from their level 3 metastable state to an intermediate level 4 terminal state. Many wavelengths of light are created in the helium-neon gas mixture, but selective filtering provides a beam with a nearly pure wavelength of 632.8 nm (some He-Ne tubes operate at 1523 nm and 543 nm).*

THE ROLE OF THE BALLAST RESISTOR

Laser tubes operate at voltages between 1.2 and 3.0 kV (though some are higher or lower), but their current requirements are much more stringent. Most tubes in the 1- to 5-mW power output range are designed to consume between 3.5 and 5 milliamps of current. If consumption is any less, the tube won't fire; any more and it could be damaged.

A helium-neon laser tube uses a ballast resistor to limit its current and to provide stable electrical discharge. The value of the ballast resistor varies depending on the tube and power output, but it's generally between 47kΩ and 230kΩ. The ballast resistor is usually placed close to the anode of the tube to minimize anode capacitance and to provide more stable operation.

You can safely operate most lasers between the 3.5- to 7-mA band gap recommended for He-Ne tubes, but the lower the current, the better. The tube will last longer and the power supply will operate more efficiently. If you have a choice of ballast resistor values, choose the highest one you can that fires the tube and keeps it running. If the tube sputters or blinks on and off, it might be a sign that it's not receiving enough current. Lower the value of ballast resistor and try again.

Commercially made power supplies designed for use with cylindrical head lasers are typically engineered *without* a ballast resistor. Rather, the resistor is housed in the end cap of the aluminum laser head—never forget this. If you use the power supply with a bare tube that lacks a ballast resistor, you can permanently ruin the laser. If in doubt, add the ballast resistor; you can always take it out later.

BUYING AND TESTING HE-NE TUBES

Not all helium-neon laser tubes are created equally. Apart from size and output power, tubes vary by their construction, reliability, and beam quality. After buying an He-Ne tube, you should always test it; return the tube if it doesn't work or if its quality is inferior.

Should you need a laser for a specific application that requires precision or a great deal of reliability, you might be better off buying a new, certified tube. The tube will come with a warranty and certification of power output. Because you're the only one handling the laser, you won't be bothered with such headaches as misaligned or chipped mirrors—unless you misalign or chip the mirrors yourself!

The documentation—a sort of pedigree that comes with the best tubes—will specify actual output, often within a tenth of a mW. For example, the tube may be rated by the manufacturer at 1 mW. That's the nominal output figure, but the documentation for the particular tube you receive might state the output was tested at exactly 1.3 mW. If you purchase many new tubes, you'll quickly discover that no two lasers are the same. Even though manufacturing tolerances are extremely tight, slight variations still exist. Much of this variation is due to mirror alignment.

Visual Inspection

The first step in establishing the quality of the tube is to inspect it visually. If the tube is used, be on the lookout for scratched, broken, or marred mirrors. Check both mirrors and use lens tissue and pure alcohol to clean them. Gently shake the tube and listen for loose components. Most tubes have a metal spider that holds the bore in place, and though it may rattle a bit, the spider should not be excessively loose. If the tube is an old one, it might have externally mounted mirrors (with most He-Ne tubes, the mirrors are mounted within the glass envelope). Carefully check the mirrors and mounts for loose components and scratches.

Inspect the tube carefully under a bright light and look for hairline cracks. If the tube is encased in a housing that contains a shutter, be sure the shutter opens when the laser is turned on (open the shutter manually if it is not electrically controlled).

Checking Laser Operation

After inspection, connect the tube to a suitable power supply. Be sure to read the section below if you are unsure how to connect the tube to the supply. Place the tube behind a clear plastic shield or temporarily cover it with a piece of cardboard. Flick the power supply on and watch for the beam. Listen for a sputtering or cracking sound; immediately turn off the power supply if you hear any anomalies. Point the laser toward a wall. If the laser is working properly, the beam comes out one end only and the beam spot is solid and well-defined.

Although one of the mirrors in an He-Ne tube should be completely reflective, they aren't always so. Occasionally, the totally reflective mirror allows a small amount of light to pass through and you see a weak beam coming out the back end (this is especially true if the mirror is not precisely aligned). Usually, this poses no serious problem unless the coating on the mirror is excessively weak or damaged or if the mirrors are seriously out of alignment.

All lasers exhibit satellite beams—small, low-powered spots that appear off to the side of the main spot. In most cases, the main beam and satellites are centered within one another, so you see just one spot. But slight variations and adjustment of the mirrors can cause the satellites to wander off axis. This can be unsightly, so if it matters enough to you, you might want to choose a tube that has one solid beam.

The satellites are caused by internal reflection from the output (partially reflective) mirror. The main and satellite beams are out of phase to one another because their path lengths are different. Some applications, such as advanced holography or interferometry, require you to separate the satellite beams from the main beam. You can do so with a spatial filter or metal flap.

When projected against a white or lightly colored wall, the beam from the laser might "bloom" with a visible, lightly colored ghost or halo. The ghost, which generally has a blue cast to it, sometimes makes it difficult to see the shape of the beam itself. Wearing a pair of green safety goggles helps to reduce the ghost, allowing you to see just the beam spot. You can also tack a piece of flat black or dark green paper on the wall and project the beam at it to reduce the effects of the ghost.

With just the beam itself cast on the wall or paper, inspect it for any irregularities. It should be perfectly round. If there are smears, try cleaning the output mirror and try again. Note that the beam will show nulls (dark spots) if the tube is not operating in TEM_{00} mode, as described above. These cannot be eliminated by cleaning.

Should the tube start but no beam comes out, check to be sure that nothing is blocking the exit mirror (or, is the laser an infrared type?). Clean it and try again. If the beam still isn't visible, the mirrors might be out of alignment. You cannot re-align mirrors that are fused to the glass tube unless the tube has alignment rings or wedges—and then the process depends largely on trial and error (and usually results in error).

Each ring is equipped with three or four hex screws. You readjust the mirrors by tightening and loosening one or all of the screws. Use a hex wrench that is *completely insulated* with high-voltage tape and take care not to touch any exposed metal parts. With the laser on, watch the output mirror and try loosening one of the screws. Does a beam appear? Keep trying, noting how you adjust each screw.

You might be able to make the beam appear by pressing down slightly on the wrench. That stresses the ends of the tube and can bring the mirrors into partial alignment. The idea is to work slowly and note the positions of the mirrors at each adjustment interval. Alignment becomes very difficult if both mirrors are out of whack.

If the tube doesn't ignite at all, check the power supply and connections. Try a known good tube if you have one. The tube still doesn't light? The problem could be caused by:

★ *Bad tube*. The tube is "gassed out," has a hairline crack, or is just plain broken.

★ *Power supply too weak*. The tube may require more current or voltage than the levels provided by the power supply.

★ *Insulating coating or broken connection on terminal*. New and stored tubes may have an insulating coating on the terminals. Be sure to clean the terminals thoroughly. A broken lead can be mended by soldering on a new wire.

68

Table 6-1. Lab Laser Parts List

R1	470 ohm resistor
R2	80 kilohm ballast resistor, 3 to 5 watts
LED1	Light-emitting diode
S1	SPST switch
J1	¼-inch phone jack
1	He-Ne laser tube (as specified in text)
1	12 Vdc modular power supply (as specified in text)
1	Plastic project box (7¾ by 4⅜ by 2⅜ inches)
2	¾- or 1-inch pipe hangers
2	⁸⁄₃₂ by ½-inch bolts
4	⁸⁄₃₂ inch nuts
2	#8 locking washers
Misc.	Rubber grommets or weather stripping, fuse clips or springs.

All resistors are 5-10 percent tolerance, ¼-watt, unless otherwise indicated.

of making the enclosure on the large side. Unless you are specifically after miniaturization, the larger the enclosure, the better the cooling of the components within.

Place stand-offs under the power supply and large grommets to support the tube. The grommets provide ventilation space between the laser and enclosure as well as shock absorption for the tube. You can use rubber weather stripping, O-rings, or high-voltage dielectric tape for the grommets.

Electrical Connection

You must connect the high voltage anode and cathode leads to the terminals on the tube. With most tubes, the terminals are the metal mirror mounts on each end of the laser. It's decidedly a bad idea to solder the leads directly to the terminals, yet the electrical connection must be solid and stable.

One method for attaching the high-voltage leads to the tube is to use a ⅝-inch compression spring (such as Century Spring S-676), as illustrated in FIG. 6-2A. These are readily available at most hardware stores. If the spring is too long, cut it with a pair of heavy clippers. Use a file to remove the outer coating of the spring (many are plastic coated), and then solder the power lead to the spring. Slip the spring over the terminal and inch it into place. If the spring won't fit over the terminals on your laser, use a larger or smaller type. Don't use a spring that's too heavy. A lightweight compression spring maintains electrical contact without stressing the mirror mount.

Another method is to use ¼-inch fuse clips (see FIG. 6-2B). You can buy these at most any electronics store, including Radio Shack. The clip is just about the right size for most laser tube terminals and can be easily soldered to the high- voltage leads. If the clip has small indentations for holding the fuses, bend these out with a pair of needle-nose pliers. You might need to tweak the clip a bit to get it to fit around the ends of the mirror mounts.

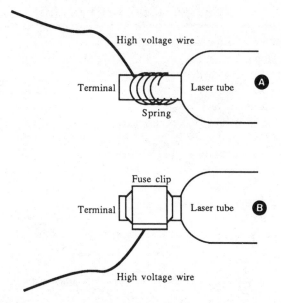

FIG. 6-2. *Two ways to attach the high-voltage leads to the laser tube terminals: by compression spring, and fuse clip. Be sure to file away any coating that might be on the spring, or electrical contact will be impaired.*

Final Assembly

Before permanently mounting the laser and power supply, note the position of the output mirror and drill a hole for the beam to escape. A ⁵⁄₁₆-inch hole provides plenty of room for mounting error and is about the size of the output mirror. You can drill a smaller hole, but you must be sure to precisely align the tube in the enclosure.

Next, drill holes in the top cover for the 12 Vdc power socket, on/off switch, and LED indicator. Use a ¼- or ⁹⁄₃₂-inch bit for the ¼-inch phone jack; measure the diameter of the switch and LED and use the proper size drill bits for each. When drilling plastic, start with a small pilot hole, then enlarge it with bigger drill bits until the hole is the proper size. Place a block of wood behind the plastic to prevent chipping.

You can use a metal enclosure only if the power supply is completely sealed and all interconnections are insulated (this includes the leads that connect to the terminals on the tube). Use heat-shrinkable tubing rated for at least 20 kV, high-voltage heat-shrinkable dielectric tape, or high-voltage putty. You might also want to coat the inside of the enclosure with a non-conductive paint. I've had great success using brush-on plastic coating, the stuff designed as a covering for tool handles. Many plastics outlets sell this coating, which is available in brush- and dip-on forms.

Solder the connections between all components, as shown in the schematic in FIG. 6-3, and mount them in the enclosure. Avoid long wire lengths; keep all lead lengths as short as possible. Place the ballast resistor close to the anode terminal of the tube. Be sure that no leads interfere with the output beam of the laser. You might want to tie down the leads to keep them from wandering about in the enclosure. Use plastic tie wraps and secure the leads to the inside of the enclosure with a small piece of double-sided foam tape. Keep low-voltage and high-voltage leads separate. Finish the lab laser

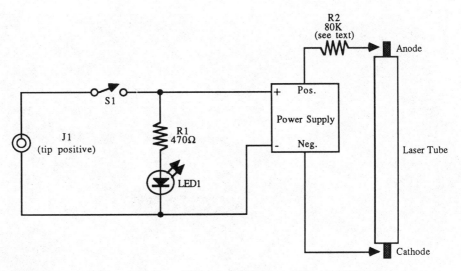

FIG. 6-3. *Wiring diagram for the lab laser. Be careful of the very high voltages present at the terminals of the laser and the output of the power supply.*

by adding rubber feet to the bottom of the enclosure. The finished laser is shown in FIG. 6-4A and B.

Power Supply

Power for the lab laser comes from a 12 Vdc battery pack or a 12 Vdc adapter/ battery eliminator. The battery or ac adapter must provide at least 350 mA current, or more, depending on the tube and high-voltage power supply used. Construction details for a battery pack and charger/adapter, suitable for use with the all-in-one lab laser, appear in Chapter 21.

BUILD AN ENCLOSED LASER HEAD

The all-in-one lab laser doesn't lend itself to hand-held portability. A bare laser tube can be easily shielded in a pipe that is suitable for hand-held use, or you can even mount one on an optical bench. You can use metal or plastic for the laser enclosure. Metal conducts heat more readily and is recommended if you plan on keeping the laser turned on for long periods of time. However, a plastic enclosure provides insulation against the high-voltage potentials present at the terminals of the laser.

The laser head used in this project is designed around a 2- mW Melles Griot plasma tube. The tube measures 1.45 inches in diameter by 10.6 inches in length. A piece of 1½-inch PVC pipe, cut to a length of 12 inches, serves as the enclosure. The pipe is capped off on both ends and a removable plug is attached to the enclosure to provide a shutter—a guard against accidental exposure to the beam.

The power supply pack, a separate component to the laser head, is designed to run off 117 Vac and includes a fuse, key switch, and pilot lamp. The design of this laser incorporates many of the safeguards required by the CDRH for a commercially sold Class II or Class IIIa device.

FIG. 6-4. *The completed lab laser: (A) With the cover removed, showing the mounting of the components, and (B) with cover in place.*

Table 6-2. Cylindrical Laser Head Parts List

R1	80 kilohm ballast resistor (3 to 5 watts)
1	He-Ne laser tube (as specified in text)
1	12-inches of 1½-inch, schedule 40 PVC
2	1½-inch plastic test plugs
1	1½-inch PVC end cap
Misc.	O-rings or grommets, fuse clips or springs, silicone sealant, 24-inch length of miniature plastic or metal chain, two eyelets for chain, two ¼-inch grommets.

Building the Laser Head

Refer to TABLE 6-2 for a parts list. FIGURE 6-5 shows the construction details of the laser head. Cut a piece of 1½-inch schedule 40 PVC to 12 inches. Make sure the cuts on both ends are square. Remove the rough edges with a file or fine-grit sand paper (300-grit wet/dry paper, used dry, is a good choice). Drill a hole, using a number 48 bit, a distance of ⅝-inch from the back end of the tube. This hole is for attaching the eyelet used to secure the protective cap chain (see below).

Solder the high-voltage leads to pair of springs or fuse clips, as detailed above. Insert a ballast resistor (80 kΩ, or more) between the anode terminal and anode lead, as shown in FIG. 6-6, and wrap the connections in heat-shrinkable tubing. Waft a lighter or match under the tubing or use a heat gun to shrink it around the resistor and other connections.

Thread the wires through a grommet and poke the grommet inside a ¼-inch hole drilled in the center of a 1½-inch test plug. Secure the grommet and resistor by applying a layer of silicone rubber over the inside surface of the test plug and let dry.

Once the sealant has set, attach the springs or fuse clips to the terminals on the laser tube. Wrap O-rings, electrical tape, or rubber bands around the tube, as shown in FIG. 6-7, and insert the tube in the PVC pipe. The output mirror should be

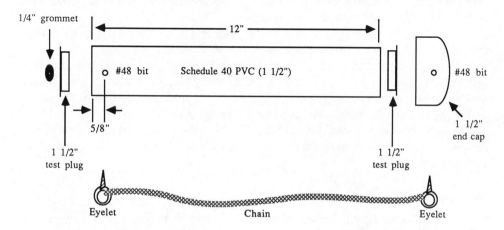

FIG. 6-5. *Construction details for the cylindrical laser head. Adjust pipe diameter and length to accommodate the exact dimensions of the tube you are using.*

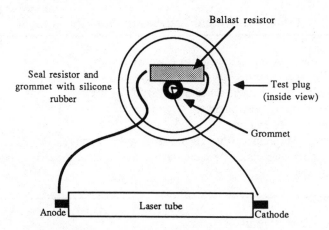

FIG. 6-6. *After soldering the leads to the ballast resistor, secure the resistor to the test plug with silicone rubber sealant. For a professional look, feed the high-voltage leads through a grommet mounted in the center of the plug. Be sure to use high-voltage dielectric wire (rated 10 kV or more) or you can receive a bad shock.*

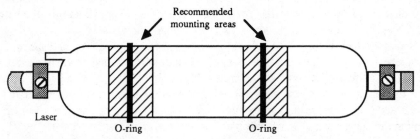

FIG. 6-7. *Recommended mounting areas on a bare glass He-Ne tube. You can use large O-rings or grommets as shock absorbers (shock absorption reduces "microphonic noise" that can impair the coherent operation of the laser).*

approximately ½-inch from the end of the tube. Secure the test plug to the other end of the tube with a dab of all-purpose adhesive.

Next, drill a ¼-inch hole in the center of another test plug and mount it on the output end of the tube. Secure the plug with glue (not PVC solvent cement).

Construct the protective plug using a 1½-inch PVC end cap. Secure it to the tube using a miniature eyelet and 24-inch length of lightweight metal or plastic chain. Insert another eyelet in the small hole previously drilled in the PVC pipe. Secure the other end of the chain to the eyelet. Be sure that the eyelet doesn't interfere with the tube, ballast resistor, or high-voltage leads. If the threads of the eyelet contact internal parts, cut the shaft of the eyelet so that it doesn't protrude inside the pipe. Apply glue to hold the eyelet in place.

Building the Power Pack

Refer to TABLE 6-3 for the parts list for the power pack. The power pack uses a commercially available 117 Vac high-voltage laser power supply. You can use a dc power supply if you need a completely portable and self-contained laser. Many new and surplus

Table 6-3. Laser Head Power Supply Parts List

S1	Key switch
L1	Neon lamp (with dropping resistor)
1	Plastic project box (6½ by 3¾ by 2 inches)
Misc.	Fuse holder, ac plug and cord, two ¼-inch grommets

ac power supplies are modular, sealed in a box measuring 1⅜-by-4¼-by-3⁵⁄₁₆-inches. The power supply may be equipped with an Alden high-voltage connector or flying leads (tinned pigtail leads, with no connector). If the supply has an Alden connector, cut it off and save it for future use—female Alden connectors are hard to find. Use the two or three holes in the module to mount the supply in a 6½-by-3¾-by-2-inch plastic experimenter's box (you can use any plastic box that is large enough for the power supply, switch, and other components). *Do not* drill holes in the module—you'll undoubtedly drill through the components sealed inside.

Follow the drilling guide shown in FIG. 6-8, and drill holes in the enclosure for the power cord (allowing extra for the grommet), indicator lamp, and key switch. Use a rotary

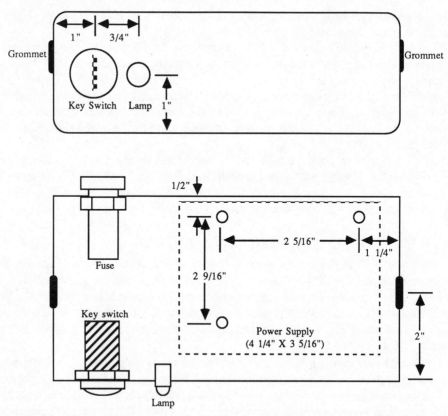

FIG. 6-8. *Layout and drilling guide for the laser head power supply. Drill holes large enough to accommodate the fuse holder, key switch, and neon lamp that you are using. The drilling guide assumes you are using a commercially made modular power supply (dimensions indicated in the figure).*

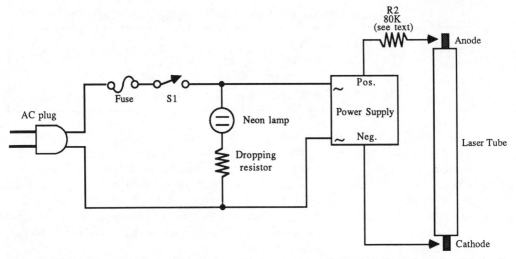

FIG. 6-9. *Wiring diagram for the laser head power supply. Beware of potentially lethal voltages throughout this circuit.*

rasp or countersink bit for the larger holes. Drill small pilot holes first to avoid chipping or cracking the plastic with large bits.

Use 16-gauge wire and solder the components as shown in FIG. 6-9. Keep wire lengths as short as possible. After soldering is complete, mount the components in the enclosure and route the grommet and power cord in the appropriate hole. Wrap all high-voltage leads in 20 kV heat-shrinkable tubing or tape and apply high-voltage putty to eliminate arcing. You may also insulate the wiring using clear aquarium tubing. Double check your work (use a meter as a safety measure) and insert a 2-amp fast-acting fuse in the fuse holder.

Test the laser by plugging in the power supply and turning the key switch to the ON position. The indicator light should turn on, and within 3 to 5 seconds, the laser should fire. If the indicator lamp and laser do not turn on, it could indicate a wiring problem or blown fuse. Recheck your work to make sure the wiring is correct.

BUILD A COMPACT LASER BREADBOARD

A breadboard allows you to experiment with lasers and other optical components without elaborate mounting and construction. The laser breadboard presented here is a semi-permanent mount for a commercial- or home-made cylindrical laser head and serves as a "head-end" for the optical breadboards detailed in the next chapter. The laser and optical breadboards are constructed with the same inexpensive materials, allowing you to mix and match as you desire.

Follow the construction details shown in FIG. 6-10 (parts list in TABLE 6-4). Cut a piece of ¼-inch pegboard to 6- by 18-inches. Sand the front, back, and edges and spray on a coat or two of clear lacquer (this seals the wood). Cut lengths of 2-by-2-inch lumber and use nails or screws to attach the wood to the pegboard. You may wish to substitute ⁴¹⁄₆₄-by-½-by-¹⁄₁₆-inch aluminum channel for the 2-by-2 lumber. Refer to the next chapter on how to use aluminum channel stock.

80

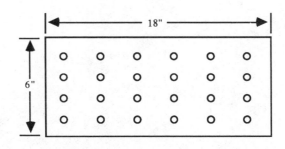

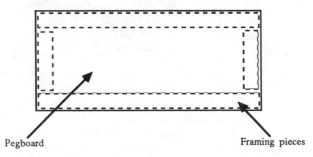

Pegboard   Framing pieces

FIG. 6-10. *A simple optical breadboard can be constructed using an 18- by 6-inch piece of ¼-inch Masonite pegboard and lengths of 2-by-2 framing lumber. Cut the lumber as indicated in the parts list in* TABLE 6-4.

Cylindrical laser heads mount easily in a clamp made with PVC pipe. Cut a length of 2-inch PVC to 3 inches. Next, carefully split the pipe in half down the middle using a hacksaw or table saw. Finish all edges with a file. Use the holes already in the pegboard as a template, and drill two holes in each PVC half. Next, mount the pipe halves in the breadboard. Use ⁸⁄₃₂-by-¾-inch bolts, ⁸⁄₃₂ nuts, and washers. Before tightening the hardware, slip the loose end of a 3-inch adjustable hose clamp under each piece of pipe. Tighten the hardware and close the clamp. Mount small rubber feet on top of the four bolt heads, then slide the laser head into the mount, as shown in FIG. 6-11. Tighten the clamps.

Table 6-4. Parts List for Mini Optical Breadboard

1	18- by 6-inch sheet of ¼-inch pegboard
2	18-inch length of 2-by-2-inch framing lumber
2	3-inch length of 2-by-2-inch framing lumber
2	3-inch lengths of 2-inch schedule 40 PVC (cut lengthwise)
4	⁸⁄₃₂ by ¾-inch bolts, nuts, and washers
2	3-inch (approx.) adjustable hose clamps
Misc.	Nails or screws for securing pegboard to framing lumber, rubber feet or rubber weather stripping

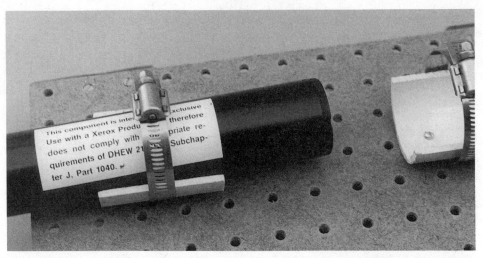

The text on the label in the image reads:

This component is inte...
Use with a Xerox Produ...
does not comply with ...
quirements of DHEW 21...
ter J, Part 1040.

...xclusive
therefore
priate re-
Subchap-

FIG. 6-11. *Secure the laser on the breadboard using lengths of 2-inch PVC cut lengthwise and fastened to the pegboard with ¹⁰/₂₄ hardware. The laser can be lashed in place with an adjustable car radiator hose clamp.*

You can mount the power supply in a number of locations, depending on its size and design. Sealed, modular power supplies can be mounted on top of the breadboard next to the laser. If it's small enough, you can tuck the power supply underneath the breadboard. Unsealed supplies should be placed in a protective housing and can be mounted on the breadboard or simply placed along side of it.

7

Constructing an
Optical Bench

An optical bench is a device that allows you to experiment with lasers and optics by mounting them securely on a rigid base. Components can be easily added or deleted, using any of a number of fastening systems, including nuts and bolts, magnets, sand, styrofoam, or even clay. The optical bench is the equivalent of the solderless breadboard used in electronics. Once you are satisfied that your project works, you can disassemble it and start on another project or re-assemble the components in a permanent housing.

There are a number of useful design approaches to optical benches in this chapter. You can make an optical bench of just about any size, up to a practical limit of 4 by 8 feet. Projects that follow show how to build a 2-by-4 foot bench as well as a 4-by-4 foot model. Construction materials are cheap and easy to get, and you're free to use more exotic materials if you desire.

You'll also learn how to construct components for an optical breadboard system. Small, Lego-like blocks can combine to facilitate just about any design. The mounting of lenses, mirrors, and other optical parts is tricky business; this chapter includes several affordable approaches that make the job easier.

BASIC 2-BY-4-FOOT OPTICAL BENCH

The basic 2-by-4-foot optical bench is made completely of wood. Using the parts indicated in TABLE 7-1, start with a 2-by-4-foot chunk of ¼-inch pegboard.

After sizing the pegboard, sand the edges as well as the front and back (the back will have a coarse finish and the front will be smooth). Seal the wood by spraying or brushing on one or more coats of clear lacquer or enamel. Let dry completely, then proceed to the next step.

Table 7-1. 2-by-4-Foot Optical Bench Parts List

1	2-by-4-foot, ¼-inch pegboard
2	48-inch lengths of 2-by-2-inch framing lumber
2	18-inch lengths of 2-by-2-inch framing lumber
Misc.	Nails or screws to secure pegboard to lumber, weather stripping

Now for the frame of the bench. Cut two 4-foot lengths of 2-by-2 lumber. Use nails or screws and attach the lumber pieces to the long sides of the pegboard. Next, cut two 18-inch lengths of 2-by-2 lumber and secure them to the ends. Center the 18-inch lengths so that there is an even gap on either side. You can use just about any size lumber for the frame, but you should be consistent in case you build more optical benches. By using the same framing lumber, you maintain a constant height for each bench. Several benches can be used together on a large table.

While you can use the bench as is, it's a good idea to paint it with flat black paint. The black paint helps cut down light scatter and also imparts a more professional look. Rubber feet (for electronics projects) or rubber or foam weather stripping make for good cushions and shock absorbers, and they prevent marring desks and tabletops. Apply the rubber to the underside of the frame.

ENHANCED 4-BY-4-FOOT OPTICAL BENCH

The enhanced 4-by-4-foot optical bench is also made with ¼-inch pegboard, but the frame is of all-metal construction. You may use aluminum or steel shelving standards or ⁴¹⁄₆₄-by-½-by-¹⁄₁₆-inch extruded aluminum channel for the frame.

The parts list for the 4-by-4-foot bench is included in TABLE 7-2. Use a hacksaw (with a fine-tooth blade) to cut the metal to size as shown in FIG. 7-1. Be sure to cut each piece with the proper 45-degree miter. Use a miter box and C-clamps for best results. After cutting, remove the flash and rough edges from the ends of the metal with a file. Secure the framing pieces to the underside of the pegboard using ⁸⁄₃₂-by-½-inch hardware. A diagram of the completed bench is shown in FIG. 7-2.

Because of the size of the pegboard sheet, it might buckle in the middle under the weight of a heavy object. If this is a problem, add one or more reinforcing struts to the underside of the bench. An extra piece of shelving (standard or aluminum channel) can be mounted down the center of the bench. Secure the extra piece to the frame using ½-inch angle-iron brackets (available at the hardware store).

Table 7-2. 4-by-4-Foot Optical Bench Parts List

1	4-by-4-foot, ¼-inch pegboard
4	48-inch lengths of ⁴¹⁄₆₄- by ½- by ¹⁄₁₆-inch aluminum channel stock
4	1½- by ⅜-inch flat corner irons
8	⁸⁄₃₂ by ½-inch bolts, nut, lock washers
Misc.	Weather stripping or rubber feet for bottom of channel stock

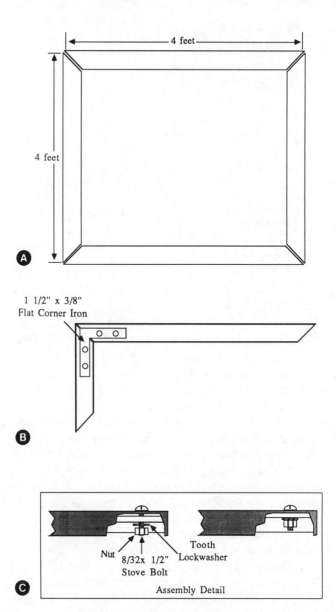

FIG. 7-1. *(A) Cutting guide for the aluminum channel used as a frame for the 4-by-4-foot optical bench. Be sure to miter the ends as shown. (B) Use 1½-by ⅛-inch flat corner irons to secure the ends. (C) Assembly detail for the framing pieces and corner irons showing ⁵⁄₃₂ hardware.*

Note: when drilling through metal, use only a new or sharpened bit. The drill motor should turn slowly—less than 1,000 rpm. Exert only enough pressure to bite into the metal, not enough to bend the metal or bit while drilling. Remove the flash around the drilled hole with a file.

As with the smaller version of the optical bench, paint this one a flat black to reduce light reflections. Also be sure to paint over the shiny heads of the machine screws.

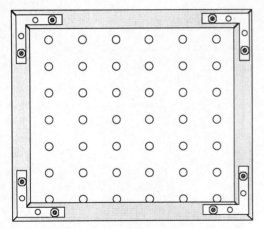

FIG. 7-2. *The underside of the optical bench, showing aluminum channel framing pieces, corner irons, and mounting hardware.*

Bottom View

Adding a Sheet-Metal Top

The pegboard optical bench is suitable for most hobbyist applications. The wood is sturdy, absorbs many vibrations, and is easy to drill. However, experiments that require greater precision need an all-metal optical bench. One can be constructed using medium- to thin-gauge aluminum, steel, or regular sheet metal. Large sheets or plates of metal are not routinely available at a hardware store, so look in the Yellow Pages under headings such as aluminum, metal specialties, sheet metal work, and steel.

Many sheet-metal shops work with thin-gauge metal—20- to 24-gauge or thinner. This is too thin to be used alone, but you can laminate it over the pegboard. The thickness of plate aluminum and steel is most commonly listed in inches, not gauge. A ⅛- or ³⁄₁₆-inch thick piece of aluminum or steel does nicely as the top of the optical bench. Steel is cheaper, but aluminum is easier to work with. The average price for a 2-by-4-foot piece of ³⁄₁₆-inch hot roll steel is about $25 to $30. The same size plate in un-anodized aluminum is approximately three times as much.

You might be lucky and find an outlet that sells pre-drilled stock. More than likely, however, you have to do the work yourself if you want the convenience of pre-drilled holes in your optical bench. Use a large carpenter's square to lay out a matrix of lines spaced either ½- or 1-inch apart (I prefer holes at ½-inch centers, but obviously this requires you to drill four times as many holes).

You might find it easier to place masking tape on the metal and mark the lines with a pencil. Use a carbide-tipped bit (#19 bit for ⁸⁄₃₂ hardware) and heavy-duty drill motor. A drilling alignment tool, available as an option for many brands of drill motors, helps to make perpendicular holes.

Using Plastic Instead of Metal

Another approach to the benchtop is acrylic plastic (Plexiglas, Lexan, and so forth). The advantages of plastic are that it is extremely strong for its weight and is easier to work with than metal. A steel top adds considerable weight to the bench (which in some cases is desirable), while a plastic top adds little weight.

A 2-by-4-foot piece of ³⁄₁₆- or ¼-inch acrylic plastic costs $6 to $8; you can buy sheets of plastic at most plastic fabricator outlets. Have them cut the plastic to size and finish the edges. You can drill through the plastic for the mounting holes using a regular bit, but better results are obtained when using a special plastic/glass drill bit.

USING OPTICAL BENCHES

Because pegboard stock already has holes, you probably won't need to drill new ones to mount the laser and other optical components. The holes are pre-drilled with better-than-average accuracy so you can use them for alignment. You can place a series of mirrors and lenses along one row of holes, for example, and they will all line up to one another.

There might be occasion, however, to drill new holes in the pegboard. Extra holes (⁵⁄₃₂-inch or #19 bit for ⁸⁄₃₂ hardware) should be ¼ inch or more away from existing holes, or the bit might slip while you are drilling. Schemes for mounting the laser and optical components appear later in this chapter.

As you complete a project or make changes to an existing experiment, keep notes and record the placement of all components. You might want to mark the rows and columns of holes as a reference for indicating the location of components. For example, you can mark all the rows with a letter and all columns with a number. A mirror placed at the intersection of C6 could be marked on a piece of graph paper. You can develop your own shorthand for identifying various optical components, but here are some suggestions:

★ S-FSM—small front-surface mirror
★ L-FSM—large front-surface mirror
★ P-BS—plate beam splitter
★ C-BS—Cube beam splitter
★ RAP—right-angle prism
★ EQP—equilateral prism
★ SPP—special prism
★ POL—polarizer
★ EXP—beam expander
★ SPA—spatial filter
★ L(XXX)—Lens (with type, size, and focal length)

You might also want to provide shorthand notation for any special components you use, such as single and double slits, diffraction gratings, optical fibers and fiber couplers, diode lasers, LEDs, sensors, and more.

OPTICAL BREADBOARD COMPONENTS

Small versions of the optical benches described above make perfect optical breadboards. Each breadboard, measuring perhaps 8- by 12-inches, can contain a complete optical sub-assembly, such as a laser and power supply (as described in the previous chapter), beam expander, beam director or splitter, sensing element, or lens array. You can construct a breadboard for each sub-assembly you commonly use. For example, if you often experiment with lasers and fiberoptics, one breadboard would consist of a laser/power supply and the other a fiberoptic coupling, cable, and sensor.

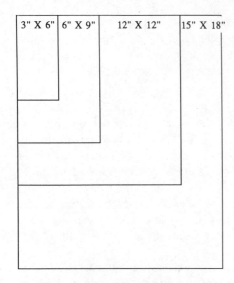

| 3" X 6" | 6" X 9" | 12" X 12" | 15" X 18" |

FIG. 7-3. *Suggestions for possible sizes for optical breadboards, but you can make your breadboards any size you wish.*

Construct the small breadboards in the same fashion as the larger optical benches detailed earlier in this chapter. Use the same materials for each breadboard to maintain consistency. Several sizes of breadboards are shown in FIG. 7-3. Use the scrap from the construction of the optical bench for the short lengths of framing pieces.

To use the breadboard pieces, lay them out like building blocks on a large, flat surface. The dining room table is a good choice, but be sure to place a drop cloth or large piece of paper on the table surface to prevent marring and scratching. You can also add rubber feet, rubber weather stripping, or foam weather stripping to the bottom of the breadboard pieces.

Lay out the pieces in the orientation you require for the project. If you've used paper as a drop cloth, you can draw on it to mark the edges of the breadboards. This is helpful in case you want to repeat an experiment and need to replace the breadboard pieces in the same spot.

LASER AND OPTICS MOUNTS

The optical bench or breadboard serves as a universal surface for mounting lasers and various optical components. The type of mounting you use for your laser and optics depends on their individual design, and most parts can be successfully secured to a bench or breadboard using one or more of the following techniques. Keep in mind that you might need to adjust the dimensions and design for each mounting configuration to conform to your particular components.

Laser Mounts

One of the most versatile yet easiest to construct laser mounts uses a piece of 2-inch (inside diameter) PVC cut in half lengthwise. Building plans are provided in Chapter 6. The mount is designed for use with cylindrical laser heads, either commercially-made versions or ones you build yourself using bare laser tubes.

Another approach to laser mounting is shown in FIG. 7-4. A pipe hanger, designed for electrical and plumbing pipe, secures a bare laser tube to the optical bench. This

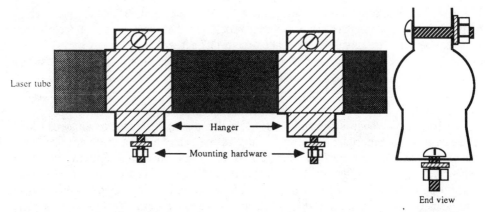

FIG. 7-4. *Cylindrical laser heads can be secured to optical benches and breadboards using pipe hangers. Choose a hanger large enough to accommodate the diameter of the laser, and avoid overtightening.*

approach is not recommended unless you shield the tube with an insulating box or cover the anode and cathode terminals on the laser with high-voltage putty or high-voltage heat-shrinkable tubing.

Pipe hangers come in various sizes to accommodate pipe from ½-inch to 2½-inches (outside dimension is approximately ¾ inch to 3 inches). Get a size large enough to easily fit the tube. Add a few layers of plastic tape around the tube where the hanger touches. *Do not* overtighten the hanger, or you run the risk of cracking or breaking the tube! You may use bigger hangers to mount cylindrical laser heads. Most commercially manufactured cylindrical heads measure 1.74 inches in diameter and can be successfully mounted using a 1½- or 2-inch hanger. All-in-one lab lasers (commercial or home-built) don't usually need mounting, because their housing lets you place them just about anywhere, yet some projects require you to secure the laser to the optical bench so that it doesn't move. Use straps or wood blocks to hold the laser in place, or drill mounting holes in the housing and attach the laser to the bench using nuts and bolts. The latter technique is not recommended if you are using a commercially made lab laser, because any modification voids its CDRH certification. Your own home-built lab laser (as detailed in Chapter 6) can be modified by locating a spot inside where the hardware will not interfere with the tube or power supply.

Lens Mounts

Lenses present a problem to the laser experimenter because they come in all sizes and shapes. Unless you are very careful about which lenses you buy (and purchase them new from prime sources), you will have little choice over the exact diameter of lenses

Table 7-3. Tube Hanger Laser Mount System Parts List

2	Plumbing pipe hanger (see text for diameters)
2	$^{10}/_{24}$ by ¾- or 1-inch carriage bolts, nuts, lock washers
2	$^{8}/_{32}$ by ½-inch bolts, nuts, and lock washers

you use. That makes it hard to build lens mounts using tubing and retaining rings, which come in only a few standard sizes. Another problem is that many lenses suitable for laser experiments are small, which makes them difficult to handle.

With a bit of ingenuity, however, you can make simple lens holders for most any size lens you encounter. The basic ingredient is patience and a clean working environment. As much as possible, handle lenses by the edges only and work in a well-lit, clean area—free from dust, cigarette smoke, and other contaminates.

A basic lens mount is shown in FIG. 7-5, and is useful for lenses as small as 4 mm to as large as 30 mm. Start by cutting a ledge in one end of a piece of ¾-inch PVC pipe coupling. Use a fine-toothed hacksaw to make two right-angle cuts. Smooth the rough edges with a file or wet/dry sandpaper (used dry). Next, use a caliper or accurate steel rule to measure the diameter of the lens. Choose a drill bit just slightly smaller than the diameter of the lens, and drill a hole in the remaining stub on the pipe coupler. Remember: you can always enlarge a hole to accommodate the lens, but you can't make it smaller.

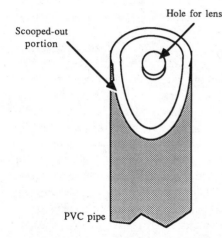

Hole for lens

Scooped-out portion

PVC pipe

FIG. 7-5. *One of the best (and easiest) ways to mount a lens is inside one wall of a piece of PVC pipe or coupler. You can make the hole the same size as the lens (for a friction fit) or slightly smaller (glue the lens on the outside wall of the pipe using all-purpose adhesive).*

Press the lens into the hole. Don't exert excessive pressure or you may crack the lens. If the hole is too small, use a larger bit or enlarge the hole with a fine rat-tail file (the kind designed for model building works nicely). After careful drilling and filing, you should be able to pop the lens into the hole but still have enough friction-fit to keep it in place. The lens is not mounted permanently, so it can be removed if you have another use for it. Clean the lens using an approved lens cleaner. Refer to Chapter 3 for details on lens care and cleaning.

The PVC coupling holder can be mounted on the optical bench in a number of ways. Several methods are shown in FIG. 7-6. The hole for the thumbscrew does not require tapping but tapping makes the job easier. Use a ⁴⁄40 or ⁶⁄32 thumbscrew or regular pan-head machine bolt as the set screw. The hardware used for mounting the holder on the bench is a ¹⁰⁄24- or ¼-inch 20 carriage bolt, secured in place with a threaded rod coupler (cut the coupler if it is too long for the PVC holder). You can adjust the height and angle of the lens by loosening the thumbscrew and adjusting the holder.

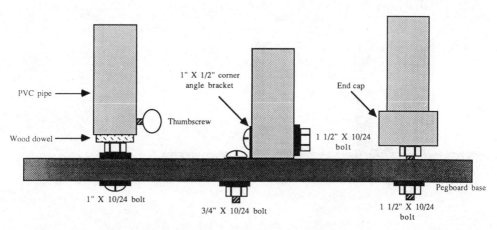

FIG. 7-6. *Various ways to attach PVC pipe to a pegboard optical bench. If the bench is ferrous metal (steel, not aluminum), you can build a magnetic base by gluing a magnet to the bottom of the pipe.*

White PVC plastic can cause light to scatter, so you might want to coat the lens holders with flat black paint. The small spray paints sold by Testor (and available wherever model supplies are sold) adhere well to PVC.

The PVC pipe can be used simply as a holder for mirrors, beam splitters, and other components. A length of 1- or 1¼-inch PVC, pressed in a vise, can be used to hold thin Masonite board or acrylic plastic. You can mount mirrors, lenses, plate beam splitters, and other components on the Masonite. Cut the Masonite to a width just slightly larger than the inside diameter of the PVC (1¼-inches for schedule 40 1¼-inch PVC).Clamp the pipe in a vise until it deforms into an ellipse. Stick the Masonite board in the pipe and slowly release the vise. The pipe will spring back into shape, holding the Masonite. You can also directly mount thick plate beam splitters and mirrors into the PVC. Don't try it with thin optics, or they could shatter as the PVC springs back into shape.

A way to mount lenses of almost any size is to tack them on a piece of ⅛-inch acrylic plastic. The acrylic pieces needn't be much larger than the lens itself, but you should allow room for mounting hardware. Black plastic with a matte finish is the all-around best choice, although you can always coat the plastic with black matte paint.

Drill a hole in the plastic just smaller than the lens, leaving enough room for the edges of the lens to make contact with the plastic. After drilling, place the lens directly over the hole. Apply two or three small dabs of all-purpose adhesive (such as Duco cement) to the outside edges of the lens. Use a syringe applicator or toothpick to dab on the adhesive.

Be sure that no "strings" of the cement cover or come in contact with the usable area of the lens. If you make a mistake, immediately remove the lens, wash it in water, and clean the lens with approved cleaner. Although you can remove undried cement with chemicals such as acetone and lacquer thinner, these compounds could have an adverse effect on lens coatings.

Some lenses are square as opposed to round and can be mounted on the top of a short length of PVC pipe or small piece of ⅛-inch acrylic plastic. Be sure to mount the lens perpendicular to the plastic, or the laser beam might not follow the path you want.

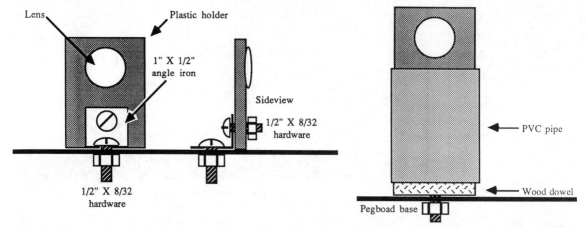

FIG. 7-7. *Two ways to mount lenses to acrylic plastic and how to secure the plastic to the optical bench. You can use angle irons (available at the hardware store) or insert the plastic into a short length of PVC pipe.*

Let the adhesive dry completely, then mount the lens and holder as shown in FIG. 7-7. The illustration provides a number of mounting techniques. In one, the holder is mounted to a pipe, which is attached to the shaft of a bolt. The length of the bolt depends on the size of the holder and lens but is generally between 1 and 3 inches long. You can alter the height of the bolt by adjusting the mounting nuts, as illustrated in FIG. 7-8. Parts lists for the lens-mounting systems are included in TABLE 7-4.

Round lenses measuring between 21 and 26 mm in diameter can be inserted inside ¾-inch PVC coupling. The lenses are held in place with thin pieces of ¾-inch pipe. Follow the instructions shown in FIG. 7-9. Cut a ¾-inch slip coupling in half, discarding the

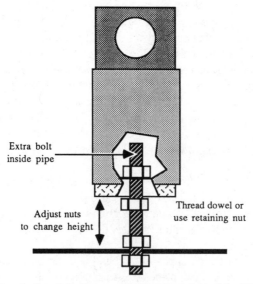

FIG. 7-8. *A set of nuts and a long bolt allow you to adjust the height of the PVC pipe in relation to the surface of the optical bench.*

Table 7-4. Lens Holders and Mounts Parts List

For PVC Lens Mount/Thumbscrew Adjustable Height (Fig. 7-6)
1   ¾- or 1-inch schedule 40 PVC pipe, length 2 to 5 inches
1   2- to 4-inch length of ¾- or 1-inch wood dowel (trim to fit in pipe)
1   $8/32$ machine thumbscrew
1   $10/24$ by 1-inch bolt, nut
2   #10 flat washers

For PVC Lens Mount/Angle Bracket Bench Mount (Fig. 7-6)
1   ¾- or 1-inch schedule 40 PVC pipe, length 2 to 5 inches
1   $10/24$ by 1½-inch bolt, nut, washer
1   $10/24$ by ¾-inch bolt, nut, washer
1   1- by-½-inch corner angle bracket

For PVC Lens Mount/Bolt Adjustable Height (Fig. 7-6)
1   ¾- or 1-inch schedule 40 PVC pipe, length 2 to 5 inches
1   ¾- or 1-inch PVC end cap
1   $10/24$ by 1½-inch bolt
4   $10/24$ nuts, flat washers

For Simple Lens Holder Only (Fig. 7-7)
1   ½- by ¾-inch, ⅛-inch-thick acrylic plastic (adjust size to accommodate lens)

For Simple Lens Holder and Bench Mount (Fig. 7-7)
1   ½- by ¾-inch, ⅛-inch-thick acrylic plastic (adjust size to accommodate lens)
1   1-by-½-inch corner angle bracket
2   $8/32$ by ½-inch bolts, nuts, split lock washers

For PVC Lens Holder and Mount/Adjustable Height (Fig. 7-8)
1   ½- by ¾-inch, ⅛-inch-thick acrylic plastic (adjust size to accommodate lens)
1   ¾- or 1-inch schedule 40 PVC pipe, length 2 to 5 inches
1   2- to 4-inch length of ¾- or 1-inch wood dowel (trim to fit in pipe)
1   2- to 4-inch length $8/32$ threaded rod
4   $8/32$ nuts, flat washers

innermost portion to avoid the raised pipe-end stops. Thoroughly smooth the ends with a file, fine-grit sandpaper, or grinding wheel. Next, cut two ⅛- to 3/16-inch pieces of ¾-inch PVC pipe, sand the edges, and insert one into the coupling (the inside diameter of the coupling is tapered, so the pipe will only fit one way). Drop in the lens, center it, and press in the other small piece of pipe.

You can use the mounted lens in a number of ways, including cementing it to a PVC holder (using PVC solvent cement) or using it in the optical rail system described later

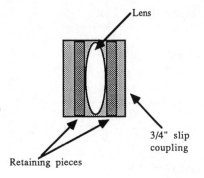

Lens

FIG. 7-9. *Lenses can be mounted inside a ¾-inch PVC clip coupling using the scheme shown in this diagram. The retaining pieces are small slivers of ¾-inch PVC. Paint all parts flat black to prevent light scatter.*

3/4" slip coupling

Retaining pieces

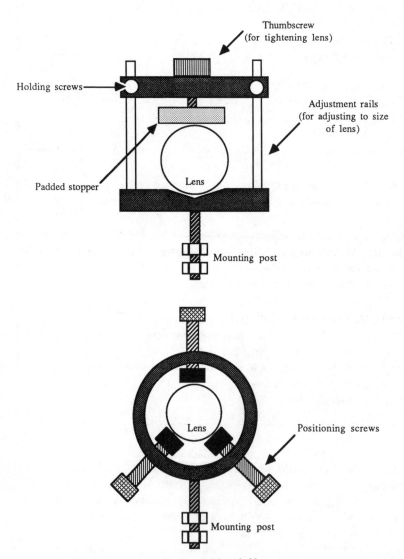

Thumbscrew
(for tightening lens)

Holding screws

Adjustment rails
(for adjusting to size
of lens)

Padded stopper

Lens

Mounting post

Lens

Positioning screws

Mounting post

FIG. 7-10. *Two ways to construct advanced lens holders.*

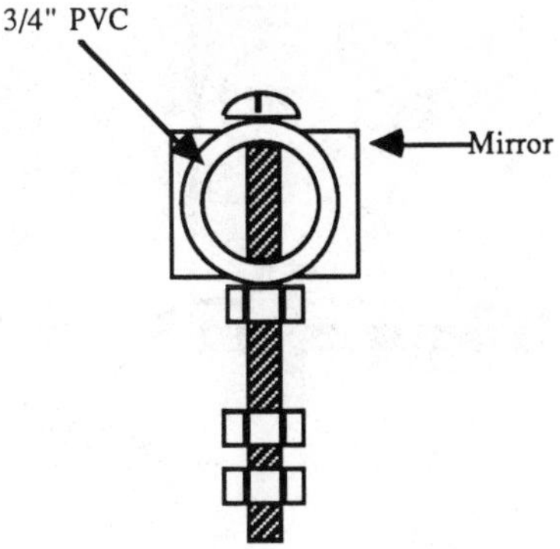

3/4" PVC

FIG. 7-11. *Build a simple mirror mount by gluing it to a short length of ¾-inch PVC. Mount the PVC to the optical bench with a nut and bolts, and adjust the height and angle of the mirror by turning the bolts.*

Mirror

in this chapter. As mentioned before, the white PVC pipe can cause light scattering. Reduce the scatter by painting the coupling and pipe pieces flat black. If you must use smaller lenses, ½-inch PVC pipe couplers accommodate optics between 16 and 21 mm in diameter.

Advanced, adjustable lens holders are shown in FIG. 7-10. These can be made using metal or acrylic plastic and require some precision work on your part. The holders are adjustable and can be used with most any size of circular lenses. On one, change the size by loosening the holding screws and sliding the top along the rails. Clamp the lens in place by turning the top screw. Delete the rounding on the bottom when using cylindrical or square lenses.

Mirror Mounts

Front-surface mirrors are often used to redirect the beam or to increase or decrease its height in relation to the surface of the bench. A simple mirror mount, using just about any thickness and size of mirror, is shown in FIG. 7-11. Here, ¾-inch PVC pipe is cut into short ½-inch sections with a mirror glued onto one end and a hole drilled through the top and bottom. Mounting hardware, such as $^{10}/_{24}$- or ¼-inch 20 (detailed in TABLE 7-5), is used to secure the pipe to the optical bench. You can swivel the mirror right and left by loosening the retaining nut. By attaching the mirror/pipe to an angle-iron

Table 7-5. Mirror Mounts Parts List

For Stem PVC Pipe Mount (Fig. 7-11)	
1	1- to 4-inch length of $^{10}/_{24}$ machine bolt
1	½-inch length of ¾-inch schedule 40 PVC pipe
3	$^{10}/_{24}$ nuts, flat washers
For Swivel PVC Pipe Mount (Fig. 7-12)	
1	$^{10}/_{24}$ by 1½-inch bolt, washer, nut
1	½-inch length of ¾-inch schedule 40 PVC pipe
1	$^{8}/_{32}$ by ½-inch bolt, washer nut

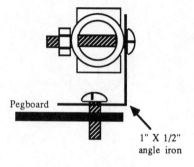

FIG. 7-12. *A swivel mirror mount (allowing 2 degrees of freedom) can be built using PVC pipe and an angle iron. Choose the right size of angle iron to fit the application.*

Pegboard

1" X 1/2"
angle iron

bracket, as illustrated in FIG. 7-12, you can provide two axes of freedom—right and left, and up and down.

Because you might never know what you'll need to complete an optical experiment, build several mounted mirrors in the following configurations (note that "small" means approximately ½-inch square and "large" means 1-inch square).

⭐ Small mirror mounted to ¹⁰⁄₂₄-by-1-inch bolt (bench hugger)
⭐ Small mirror mounted to ¹⁰⁄₂₄-by-2½-inch bolt
⭐ Large mirror mounted to ¼-inch 20 by 3-inch bolt
⭐ Small mirror mounted to ½-inch angle-iron bracket and ¹⁰⁄₂₄-by-1-inch bolt
⭐ Small mirror mounted to ½-inch angle-iron bracket and ¹⁰⁄₂₄-by 2-inch bolt
⭐ Large mirror mounted to 1½-inch angle-iron bracket and ¼-inch 20 by 3-inch bolt

You can make other mounts by using different mirrors, bolt lengths, and angle-iron sizes, as needed. Be sure that the mirror is small enough to clear any hardware. For instance, an overly large mirror might interfere with the angle bracket.

Commercially made mirror mounts are occasionally available from surplus sources. The components are optically flat, dielectric mirrors often designed for use with helium-neon lasers. The mounts can be secured to the bench in a variety of ways including angle irons, hold-down brackets, and clamps. Many of the mounts come with holes pre-drilled and pre-tapped at ½- or 1-inch centers. The holes can be used to facilitate quick and easy mounting to the bench.

The major benefit of commercially made mirror mounts is their precision. Many have fine-threaded screws for adjusting the inclination or angle of the mirror. If you're handy with tools, you can fashion your own precision mirror mounts from aluminum or plastic. Best results are obtained when you use a precision metal-working lathe or mill.

Another method of mounting mirrors, popular among holographers, is depicted in FIG. 7-13. Here a mirror is cemented inside a nook sawed into a piece of 1¼-inch PVC pipe. The pipe can be mounted on a stem, placed in sand (see below and in Chapters 17 and 18), or simply balanced on the bench. You can mount extra large mirrors on the pipe (use bigger pipe) and the pipe pieces can be any length to suit your requirements.

One method of securing the pipe is to use a dowel and short piece of PVC. Secure the dowel inside a short pipe end (¾- to 1-inch long). This pipe holds the mirror. A thumbscrew set in another length of pipe (1 to 4 inches) lets you adjust the height of the mirror. The bottom pipe piece can be secured to the bench using hardware or a

96

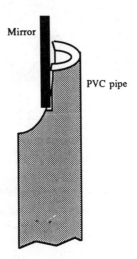

Mirror

PVC pipe

FIG. 7-13. *A mounting method popular with holographers is to glue the mirror onto a piece of scooped-out PVC pipe.*

magnet. The magnet obviously requires a metal benchtop. You can add metal to a wood benchtop with pieces of galvanized sheet metal. Small squares (up to about 6 by 12 inches) are available at many better hardware stores. The sheet metal comes with holes pre-drilled and can be cut to size with a hacksaw.

Cement the mirror in the pipe using gap-filling glue (Duco works well, and the mirror can be removed later if necessary). On occasion, you may want to make the mirror mount temporary, for example, if you have one extra-nice mirror and don't want to commit it to any one type of mount. You can temporarily secure the mirror to the pipe using florist or modeling clay. Build up a bead of clay around the pipe and press the mirror into place. You can vary the angle of the mirror by adding extra clay to one edge.

Mounting Prisms, Beam Splitters, and Other Optics

Prisms and beam splitters require unusual mounting techniques. A porro prism, often found in binoculars and a common find on the surplus market, can be used as laser retroreflectors (or corner cubes). The prisms act as mirrors where the axis of the outgoing beam is offset, yet parallel, from the incoming beam. You'll find plenty of uses for porro prisms, including using them to change the height of the beam over the bench.

The porro prism is easy to mount on a small piece of ⅛-inch acrylic plastic. Cut the plastic to accommodate the size of prism you have. Secure the plastic holder on a sliding post mount (as described above for mirror mounting) and you have control over the height and angle of the prism.

A prism table is a device that acts as a vise to clamp the component in place. The table can be used with most any type or shape of prism. A basic design is shown in FIG. 7-14; the parts list is provided in TABLE 7-6. Construct the table using small aluminum, brass, or plastic pieces. The thumbscrew is a ⁸⁄₃₂-by 1-inch bolt; the top of the table is threaded to accommodate the bolt.

Beam splitters require an open area for both reflected and refracted beams. Cube beam splitters can often be used in prism tables. Alternatively, you can build a number of different mounts suitable for both cube and plate beam splitters, as shown in FIG. 7-15. The idea is to not block the back or exit surface(s) of the component.

97

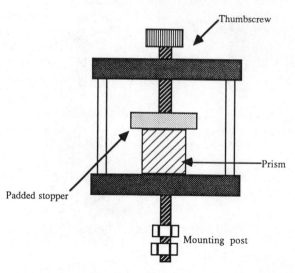

FIG. 7-14. *A prism or cube beam splitter table can be constructed using ordinary hardware and acrylic plastic. Secure the component by tightening the thumbscrew.*

Filters can be secured using adjustable lens mounts, as shown above. Commercially available filter holders designed for photographic applications (camera and darkroom) can also be used. Filters can be mounted on plastic or, if small enough, held in place inside a PVC pipe.

ADVANCED OPTICAL SYSTEM DESIGN

The projects in this section deal with advanced optical system design. Included are plans on building an optical rack for testing lenses and their effect on the laser beam, as well as motorizing mirrors, beam splitters, and other components for remote control or computerized applications. You can adopt these plans for any number of different optical system requirements such as laser light shows, laboratory experiments, laser rangefinders, and more.

Building an Optical Rack

An optical rack provides a simple means of experimenting with optical components such as lenses and filters. A common optical rack used by hobbyists is the meter or

Table 7-6. Prism Table Parts List

2	3-inch lengths of ¼-by-¼-inch acrylic plastic square extruded rods
2	3-inch lengths of ³⁄₁₆-inch-diameter acrylic plastic round extruded rods
1	¹⁰⁄₂₄ by 2-inch thumbscrew with knurled knob
1	¹⁰⁄₂₄ by 1½-inch flat heat machine bolt (for mounting post)
2	¹⁰⁄₂₄ nuts, washers
1	½-by-½-inch padded rubber

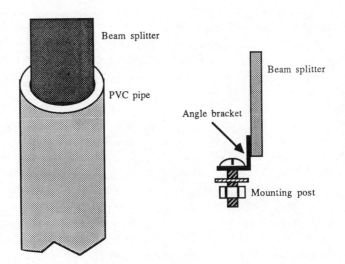

Beam splitter

PVC pipe

Beam splitter

Angle bracket

Mounting post

FIG. 7-15. *Plate beam splitters can be attached to an optical bench by inserting them into a length of PVC or by gluing them onto a small angle bracket. Be sure you can adjust the angle of the beam splitter and the transmitted and reflected beams are not obscured.*

yardstick. You place various clamps along the length of the stick to mount optical components. Meter sticks and their clamps, such as the set shown in FIG. 7-16, are available from a number of sources, including Edmund Scientific. The clamps hold only lenses of certain sizes (usually 1 inch or 25 mm), so be sure you get lenses to match.

Another approach to the optical rack is shown in FIG. 7-17. Here, aluminum channel is used as a "trough" for the optical components. The lenses are mounted in tubes such

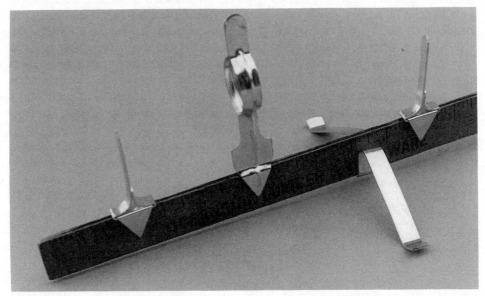

FIG. 7-16. *A yardstick with inexpensive optical components attached. The stick provides an easy way to align optics and measure their distances.*

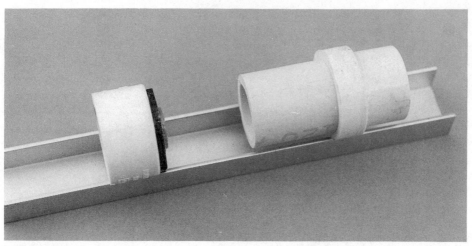

FIG. 7-17. *A homemade optical rail can be constructed using an aluminum channel and lenses mounted inside PVC pipe.*

as PVC or brass pipe. Each lens is centered in the tube and the outside dimension of the pipe is the same for all components. That means the light will pass through the optical center of each lens. You may use smaller or larger tubes for some components, but you must adjust the optical axis with shims or blocks.

Cut a piece of $^{57}/_{64}$-by-$^{9}/_{16}$-by-$^{1}/_{16}$-inch aluminum channel stock to whatever length you need for the rack. A 1-to-3-foot length should be sufficient. Mount the lenses (and/or filters) in PVC pipe, as explained in the lens-mounting section previously in this chapter. With ¾-inch PVC, you are limited to using lenses that measure between 21 mm and 26 mm in diameter. That comprises a large and popular (not to mention inexpensive) group of lenses, so you should have no trouble designing most any optical system of your choice. When using ½-inch PVC, you can use lenses with diameters between 16 mm and 21 mm.

A sample lens layout using the optical rack is shown in FIG. 7-18. A parts list is provided in TABLE 7-7. The two lenses together comprise a beam expander, which is a common optical system in laser experiments. The laser beam is first diverged using a plano-concave or double-concave lens and is then collimated using a plano-convex or double-convex lens. The focal length of the plano-convex (or double-convex) lens deter-

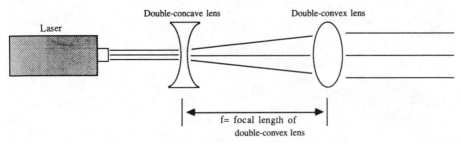

FIG. 7-18. *A double-concave lens and a double-convex lens make a laser beam expander. Adjust the distance between lenses to match the focal length of the double-convex lens.*

100

Table 7-7. Optical Rack Beam Collimator Parts List

1	1-3 foot length of $^{57}/_{64}$- by $^{9}/_{16}$- by $^{1}/_{16}$-inch extruded aluminum channel stock
1	Double-convex lens, in $^{3}/_{4}$-inch PVC holder
1	Double-concave lens, in $^{3}/_{4}$-inch PVC holder

mines whether the beam will be collimated for focusing to a point. You can experiment with different lenses and distances until you get the results you want.

A good way to use the rack is to calculate on paper the effects of the lenses in the system. Then see if your figures hold true. Armed with some basic optical math (see Appendix B for a list of books that provide formulas), you should be able to compute the result of just about any two- or three-lens combination.

R/C TRANSMITTER/RECEIVER

Servo motors designed for use in model airplanes and cars can be used as remote-control beam-steering units. Perhaps the easiest way to use the motors is with their intended receiver and transmitter. Unless you need to control many mirrors or other optical components, a two- or three-channel transmitter should do the job adequately. That allows you direct control over two or three servo motors. More sophisticated radio control (R/C) transmitters exist, of course, but they can be expensive. Fortunately, several Korean companies have recently joined the R/C market and offer low-cost

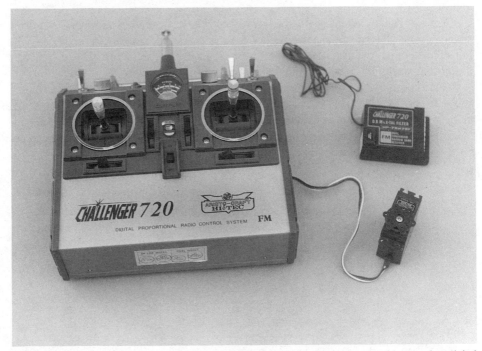

FIG. 7-19. *A complete R/C transmitter/receiver/servo setup. This transmitter provides complete digital proportional control of four independent servos (additional channels provide "on/off" functions only).*

Table 7-8. Beam Steering Servo Parts List

1	Four-channel (or more) R/C transmitter, with batteries
1	R/C receiver, with batteries
4	R/C servos
4	Plastic bell cranks and angle brackets (for mounting mirrors to servos)
4	First-surface mirrors (approx. ½-inch square, or size as needed)

alternatives to the more expensive brands. While these might not be as well made as the name brands, they provide similar functionality.

FIGURE 7-19 shows a six-channel R/C transmitter and companion receiver. The servo motors, which contain a small dc motor, feedback potentiometer, and circuit, connect to the receiver. This transmitter is designed for model aircraft use, so some of its channels are dedicated to special purposes, such as raising and lowering the landing gear. These channels are effectively on or off, without intermediate steps, as are the aileron and rudder controls. Even with the extra channels, only three or four can be adequately used for optical bench beam-steering.

The transmitter and receiver operate on battery power. The servos are designed for use in many types of airplane fuselages, so a variety of mounting hardware is included. If you don't find what you need in the parts included with the R/C transmitter and receiver, you can always take a trip to the hobby store and get more. A well-equipped hobby store, particularly those that specialize in R/C components, will stock everything you need.

Mirrors can be mounted to the servo motors in a number of ways. Gap-filling cyanoacrylate glue can be used to secure the mirror to the various plastic pieces. The hubs on servo motors are interchangeable by removing the set screw. If you can't find a hub that works for you, you can fashion your own using metal or plastic.

Control the motors by turning on the transmitter and receiver. Rotate the control sticks as necessary to move the motors. If the transmitter is set up correctly, the servo motors should return to their midway position when the control sticks are centered. If this is not the case, adjust the trimmer pots located on the transmitter. A parts list for a typical servo-controlled system is included in TABLE 7-8.

8

Laser Optics Experiments

You can use lasers for almost any experiments customarily associated with a non-coherent or white light source. For example, laser light refracts and reflects the same as ordinary light, allowing you to use a laser for experimenting with or testing the effects of refraction and reflection.

Moreover, laser light provides some distinct advantages over ordinary white light sources. The light from a laser—such as a helium-neon tube—is highly collimated. The pencil-thin beam is easily controllable without the use of supplementary lenses. That makes routine optics experiments much easier to perform.

Because laser light is comprised of so much intense, compact illumination, you can readily use it to demonstrate or teach the effects of diffraction, total internal reflection, and interference. When using ordinary light sources, these topics remain abstract and hard to comprehend because they are difficult to show. But with laser light, the effects are clearly visible. You can readily see the effects of diffraction and other optical phenomena using simple components and setups.

This chapter shows how to conduct many fascinating experiments using visible laser light. While you don't need to complete each experiment, you should try a handful and perhaps expand on one or more for a more in-depth study. For example, the effects of polarization by reflection is a fascinating topic that you can easily develop into a science fair project.

EXPERIMENTING WITH REFRACTION

Recall that refraction is the bending of light as it passes from one density to another. Light bends away from line normal at a dense-to-rare transition; light bends toward line

Table 8-1. Parts List for Refraction Experiments

1	Laser
1	Plate glass (approx. 2 inches square)
1	Plastic sheet or block
1	Clear glass or plastic container to hold water, mineral oil, isopropyl alcohol, glycerin, etc.
1	Plastic protractor
1	Equilateral prism

normal at a rare-to-dense transition. TABLE 8-1 lists the materials you'll need to carry out the experiments in this section.

Refraction Through Glass Plate

You can most readily see the effects of refraction by placing a piece of clear plate glass in front of the laser. The beam will be displaced some noticeable distance. Try this experiment: place a laser at one of the room so that the beam strikes a back wall or screen. Tape a piece of paper to the wall and lightly mark the location of the beam with a pencil.

Next, have someone place a small piece of regular window glass in front of the laser. Position the glass so that it is canted at an angle to the beam. The spot on the paper should shift. Mark the new location.

If you know the distance between the glass and wall as well as the distance between the two spots, you can calculate the angle of deflection (see FIG. 8-1). (With that angle, you can then compute the refractive index of the glass; consult any book on basic

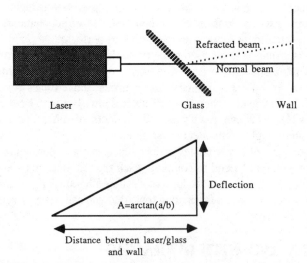

FIG. 8-1. *Measuring the deflection of the beam enables you to calculate the refractive index of glass, lenses, and other optical components.*

104

trigonometry for the formula.) The figure shows how the distance between (A)—glass to wall—and (B)—spot to spot, or deflection—are the equivalent of the adjacent and opposite sides of a right triangle. The illustration shows the particular formula to use to find the angle of deflection: $A = \arctan(a/b)$. Use trig tables or a scientific calculator to solve for it.

Try the experiment with different types of glass, plastic, and other transparent materials. Move the paper each time you try a new material. Note that not all glass is made of the same substances, and refractive indexes can vary. For example, one piece of glass may have a refractive index of 1.55 and another may have a refractive index of 1.72. The two may look identical on the outside but refract light differently. The higher the index of refraction, the greater the distance between the two beams.

Refraction Through Water

Bears, eagles, and others of the animal kingdom know a lot about refraction; how else could they catch fish out of lakes and streams with such precision! Because water is more dense than air, it has a higher index of refraction. Images in water not only look closer, but they appear at a different place than they really are. If you don't take refraction into account, you'll come up with nothing after throwing your spear into the water.

One easy way to demonstrate refraction in water is to place a narrow stick, like a dowel, in a jar. Fill the jar with water and the stick seems to bend. If the water is a little cloudy (as it often is after coming out of the tap), you can demonstrate the actual refraction of light with a laser. Fill the jar with water. Position the laser above the jar but at a 10 to 15 degree angle to the water line. Turn off the room lights so that the beam can be more readily seen in mid-air (chalk dust or smoke helps bring it out). You'll see the shaft of light from the laser ''magically'' bend when it strikes the water.

By dipping a protractor in the water, you can visually measure the angles and compute the index of refraction using the classic formula $n = \sin i / \sin r$. That is, the index of refraction is equal to the sin of the angle of incidence over the sin of the angle of refraction (this is often referred to as Snell's Law). Repeat the experiment with other liquids, including glycerin, alcohol, and mineral oil. Correlate the refractive indexes of these fluids with the glass and plastic from above (you can use Snell's Law for these materials, too).

Refraction Through Prism

Equilateral prisms are most often used to disperse white light (break up the light into its individual component colors). The light from a helium-neon laser is already at a specific wavelength, so it cannot be further dispersed. However, prisms can be used to show the effects of refraction and how the light can be diverted from its original path.

To demonstrate refraction in a prism, place the prism at the edge of a table. Aim the laser up toward the prism at a 45 degree angle. The beam should strike one side and then refract so that the light exits almost parallel to the surface of the table. For best results, you'll need to clamp the laser in place so that the beam doesn't wander around.

EXPERIMENTING WITH REFLECTION

The law of reflection is simple and straight-forward: the angle of reflectance, in relation to line normal, is equal but opposite to the angle of incidence. As an example, light incident

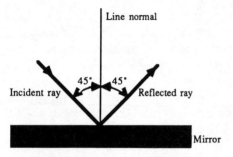

FIG. 8-2. *The law of reflection.*

on a mirror at a 45-degree angle, as shown in FIG. 8-2, will also bounce off the mirror at a 45-degree angle. The total amount of deflection between incident and reflected light will be 90 degrees. Materials needed for conducting the experiments in this section are found in TABLE 8-2.

Front-Surface Versus Rear-Surface Mirrors

Light is reflected off almost any surface, including plain glass. An ordinary mirror consists of a piece of glass backed with silver or aluminum. Light is reflected not only off the shiny silver backing but the glass itself. You can readily see this effect by shining a laser at an ordinary rear-surface mirror. With a relatively large distance between mirror and wall, you will note two distinct spots. The bright spot is the laser beam reflected off the silver backing; the dim spot is the beam reflected off the glass itself.

Now repeat the experiment with a front-surface mirror. Because the highly reflective material is applied to the front of the glass, the light is reflected just once. Only a single spot appears on the wall (see FIG. 8-3).

Total Internal Reflection

When light passes through a dense medium toward a less dense medium, it is refracted into the second medium if the angle of incidence is not too great. If the angle of incidence is increased to what is called the *critical angle of incidence*, the light no longer exits the first medium but is totally reflected back into it. This phenomenon, called *total internal reflection*, is what makes certain kinds of prisms and optical fibers work.

Consider the right-angle prism shown in FIG. 8-4. Light entering one face is totally reflected at the glass-air boundary at the hypotenuse. The reflected light emerges out of the adjacent face of the prism. Depending on how you tilt the prism with respect to the light source, you can get it to reflect internally in other ways.

Table 8-2. Parts List for Reflection Experiments

1	Laser
1	Rear-surface mirror (approx. 1 inch square)
1	Front-surface mirror (approx. 1 inch square)
1	Right-angle prism
1	Large paper sheet (such as 11 by 17 inches) to make cylinder
1	Long lens (for laser scanning)

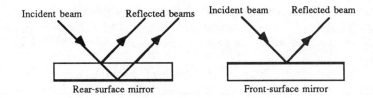

Rear-surface mirror   Front-surface mirror

FIG. 8-3. *A rear-surface mirror creates a ghost beam along with the main beam; a front-surface mirror reflects just one beam.*

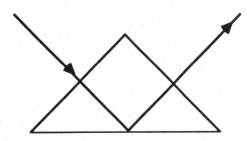

FIG. 8-4. *A right-angle prism allows you to deflect and direct light in a number of interesting ways.*

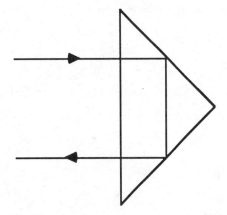

    The main advantage of total internal reflection is that it is more efficient than even the most highly reflective substances. Reflection is almost 100 percent, even in optics of marginal quality. Another advantage is that the reflective portion of the prism is internal, not external. That lessens the chance of damage caused by dust and other contaminants.

    You can readily experiment with total internal reflection (TIR) using just about any prism. A right-angle or equilateral prism works best. Shine the beam of a laser in one side of the prism. Now rotate the prism and note that at some angles, the beam is refracted out of the glass without bouncing around inside. At other angles, the light is reflected once and maybe even twice before exiting again.

    If the prism is poorly made—the glass has many impurities, for example—you can even see the laser beam coursing around inside the prism. This effect is best seen when the room lights are turned low or off. To see the effect of total internal reflection and refraction in a prism, make a tube by wrapping a large piece of white construction paper

107

or lightweight cardboard around the prism. Poke a hole for the light beam to go through and turn on the laser. If the paper is thin enough, you can see the laser beam striking it (looking from the outside). If the paper is thick, you'll have to look down the tube to see the beam on the inside walls.

Rotate the prism (or the laser and tube) and note how the beam is either refracted or internally reflected. Keep a notebook handy and jot down the effects as the prism is rotated. Use a protractor if you want to record the exact angles of incidence.

Another experiment in total internal reflection uses a lens. A long, single-axis lens (the kind shown in FIG. 8-5 like those used for laser scanning) works the best. Shine the light through the lens in the normal manner. Depending on the type of lens used, the beam will refract into an oval or slit. Now cant the lens at an oblique angle to the laser beam, as shown in the figure. If you hold the lens just right, you should see the laser beam internally reflected several times before exiting the other end. You can see the internal reflections much more clearly if the room lights are off.

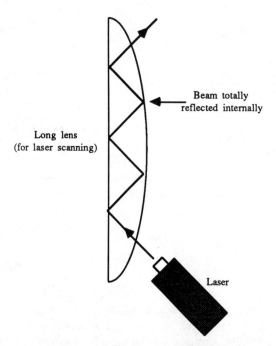

FIG. 8-5. *A long beam-spreading lens (or something similar) can be used to readily demonstrate total internal reflection. A solid glass rod can also be used.*

EXPERIMENTING WITH DIFFRACTION

TABLE 8-3 provides a parts list for the diffraction experiments that follow. Diffraction is a rather complex subject, thoroughly investigated by Thomas Young in 1801. Young's aim was to prove (or disprove) that light traveled in waves, and were carried along as particles in some invisible matter (the early physicists called this invisible matter the *ether*). If light was really made of waves, those waves would break up into smaller waves (or secondary wavelets) when passed through a very small opening (according to Huygens'

Table 8-3. Parts List for Diffraction Experiments

For Single Slit

2	Utility knife razor blades
1	2-by-3-inch acrylic plastic (⅛-inch thick)
2	6/32 by ½-inch bolts, nuts, washers

For Double Slit

2	Utility knife razor blades
1	2-by-3-inch acrylic plastic (⅛-inch thick)
1	Tungsten, wire, or other thin filament
4	6/32 by ½-inch bolts, nuts, washers

For Spectroscope

1	1-inch-diameter, 6-inch-long cardboard tube
1	Single slit (in cardboard or razor blade, as above)
1	Diffraction grating

General Experiments

1	Laser
1	Single slit
1	Double slit
1	Screen
1	Transmission diffraction grating (1-inch square or larger)

principle). The same effect occurs when water passes through a small opening. One wave striking the opening turns into many, smaller waves on the other side.

Young used a small pinhole to test his lightwave theory technique. Light exiting the pinhole would be *diffracted* into many small wavelets. Those wavelets would act to constructively or destructively interfere with one another when they met at a lightly colored screen. *Constructive interference* is when the phase of the waves are closely matched—the peaks and valleys coincide. The two waves combine with one another and their light intensities are added together. *Destructive interference* is when the phase of the waves are not in tandem. A peak may coincide with a valley, and the two waves act to cancel each other out.

What Young saw convinced him (and many others) that light was really made up of waves. Young saw a pattern of bright and dark bands on the viewing screen. The brightest and biggest band was in the middle, flanked by alternating light and dark bands. Bright bands meant that the waves met there on the screen constructively. Dark bands denoted destructive interference.

Making Your Own Diffraction Apparatus

Young's original experiments have been repeated in many school and industrial laboratories. Monochromatic light sources are the best choice when experimenting with diffraction. A laser is the perfect tool not only because its light is highly monochromatic but that its beam is directional and well-defined.

Use a slit instead of using a pinhole for your diffraction experiments. You can make a high-precision slit using two new razor blades, as shown in FIG. 8-6. You can use almost any type of blade, but be sure that they are new and the cutting surfaces aren't nicked. Mount the razor blades on a small piece of plastic by drilling a hole in the metal (if there isn't already one) and securing it to the plastic with ⁵⁄₃₂ or smaller hardware. Adjust the space between the blades using an automotive spark gap gauge. A gap of less than 0.040 inches is sufficient.

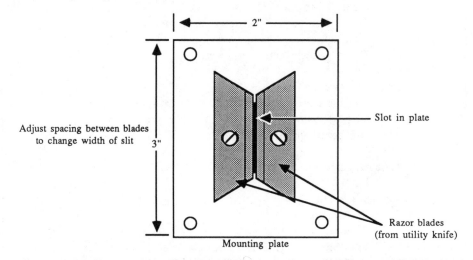

FIG. 8-6. *A single slit made by two razor blades butted close together. Mount the blades on a piece of ⅛-inch arylic plastic. Drill or cut a large slot in the center of the plastic for the beam to pass through.*

Shine the laser beam through the gap and onto a nearby wall or screen (distance between blades and wall: 2 to 3 feet). You should see a pattern of bright and dark bands. If nothing appears, check the gap and be sure that the laser beam is properly directed between the two blades. If the beam is too narrow, expand it a little with a lens. Note that the center of the diffraction pattern is the brightest. This is the *zero order fringe*.

If the fringe pattern is difficult to see, increase the distance between the blades and the screen. As the distance is increased, the pattern gets bigger. If the screen is 10 feet or more away (depending on the gap between the razor blades), the light-to-dark transitions of the fringes can be manually counted. Experiment with the gap between the razor blades. Note that, contrary to what you might think, the fringes become closer together with wider gaps. You achieve the largest spacing between fringes with the smallest possible gap.

Experiments with Double Slits

You can perform many useful metrological (measurement) experiments using a double slit. Follow the arrangement shown in FIG. 8-7. Place a single slit in front of a laser so that it is diffracted into secondary wavelets. Then place a double slit in the light path. The fringes that appear on the screen will be an interference/diffraction pattern similar to the fringes seen in the single-slit experiment above. With the double slit, however,

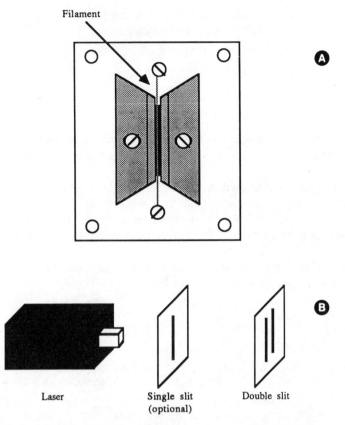

FIG. 8-7. *A double slit using a small filament (I used 30-gauge wire-wrap wire. Stretch the filament between the blades; adjust the gap between the blades so that it is even on both sides of the filament. (A) Details of the double slit. (B) arrangement of laser, single slit, and double slit for experiments to follow.*

it's possible to perform such things as measuring the wavelength of light, measuring the speed of light, or measuring the distance between the double slit and the screen.

Make a double slit by opening the gap between two razor blades and inserting a thin strip of tungsten filament. Secure the filament using miniature watch or camera screws. You can also use small gauge wire (30 gauge or higher) or a strand of hair (human hair measures about 100 micrometers across). Measure the filament or wire with a micrometer. Next, adjust the spacing between the razor blades so the gap is less than a millimeter or so across. The filament should be positioned in the middle of the gap.

Another more accurate method is to use a *Ronchi ruling*, which is a precision-made optical component intended for testing the flatness (or equal curvature) of lenses, mirrors, and glass. The ruling is made by scribing lines in a piece of glass. The spacing between the lines is carefully controlled. Typical Ronchi rulings come with 50, 100, and 200 lines per inch. At 50 lines per inch, the distance between gaps is 0.02 inch, or 0.508mm.

With the distance between gaps known, you can now perform some experiments. Set up the double slit precisely 1 meter (1,000 mm) from a white or frosted glass screen. Turn on the laser, and as accurately as possible, measure the distance between fringes. With a red helium-neon laser, the distance should be approximately 1.25 mm between fringes. If this proves too difficult to measure, increase the distance to 5 meters (5,000 mm). Now you are ready to perform some calculations.

Here is the formula for computing light wavelength.

$$w = \frac{f \times b}{d}$$

where w is the wavelength of the light, f is the distance between two bright fringes, b is the distance between the slits, and d is the distance between the slits and the viewing screen.

Measuring the fringes indicates they are 1.25 mm apart using a 50-line-per-inch Ronchi ruling. And you know that the distance between the slits is 0.508 mm. Multiply 1.25 times 0.508 and you get 0.635. Now divide the result by 1,000 (for the distance between slits and screen) and you get 0.000635. That is very close to the wavelength, in millimeters, of red He-Ne light.

You can interpolate the formula (using standard algebra) to find the distance between the slits and screen. The formula becomes:

$$d = \frac{f \times b}{w}$$

As an example, if the fringes measure 6.25 mm apart with a slit distance of .508 mm and the wavelength of the laser light is 0.000632 mm, the distance is 5,017 mm, or a little over 5 meters.

If you aren't sure of the spacing between slits, you can use the formula:

$$b = \frac{w \times d}{f}$$

The result is the spacing, in millimeters. Calculating the space between slits is handy if you use the tungsten-and-razor-blade method. Two other (less accurate) methods of making double slits include:

★ Photograph two white lines with color slide film and use the film as an aperture.
★ Scribe two lines in film, electrical tape, or aluminum foil.

You might also use this technique to measure small parts like machine pins, needles, and wire. The spacing of the fringes reveals the diameter or width of the part. Keep in mind that the best diffraction effects are obtained when the slits (and center object) have the thinnest edges possible. The fringes disappear with thicker edges.

Calculating Frequency, Wavelength, and Velocity

Knowing the wavelength of the light used in the diffraction experiments can be used to calculate the frequency of the light as well as its velocity. Use the following formulas for calculating frequency, wavelength, and velocity of light:

$$\text{velocity} = \text{frequency} \times \text{wavelength}$$

$$\text{frequency} = \frac{\text{velocity}}{\text{wavelength}}$$

$$\text{wavelength} = \frac{\text{velocity}}{\text{frequency}}$$

These figures will help you in your calculations:

- ★ Speed of light in a vacuum: 299,792.5 km/sec
- ★ Approximate speed of light in air: 299,705.6 km/sec (sea level, 30°C)
- ★ He-Ne laser light: 632.8 nm
- ★ Green line of argon laser: 514.5 nm
- ★ Blue line of argon laser: 488.0 nm

As an example, to calculate the frequency of red He-Ne laser light, take velocity (299,792.5) divided by wavelength (632.8). The result is 473.7555 terahertz.

Using Diffraction Gratings

A *diffraction grating* is a piece of metal or film that has hundreds or thousands of tiny lines scribed in its surface. The grating can be either transmissive (you can see through it) or reflective (you see light bounce off of it). Although the reflective type makes interesting-looking jewelry, it has limited use in laser experiments. A small piece (1-inch square) of transmissive diffraction grating can be used for numerous experiments. The exact number of scribes is not important for general tinkering, but one with 10,000 to 15,000 lines per inch should do nicely. Edmund Scientific and American Optical Center (see Appendix A) sell diffraction gratings and kits at reasonable cost.

The diffraction grating acts as an almost unlimited number of slits and disperses white light into its component colors. When used with laser light, a diffraction grating splits the beam and makes many sub-beams. These additional beams are the secondary wavelets that you created when experimenting with the diffraction slits detailed above. The beams are spaced far apart because the scribes in the diffraction grating are so close together.

The pattern and spacing of the beams depends on the grating. A criss-cross pattern shows a grating that has been scribed both horizontally and vertically. You can obtain the criss-cross material from special effects ''rainbow'' sunglasses sold by Edmund. Most gratings, particularly those used in compact disc players and scientific instruments, are scribed in one direction only. In that case, you see a single row of dots.

Besides breaking up the beam into many sub-beams, one interesting experiment is what might be called ''diffraction topology.'' The criss-cross rainbow glasses material shows the effect most readily. Put on the glasses and point the laser beam at a point in front of you. Tilt the glasses on your head and note that the sub-beams appear almost 3-D, as if you could reach out and grab them. Of course, they aren't there but the illusion seems real.

Now move the beam so that it strikes objects further away and closer to you. Not only does the apparent perspective of the sub-beams change, but so does the distance between the spots. The closer the object, the greater the perspective and the closer the spots are spaced to one another. Scan the laser back and forth and the perspective and distance of spots changes in such a way that you can visually see the topology of the ground and objects in front of you.

One practical application of this effect is to focus the diffracted light from the film onto a solid-state imager or video camera, then route the signal to a computer. A program running on the computer analyzes the instantaneous arrangement of the dots and correlates it to distance. If the laser beam is scanned up and down and right and left like the electron beam in a television set, the topology of an object can be plotted. The easy part is setting up the laser, diffraction grating, and video system; the hard part is writing the computer software! Anybody want to give it a try?

While you've got your hands on a diffraction grating, look at the orange gas discharge coming from around the tube. You'll be startled at all the bright, well-defined colors. Each band of light represents a wavelength created in the helium-neon mixture. The dark portions between each color represents wavelengths not produced by the gases.

If you can't readily see the lines with the diffraction grating you're using, you might have better luck with a home-built pocket spectroscope. Place a plastic or cardboard cap on the end of a 1-inch diameter tube. Saw or drill a slit in the cap, or make a gap using a pair of razor blades, as detailed earlier in this chapter. On the other end of the tube, glue on a piece of transmissive diffraction grating. Aim the slit-end of the spectroscope at the light source and view the spectra by looking at the inside of the tube, as shown in FIG. 8-8. Don't look directly at the slit.

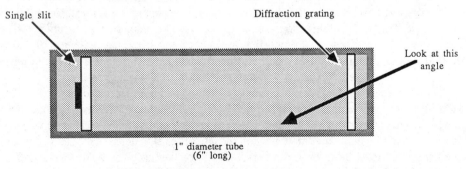

FIG. 8-8. *You can construct your own pocket spectroscope with a a short 6-inch long cardboard tube, slit (cutout or razor blade), and diffraction grating. View the diffracted light by looking at the inside wall of the tube. Rotate the tube (and grating) to increase or decrease the width of the spectra lines.*

POLARIZED LIGHT AND POLARIZING MATERIALS

When I was a kid, I learned about polarized light the same way that most other people did at the time—in ads for sunglasses. Specially made sunglasses somehow blocked glare by the magic of polarization. Not until I began experimenting with lasers did I learn the true nature of polarized light and how the sunglasses perform their tricks. Be aware that the subject of polarized light is extensive and at times complicated. The following is just a brief overview to help you understand how the experiments in this section work.

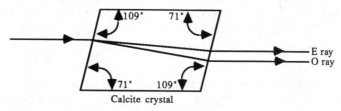

as one of television's early stars on
an With a Camera'' series.
g the '60s, Bronson worked hard at
doing a that one
that we soaring.
emorab were ''The M ificent
in 1960 ''The Great Esc e'' in
The Sanq per in 1965 ''The
ozen'' in Bron-
portunity to co-star with French
idol Alain Delon in ''Adieu
Suddenly the international world

FIG. 8-14. *Calcite exhibits birefringence, causing double images.*

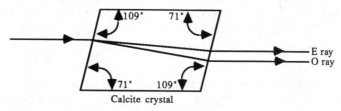

Calcite crystal

FIG. 8-15. *A single beam entering a calcite crystal is refracted into two beams—E and O rays. The refracted beams are orthogonally polarized.*

Retardation Plates

Retardation plates, typically made of very thin sheets of mica or quartz, are elements primarily used in the synthesis and analysis of light in different states of polarization. There are several types of retardation plates (also called *phase shifters*):

★ A *quarter-wave retardation plate* converts linearly polarized light into circularly polarized light, and vice versa.

★ A *half-wave retardation plate* changes the polarization plane of linearly polarized light. The angle of the plane depends on the rotation of the plate.

Both quarter- and half-wave plates find use in some types of holography, interferometry, and electro-optic modulation. They are also used in most types of audio compact disc players as one of the optical components. When coupled with a polarizing beam splitter, the quarter-wave plate prevents the returning beam, (after being reflected off the surface of the disc) from re-entering the laser diode. If this were to happen, the output of the laser would no longer be coherent. A quarter-wave plate can be similarly placed in front of a polarizing filter and laser (see FIG. 8-16) for use in the Michelson interferometer project detailed in Chapter 9. The retardation plate prevents light from reflecting back into the laser and ruining the accurate measurements possible with the interferometer.

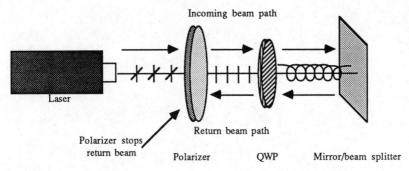

FIG. 8-16. *A quarter-wave plate, in conjunction with a polarizer, can be used to prevent a reflected beam from re-entering the laser.*

MANIPULATING THE LASER BEAM

Simple and inexpensive optics are all that's required to manipulate the diameter of the beam. You can readily focus, expand, and collimate a laser beam using just one, two, or three lenses. Here are the details. Refer to TABLE 8-5 for a parts list.

Focusing

Any positive lens (plano-convex, double-convex, positive meniscus) can be used to focus the beam of a laser to a small point. For best results, the lens should be no more than about 50 to 100 percent larger than the diameter of the beam, or the light won't be focused into the smallest possible dot. To use a particular lens to focus the light from

Table 8-5. Parts List for Beam Manipulation Experiments

1	Double-convex lens (in PVC or similar holder)
1	Double-concave lens (in PVC or similar holder)
1	Ruler (for measuring focal length)
1	Focusing screen

your laser, it might be necessary to first expand the beam to cover more area of the lens. Refrain from expanding the beam so that it fills the entire diameter of the lens.

Note that the beam is focused at the focal point of the lens, as illustrated in FIG. 8-17. The size of the spot is the smallest when the viewing screen (or other media such as the surface of a compact disc) is placed at the focal length of the lens. The spot appears out of focus at any other distance.

If you don't know the focal distance of a particular lens, you can calculate it by holding it up to a strong point source (laser, sun, etc.) and varying the distance between lens and focal plane. Measure the distance, as shown in FIG. 8-18, when the spot is the smallest.

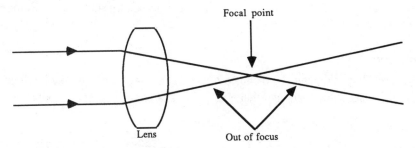

FIG. 8-17. *Positive lenses focus light to a point. The size of the beam increases at distances ahead or beyond this focal point.*

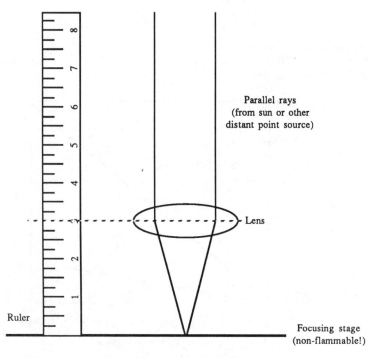

FIG. 8-18. *Measure the focal length using the setup shown here. The ruler can be marked off in inches or centimeters, as you prefer.*

Expanding

Any negative lens (plano-concave, double-concave, negative meniscus) can be used to expand the beam. An expanded beam is useful in holography, interferometry, and other applications where it is necessary to spread the thin beam of the laser into a wider area. You might also want to expand the beam to a certain diameter before focusing it with a positive lens.

The degree of beam spread depends on the focal length of the lens (remember that negative lenses have negative focal points because their focal point appears behind the lens instead of in front of it). The shorter the focal length, the greater the beam spread. If you need to cover a very wide area with the beam, you may combine two or more negative lenses. The beam is expanded each time it passes through a lens. Remember to use successively larger lenses as the beam is expanded.

Collimating

By carefully positioning the distance between the negative and positive lens, it's possible to enlarge the beam and make its light rays parallel again. A double-concave

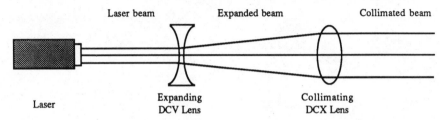

FIG. 8-19. *A common optical arrangement for lasers is the beam expander/collimator, created by coupling a negative and positive lens. Note that this arrangement is modeled after the design of a simple Galilean telescope, used in reverse.*

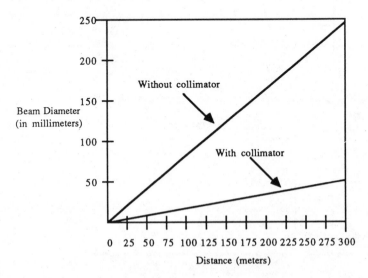

FIG. 8-20. *Beam divergence is greatly reduced over long distances by the use of an expander/collimator. This graph shows the approximate beam spread of an He-Ne laser with and without collimating optics.*

and double-convex lens positioned as shown in FIG. 8-19 make it a collimator, a useful device that can be used to enlarge the beam yet still maintain collimation and to reduce divergence over long distances. FIGURE 8-20 shows a graph that compares the divergence of the beam from a He-Ne laser with and without collimating optics. Although the collimator initially makes the beam wider, it greatly decreases divergence over long distances.

In order for the beam to be collimated, the distance between lenses A and B must be equal to the focal length of the double-convex lens B. Again, if you don't already know the focal length of this lens, determine it using a point light source and ruler. Collimation allows for the sharpest focus of the beam. Rather than simply expand the laser beam to fill the focusing lens, first expand and collimate it.

A small telescope (such as a spotting scope or rifle scope) makes a good laser beam collimator. Just reverse the scope so that the beam enters the objective and exits the eyepiece. An inexpensive ($10) sports scope can be used as an excellent collimating telescope. Mount the scope in front of the laser using clamps or brackets. Adjust for the smallest divergence by focusing the scope.

9

Build a Michelson Interferometer

Albert A. Michelson was the first U.S. citizen to win the Nobel prize for science. The 1907 award was given to him for his "precision optical instruments and the spectroscope and metrological investigations conducted herewith." Michelson was the inventor of a unique *interferometer* that opened up new vistas in the science of light and optics. With the interferometer, Michelson was able to determine—with unprecedented accuracy— the speed of light, the wavelength and frequency of light emitted by different sources, and the constancy of the speed of light through any given medium, among other feats.

Michelson's apparatus (actually a modified version of it) can be reasonably duplicated in the home shop or garage using a minimum of optical components. When finished, you'll be able to use your interferometer for many of the same experiments conducted by Michelson and others. You will learn first-hand about lightwave interference and discover some interesting and useful properties of laser light. The interferometer project presented provides an excellent springboard for a science fair project.

A SHORT HISTORY

Before launching into the construction plans for the interferometer, let's take a moment to review Michelson and his exotic contraption.

Albert A. Michelson was born in Prussia in 1872. As with many Europeans at the time, Michelson moved at the age of two with his parents to the United States. They settled in the wide expanse of Virginia City, Nevada, which was then experiencing a mining boom. Michelson's father was not a miner but a shopkeeper. He serviced the miners and local community with his dry goods store.

Even at an early age, Michelson showed great interest and adeptness in mathematics. His parents could not afford a private college but Michelson did manage to be accepted by the U.S. Naval Academy. After graduation and a short stint at sea, he accepted a job as instructor at the academy.

Michelson's foray into light and light physics came after one of his instructors asked him to prepare a demonstration of the Foucault method of measuring the speed of light. After studying the Foucault device, Michelson saw several ways that it could be made more accurate, and set about building an improved homemade version himself. In 1878 he repeated Foucault's earlier speed of light measurements and achieved results that were the most accurate to date.

A few years later, Michelson began designs on another apparatus that he believed could measure the speed of light with unmatched accuracy. Michelson also wanted to learn more about what scientists of his time called *ether*, an invisible and undetectable material that surrounded all matter, including space, planets, and stars. Though physical evidence of the ether had never been found, physicists knew it had to exist.

In the 1860s, James Clerk Maxwell had determined, on a theoretical level, that light consists of electromagnetic waves. Others postulated that if light really consists of waves, like water or sound waves, it must travel in some medium. That medium, though invisible, must be in every nook and cranny in the universe.

With a grant of about $500 from the Volta fund, Michelson began work on his first interferometer. The brass device, shown in the sketch in FIG. 9-1, consisted of two optical bench arms, each about three feet long. The arms were positioned in an asymmetrical cross, and in the apex of the cross, a beam splitter was placed. At the four ends of the arms he positioned a light source, two mirrors, and a viewing eyepiece.

In operation, a beam of light from an Argand lamp (popular in vehicles and drinking pubs as well as in the laboratory) was passed through a slit and lens, then separated by the beam splitter. The two beams were directed to a pair of fully silvered mirrors and re-directed back through the beam splitter, arriving superimposed over one another at the eyepiece.

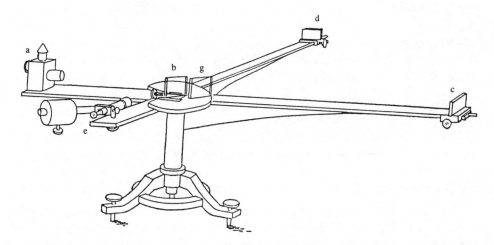

FIG. 9-1. *A preliminary sketch of the original Michelson interferometer.*

The function of Michelson's interferometer was to split the nearly monochromatic light from the lamp two ways. The two beams then traveled at right angles to one another, proceeding the same distance, and then recombined at some common converging point. As with all waves, *fringes* appeared as the two light beams recombined.

The fringing is the result of alternating reinforcement and canceling of the waves. Where the crests of two waves meet, the light is reinforced and Michelson saw a bright glow. Where the crest and valley of two waves meet, the light is canceled, causing a dark spot.

Though Michelson's device worked, it was extremely sensitive to vibration. Minute disturbances such as the changing of air temperature, the tap of a dog's tail on the floor 20 feet away, even human breathing causes the fringes to disappear. Serious experiments with the interferometer were not generally conducted until Michelson took the device to the Potsdam Astrophysical Observatory in Berlin and mounted it on the stable foundation designed for an equatorial telescope.

To test for the presence, direction, and speed of the ether, Michelson rotated the interferometer and looked for minute changes in the speed of light. If the ether existed, Michelson and others postulated, it would take longer for light waves to travel "upstream" through the medium than down or across it. By rotating the device at 90-degree intervals, Michelson hoped to find a direction where a shift in the fringes would show that the light was indeed taking longer to travel one path over the other.

Alas, Michelson could detect no differences or changes in the speed of light. Though the interferometer worked and promised many breakthroughs, he was disappointed and vowed to try again with a larger and more complex version.

It wasn't until 1905 that Albert Einstein declared that the ether did not exist and announced that light travels at the same speed in all directions. That explained why Michelson's interferometer detected no change. In the meantime, Michelson had devised several versions of his unpatented interferometer, including the interferential comparator for the standardization of the meter, a mechanical harmonic analyzer for testing the harmonic motions of interference fringes, and a stellar interferometer for measuring the size of stars. By the turn of the 20th Century, a number of firms regularly manufactured bench models of the interferometer, and it became common in laboratories around the world.

Even up to his death in 1931, Michelson kept busy trying to improve on his basic ideas. His last project was to measure, with greater accuracy then ever before, the speed of light through a mile-long vacuum tube. It's a pity that Michelson was not born later, or at least that the laser was not invented earlier. Had Michelson used a laser in his interferometer experiments, he would have realized an even finer measure of accuracy. With the benefit of hindsight, it's easy to see that modern physics owes a lot to Michelson's pioneering work. Devices and processes such as holography and the laser gyroscope are a direct outgrowth of the interferometer.

BUILDING YOUR OWN INTERFEROMETER

Michelson was plagued by the problems of weak fringes in his early interferometers because the light source he used was not entirely monochromatic, and it was certainly not coherent. With the aid of a helium-neon laser, you can construct a complete and working interferometer that displays vivid, easy-to-see fringes. The interferometer plans provided below are a modification of the Michelson apparatus. The design presented is generally referred to as the Twyman-Green interferometer, after its creators.

As with most optics experiments, a certain level of precision is required for the interferometer to work satisfactorily. However, careful construction and attention to detail should assure you of a properly working model. The design outlined below uses commonly available components and yields a device that is moderately accurate in determining the speed of light, light frequency, rate of rotation, and other observations. You might want to improve upon the basic design by adding components that provide a greater level of precision.

Component Overview

A parts lists is included in TABLE 9-1. The interferometer consists of an acrylic plastic base. Four bolts allow you to make fine adjustments in the level and position of the base. A single plano-concave or double-concave lens spreads the pencil beam of the laser to a larger spot. This allows you to see the fringes. In the center of the base is a glass plate beam splitter, positioned at a 45-degree angle to the lens. The beam splitter breaks the light into two components, directing it to two fully silvered mirrors.

One mirror is mounted on a sled that can be moved back and forth by means of a micrometer. The second mirror, placed at right angles to the first mirror, is stationary. The final component is a ground-glass viewing screen. After reflecting off the mirrors, the two beams are re-directed by the beam splitter and are projected onto the rear of the screen. When adjusted properly, the two beams exactly coincide and fringes appear. You can conveniently view the fringes on the front side of the screen.

Constructing the Base

Cut a piece of ⅜-inch acrylic plastic (Plexiglas) to 9½ by 7 inches. Keep the protective paper on the plastic until you are through with cutting and drilling to prevent chipping.

Be sure the plastic is square. Finish the edges by sanding or burnishing. Drill a series of mounting holes using the drilling guide shown in FIG. 9-2. Although there is room for some error, you should be as accurate as possible. Measure twice and use a drill press to ensure that the holes are perpendicular. Remove the protective paper from the plastic and set the base aside.

Constructing the Micrometer Sled

The micrometer sled is perhaps the most complicated component of the interferometer and will take you the longest to make. Note that the sled is optional; you don't need it if all you want to do is observe laser light fringes. The sled is required only if you wish to perform some of the more advanced experiments outlined in this chapter.

The sled consists of two 2½-inch lengths of ⁴¹⁄₆₄-by-½-by-¹⁄₁₆-inch extruded aluminum channel stock. The channel stock serves as a guide rail for the sled and is available at most hardware and building supply stores. Cut the stock with a hacksaw and miter box (to assure a perfect right-angle cut), then finish the edges with a file.

To avoid a mismatch, use the holes you drilled in the base as a template for marking the mounting holes for the channel stock. Place the stock on the base and mark the holes

Table 9-1. Michelson Interferometer Parts List

Base and Legs
1   9½-by-7-inch, ⅜-inch-thick acrylic plastic
4   ¼-inch 20 by 2-inch hex-head machine bolt
4   ¼-inch 20 nut
4   Plastic discs (for feet)

Micrometer Sled (One-Axis Translation Table)
2   2½-inch lengths of $^{41}/_{64}$ by ½-by-$^{1}/_{16}$-inch aluminum channel stock
2   1½-by-2¾-inch, ⅛-inch-thick acrylic plastic
2    $^{8}/_{32}$ by ½-inch bolts, nuts
4   #8 flat washers
2   #8 or #10 fender washers
4    $^{6}/_{32}$ by ½-inch bolt, nuts
12   #6 flat washers
2   ¼-by-¾-inch metal mending plate (one flat, one bent to make an angle bracket)
1   4-inch length $^{2}/_{56}$ threaded rod
2    $^{2}/_{56}$ locking nuts
3    $^{2}/_{56}$ brass nut (one for soldering to straight metal mending plate)
1    $^{2}/_{56}$ self-tapping wood screw
1   Plastic wheel (from R/C servo; for turning rod)

Sled Mirror
1   1¼-by-1½-inch front-surface mirror
1   Small metal bracket (Erector Set)

Side Mirror
1   1¼-by-2½-inch front-surface mirror
1   Small metal bracket (from Erector Set)
1    $^{8}/_{32}$ by ½-inch bolt, nut, flat washer

Beam Splitter
1   1-by-3-inch plate (coated) beam splitter
1   1½-inch, L-shaped metal girder (from Erector Set)
1    $^{8}/_{32}$ by ½-inch bolt, nut, flat washer

Lens
1   8 to 10 mm diameter double-concave lens, in plastic or metal holder (measuring 1-inch square)
2   2-inch flat metal strip (sheet metal, brass, or similar, from Erector Set)
1   1⅜-by-¾-inch, ⅛-inch-thick acrylic plastic
1    $^{8}/_{32}$ by ½-inch bolt, nut, flat washer

Screen
1   2½-by-3½-inch ground glass
1   1½-inch, L-shaped metal girder (Erector Set)
1    $^{8}/_{32}$ by ½-inch bolt, nut, flat washer

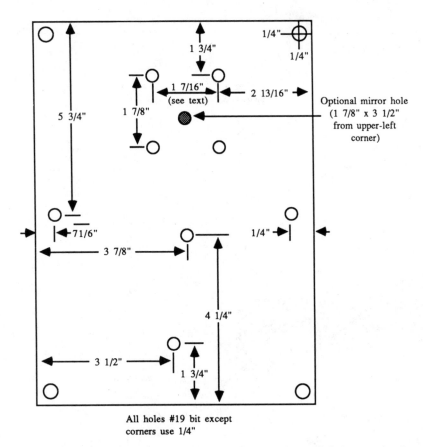

1/4"

1 3/4"

1/4"

1 7/16"
(see text)

2 13/16"

5 3/4"

1 7/8"

Optional mirror hole
(1 7/8" x 3 1/2"
from upper-left
corner)

7 1/6"

3 7/8"

1/4"

4 1/4"

3 1/2"

1 3/4"

All holes #19 bit except
corners use 1/4"

FIG. 9-2. *Drilling guide for the Michelson interferometer base. Note the gray hole for mounting the second mirror directly to the base.*

with a pencil. Drill the holes with a 5/32-inch bit. Remove the flash from the aluminum with a file.

During extrusion, the aluminum channel might develop slight ridges and imperfections along the surfaces and edges. Remove these blemishes by rubbing the top (undrilled side) and edge with a piece of 300-grit wet/dry sandpaper, used wet. The object is to make the surface as smooth as possible. Don't use a file, grinding stone, or other coarse tool, as these introduce heavy and difficult-to-remove ridges.

Set the aluminum pieces aside and construct the sled from two 1/8-inch-thick acrylic plastic pieces cut to 1½ inches by 2¾ inches. Sand and burnish the edges of the plastic to make them smooth (the exact dimensions of the plastic is not important, so you may safely remove some of it when finishing the edges). Clamp the two pieces together and drill two holes are shown in FIG. 9-3.

Finish the sled by attaching the hardware shown in FIG. 9-4. The angle bracket is a flat mending plate (¼-by-¾-inch, available at most hobby stores), bent 90 degrees at the middle. Before inserting the fender washers, select two, butt them together, and check for size. The two washers should be exactly the same size and should not have

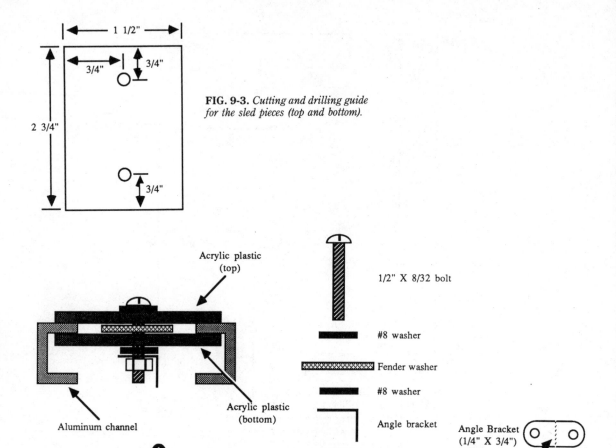

FIG. 9-3. *Cutting and drilling guide for the sled pieces (top and bottom).*

FIG. 9-4. *Use the hardware shown to assemble the sled. (A) Assembled sled (end view); (B) exploded view showing the hardware to use; (C) Bend a small, flat metal plate (available at hobby stores) to make the angle bracket.*

any ridges. Continue the selection process until you find two washers that are exactly the same size.

The mirror mount in the prototype was salvaged from an Erector Set. You can use a similar mounting bracket as long as it is sturdy yet can be bent slightly. Metal is the best choice because it prevents excessive vibration, yet heavy angle irons can't be easily bent. If you can't scrounge up a bracket from an Erector Set, check around a hobby shop for ideas. You are bound to find something.

Assemble the pieces of the sled using the ½-by-⅜₂-inch hardware as shown, but don't tighten the bolts just yet. Mount the two aluminum rails using ⁶⁄₃₂-by-½-inch bolts. Use two #6 washers stacked on top of one another as spacers. Attach nuts to each bolt but don't tighten them. A mounting assembly detail is shown in FIG. 9-5.

FIG. 9-5. *End view of the rails. Use flat washers as spacers. Be sure the two channels are parallel.*

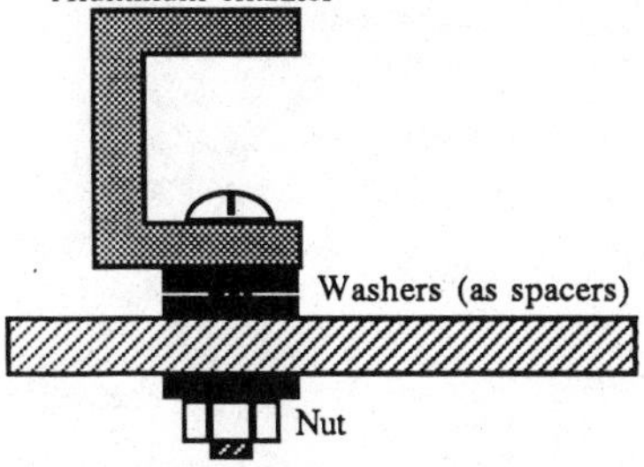

Aluminum channel

Washers (as spacers)

Base

Nut

1/2" X 6/32 bolt

Carefully insert the sled between the rails and jiggle things around until everything fits. Center the sled between the two rails and tighten the rails. Be absolutely sure that the rails are parallel or the sled won't travel properly. Once the rails are tightened in place, slip the sled back and forth until it rides evenly on both rails. Slowly tighten the nuts sandwiching the top and bottom sled pieces together.

If everything fits properly, the sled should mount on the rails with little or no side-to-side motion. If there is excessive play, loosen the nuts holding the rails and push the rails closer together, remembering to keep them parallel. After careful adjustment, you should be able to get the rails close enough together so that the sled still has freedom to move back and forth, but with no side-to-side play.

If the sled doesn't move even when the rails are pushed apart, the washers might be too thin, causing the top and bottom plastic pieces to clamp against the rails. Try slightly thicker washers. A micrometer comes in handy at this point to accurately measure the thickness of the washers. The aluminum rail should be $1/16$-inch thick, or very close to it. The washers should be just slightly thicker.

Should you find that the sled binds at one end or another, the rails are not parallel. Loosen and readjust them as necessary. If both ends of the sled are not centered between in the rails, it might move at an angle and cause considerable problems when you attempt to use the interferometer. Assuming that the plastic pieces of the sled are square, there should be equal distance between the edges of the sled and the rails. If not, loosen the two nuts holding the sled together and readjust as necessary.

The sled is now almost complete. Pull the sled out of the rails. Attach two $2/56$ locking nuts and a 4-inch length of $2/56$ threaded rod to the angle bracket on the underside of the sled, as shown in FIG. 9-6. Solder a $2/56$ brass nut to a $1/4$-by-$3/4$-inch flat mending iron (the same kind used to make the angle bracket above), being careful not to spill molten solder inside the threads. One or two small spots of solder tacked to the outside of the nut should be sufficient (it took me three tries to get it right, so don't despair if your first attempt doesn't work out).

With a small bit, drill a hole $3½$ inches from either side of the base (see FIG. 9-7). Thread the free end of the rod through the nut/mending iron, insert the sled between the rails, then secure the mending iron using a $2/56$ self-tapping screw. Finish the sled by mounting a small pulley or plastic wheel to the end of the rod. I used a $1\frac{3}{16}$-inch diameter plastic hub designed for model airplane servos. The finished sled, mounted on the base and with mirror attached (plans below), is shown in FIG. 9-8.

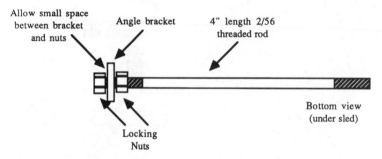

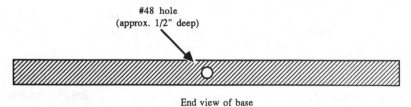

Allow small space between bracket and nuts

Angle bracket

4" length 2/56 threaded rod

Bottom view (under sled)

Locking Nuts

FIG. 9-6. *The micrometer screw is a 4-inch length of ⅖₆ threaded rod. Secure it to the angle bracket (from FIG. 9-4) with two ⅖₆ locking nuts.*

#48 hole (approx. 1/2" deep)

End view of base

FIG. 9-7. *Drilling guide for the micrometer screw.*

FIG. 9-8. *The finished sled (shown with mirror attached).*

Note that although the micrometer sled is fairly accurate and has a usable resolution of a fraction of a millimeter, it is not precise enough for some light measurement applications. Methods of improving accuracy are provided later in this chapter.

Mounting the Mirrors

One of the fully silvered mirrors attaches to the bracket on top of the sled. Use epoxy or a general-purpose adhesive to stick the mirror against the bracket. The prototype used Duco cement, which dries fairly quickly, cures overnight, and does not discolor or fog the mirrors. The mirror attached to the sled should measure approximately 1¼ wide by 1½ inches tall.

If you don't use the adjustable sled, you can mount the mirror and bracket directly to the base. The drilling template for the base shows the location for the rear mirror when no sled is used.

The second fully silvered mirror is mounted along the right edge of the base, using the same kind of bracket as the one attached to the sled. The second mirror should measure approximately 1¼ by 2½ inches. Secure the bracket for the second mirror using a ⁸⁄₃₂-by-½-inch bolt and matching hardware. Do not overtighten.

Mounting the Beam Splitter, Lens, and Viewing Screen

The beam splitter can be ordinary plate glass, but its thickness should not exceed ³⁄₃₂-inch. Internal reflections in thicker glass can cause a separate satellite beam (that beam can be removed using electrical tape, as shown in Chapter 3, but the close distances between optics make this a difficult task). The prototype used a coated, flat, glass plate measuring 1 by 3 inches. You can get by with a piece that's shorter, but stay away from beam splitters that are less than one inch wide.

A right-angle girder piece stolen from an Erector Set serves as an excellent mount for the beam splitter. Cut a piece of girder to a length of three holes wide, as shown in FIG. 9-9, and glue the beam splitter to the outside edge. When the cement is dry, attach the beam splitter and bracket to the center hole of the base using an ⁸⁄₃₂-by-½-inch bolt and hardware. Before tightening, position the beam splitter so that it is at a 45-degree angle to the sides of the base. Don't overtighten.

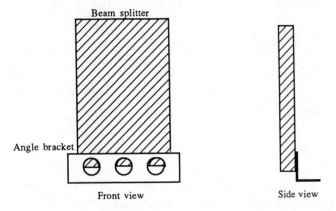

FIG. 9-9. *Mount the beam splitter on an Erector Set (or equivalent) metal angle bracket.*

The small beam from the laser must be expanded before passing through the beam splitter. A suitable choice is a plano-concave or bi-concave lens approximately 8 to 10 mm in diameter. Focal length isn't a major consideration, but you might need to choose another lens if the spot on the screen is excessively small or large.

Mount the lens using one of the techniques outlined in Chapter 7, "Constructing an Optical Bench." A metal or plastic frame is a good choice. Use flat girders from an Erector Set or similar toy, bent at right angles at the bottom. Cement the bottom of the bracket to a piece of 1⅜-by-¾-by-⅛-inch acrylic plastic. Drill a hole with a #19 bit in the center of the plastic and attach it to the base using 8/32-by-½-inch bolt and hardware.

The screen is made from a 2½-by-3½-inch piece of ground glass (plastic can also be used). Mount the glass on the same type of girder used for the beam splitter. Once the cement is dry, attach the glass to the base using an 8/32-by-½-inch bolt and matching hardware. Position the screen so that it is parallel to the front edge of the base. Once more, don't overtighten the hardware.

Attaching the Legs

Finish the interferometer by attaching the legs to the four corners. Tap the four corner holes with a ¼-inch 20 tap. Thread ¼-inch 20 with 2-inch machine bolts (threaded all the way) into the holes, and secure them into position with a ¼-inch 20 nut. You can adjust the overall height of the interferometer by loosening the nut and turning the bolt. Retighten the nut when the base is at the proper height.

Leveling discs or "feet" can be added to the tips of the bolts, as shown in FIG. 9-10. The discs used in the prototype were ¾-inch-diameter by ¼-inch-deep plastic chassis spacers. You can use just about anything for the feet in your interferometer, including plastic or metal stand-offs, acrylic discs drilled and tapped for ¼-inch 20 hardware, or plastic torriod coil cores. A visit to any well-stocked industrial surplus or supply outlet should yield several good alternatives. You can compensate for slight differences in height by turning the discs one way or the other. The finished interferometer, with sled and feet, is shown in FIG. 9-11.

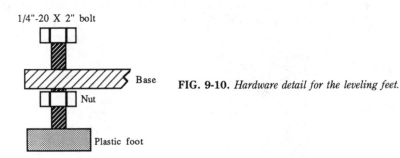

FIG. 9-10. *Hardware detail for the leveling feet.*

Adjustment and Checkout

Now comes the fun part. Set the interferometer on a hard surface, preferably on a concrete floor or sturdy table covered with carpet or foam. Even minute vibrations will upset the fringes, so it is important that the interferometer be placed on a stable platform. Thoroughly clean all of the optics, including the lens. Smudges, dirt, and

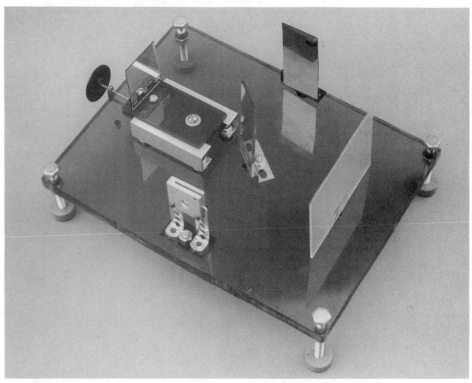

FIG. 9-11. *The finished interferometer, with all components attached and aligned.*

fingerprints may prevent the appearance of the fringes or make them extremely difficult to see.

Turn the wheel on the sled so that the distance between the mirrors and beam splitter is approximately the same (about 3 inches). Exact positioning of the sled is not important. (Obviously, if you didn't construct the sled, you can ignore this step.)

Position a laser in front of the lens and direct the beam through its center. Shim the front or back of the laser so that the beam is parallel to the base. Start out with the lens perpendicular to the left edge of the base. If the expanded beam doesn't strike the center of the beam splitter, move the laser to the right or left as needed.

As depicted in FIG. 9-12, one half of the beam should pass through the beam splitter and strike the side (say the #1) mirror. The other half of the beam should reflect off the beam splitter and strike the rear (#2) mirror. If the beam doesn't hit the #2 mirror, adjust the angle of the beam splitter as needed.

The reflected beam from the #2 mirror should pass directly through the beam splitter and hit the rear of the screen. It should appear as a round, fuzzy, red dot. The reflected beam from mirror #1 should strike against the beam splitter and also project onto the rear of the screen. Most likely, this beam will not match up with the first one. Adjust differences in horizontal spacing by rotating the #1 mirror on its mount. You can compensate for differences in vertical alignment by *carefully* bending the #1 and/or #2 mirrors up or down. Bend the brackets to stress the metal only, not the mirror, or you might break the glass.

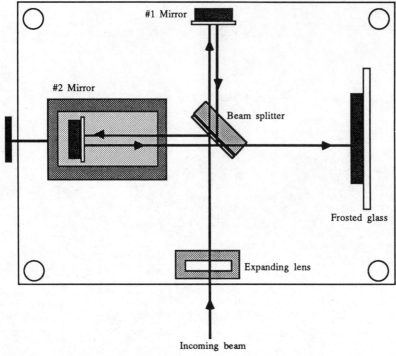

FIG. 9-12. *The beam from the laser should follow the path shown here. Be sure both reflected beams from the mirrors align perfectly at the frosted glass.*

Continue adjusting the mirrors, lens, and beam splitter until the beams coincide on the screen. It's vitally important that the two beams be as parallel as possible. Any slight incline, either horizontally or vertically, of one beam to the other will prevent the fringes from appearing.

After each adjustment, especially after bending the mirror mounts, wait 15 to 30 seconds for the vibrations in the interferometer to settle. Once you are satisfied that the beams are as parallel to one another as possible, wait a few minutes and see if the fringes appear. Although the fringes should have a concentric bullseye or ring appearance, the pattern you see may begin at the outside edges of the patch. You can center the beam with the rings simply by moving the laser. You will note that moving the laser doesn't disturb the rings nor their position on the screen. Moving the laser too much will cause the fringes to disappear.

Once the fringes show up on the screen, you can lightly tighten the optical components to the base. After tightening a nut and bolt, inspect the beam to make sure that the fringes are still visible. You'll note that even a small thump on the base of the interferometer causes the fringes to disappear. Depending on where you placed the interferometer, vibrations through the ground or table can also cause the fringes to go away. Once the vibrations settle, the fringes should reappear.

The loss of distinct fringes is a nuisance in many laser projects such as interferometry and holography, but it is handy in other applications. For example, you can detect even faint motion around a given perimeter by placing a phototransistor in front of the ground glass screen. Changes in the fringes appear as voltage or current fluctuations in the transistor, and an alarm circuit can announce a possible breach of security. You can even hear the disturbance by connecting an audio amplifier to the phototransistor.

If you built the sled, you can experiment with the effects of changing the distance of the two light paths. Slowly turn the sled wheel and watch the fringes. Even a slight movement of the wheel should cause the fringe rings to move in and out. Turning the wheel one way causes the rings to grow from the inside. Turning the wheel the other way causes the rings to shrink towards the center.

By looking at the center spot of the rings, you can move the sled in precise increments. Each change from light to dark denotes a movement of one-half of one wavelength. When using a helium-neon laser, one half a wavelength is 316.4 nanometers (632.8 nm ÷ 2), or 316.4 billionths of a meter! As you might guess, it's hard to turn the wheel so that you see only one complete light-to-dark or dark-to-light transition. The section below shows how to increase the precision of the sled so that you can use it for accurate laboratory experiments.

If the Fringes Don't Appear

In some instances, no amount of tweaking and adjustment can coax the fringes to appear. What now? If you have constructed the interferometer as described in the text and both beams of light are transversing the paths as indicated in FIG. 9-12, above, then the fringes are bound to show up sooner or later. Continue experimenting with the position of the optics until they appear.

Misalignment isn't the only cause of absent fringes. Excessive vibrations, even ones that you and your dog can't feel, can mask the fringes. Be sure that the interferometer is on a solid base. I first tried the prototype on a carpeted floor and obtained adequate but frustrating results. Things got better when I took the contraption outside and placed it on the concrete in the garage.

If you suspect excessive vibration and just can't seem to find a vibration-free spot, try planting the interferometer in a sand box. A similar type of sandbox is used for home-brew holography and is used for the same reason: to eliminate vibrations that cause the fringes to shift.

Loosely mounted optics can amplify even minute vibrations. Be sure the mirrors are securely cemented to their mounts and that the brackets are fairly tight on the base.

As you learned in Chapter 3, ''Introduction to Optics,'' air is a refracting medium. Like a lens, air causes light to bend and change speeds. If the air is moving around the interferometer, the density and therefore the index of refraction is constantly changing. Understandably, this has a dramatic impact on the appearance of fringes. Avoid placing the interferometer in a drafty place or where there may be a sudden change in temperature.

If you've just taken the interferometer from a cool to a warm atmosphere, the optics will slowly warm up and expand. This expansion, invisible to the eye, can also inhibit the appearance of the rings. Wait at least 15 to 30 minutes to acclimate the interferometer to a change in environment.

MODIFYING THE INTERFEROMETER

There are a number of ways to improve on the basic interferometer. By using a rod that has more threads per inch, you achieve greater accuracy over linear position. A quick bit of math, however, reveals that no single rod can be machined accurately enough to give a resolution good enough to turn the wheel precisely at ½-wavelength intervals.

For example, with the 2⁄56 threaded rod used, there are 56 threads per inch. Each revolution, then, is 1⁄56 of an inch, or about 0.45 mm. With practice, you can move the wheel in 1-degree increments, or 0.00125 mm at a time. Smaller hardware exists with up to 160 threads per inch, equal to about 0.159 mm per revolution, or 0.00044 mm per degree of rotation. Hardware this fine is hard to get and is fragile. In order for you to "dial" in a half wavelength of time, the sled needs to have a positional accuracy of greater than 0.0003164 mm! (316.4 nm); that's even less than you can hope for using the extremely small rods with 160 threads per inch.

An easier way to approach this accuracy is by using a gearing mechanism that reduces your movements to a snail's pace. A suitable gearbox can be obtained by salvaging the gearing mechanism of a small stepper or dc motor. Attach the control wheel to the input of the gear box; attach the output of the gearbox to the interferometer sled using a rubber band or rubber belt. The belt helps isolate the interferometer from vibration.

Using the more commonly available 2⁄56 threaded rod, you can achieve a positioning accuracy of about 0.00014 mm (140 nm) using the 16:1 gear box supplied with a typical surplus dc stepper motor (the gear ratio may differ depending on the exact model).

Even with an improved gearing system, however, the sled may still not offer sufficient resolution for some applications. There is a limit to what garage shop tinkering can do. A machinist can rebuild the sled using aluminum stock and a manual or numerically controlled mill. In addition, a number of ready-made products that do the same thing are available. Industrial manufacturers sell optical bases (called *translation stages*) that have extremely fine precision. Similar (but generally less expensive) models with micrometer adjustment bases are available at most any machinist supply outlet.

One other method is to use the works of a student's micrometer. These are available for $20 or less at many hardware stores and have a measurement accuracy of 0.001 of an inch (0.0254 mm), but you can dial in smaller amounts. With proper gearing and careful control of the knob, you can obtain far greater resolution. A 16:1 gear ratio—which you can make yourself using small plastic or brass gears pulled from a small dc motor—should provide enough accuracy to move the mirror at half-wavelength steps.

Another modification of the interferometer is removing the viewing screen. By removing the screen, you can project the fringe pattern on a wall or other surface. The larger bullseye makes viewing and counting the fringes easier. Calibrate and graduate the screen for easier measurements.

Try bouncing the fringe pattern onto a separate, larger, rear-projection (frosted glass or plastic) screen. A graduated and calibrated magnifier (such as those used in the optics and publishing trades), can then be placed directly against the screen without worry of upsetting the interferometer.

The Michelson interferometer can be used with a number of light sources. If you have other lasers that operate at different wavelengths such as argon or krypton, you can compare fringe patterns and calculate the differences in wavelengths. Both argon and krypton lasers emit several strong lines of visible light; you can separate these with a prism or dichroic filter. After separation, the beam can be sent through the lens of the device.

INTERFEROMETRIC EXPERIMENTS

While the Michelson interferometer provides a wealth of hands-on experience in optics, interference, and lasers, it's nice to be able to actually do something with the contraption. Here are some ideas.

Structural Stress

Remove mirror #2 from the slide and mount it on a wall. Position the interferometer base close to the wall but make sure the device doesn't touch the wall. Apply pressure on the wall (anywhere) and you should see a shift in the fringes. Even a brick wall under light pressure by a child's hand will show some movement.

If the stress on the wall is not too great, project the fringe pattern on a larger surface. This enables you to more accurately measure the distance of travel. Each light-to-dark or dark-to-light transition of the center bullseye in the pattern denotes a change of 316.4 nanometers. You'll find that a wood or plaster wall can bow so much that you'll spend the greater part of the evening counting fringes!

Linear Measurement

By attaching a small pointer to the sled, you can measure the size of objects with amazing accuracy. Again, each transition of the bullseye patterns marks a change of 316.4 billionths of a meter. With an extra bit of work, it's possible to locate the stage of a microscope on the base of the interferometer and sled. With the microscope, the pointer (such as a tungsten filament or even a strand of human hair) can be more easily seen than with the unaided eye.

Study Effects of Refraction

Placing any object in front of either mirror #1 or #2 causes a shift in the time it takes for light to traverse the two paths. You can study the effects of refraction in air by blowing gently through a tube. Place the end of the tube in either optical path and watch the fringes move. Try other objects like lenses, smoke, and water.

Similarly, you can explain the shimmer of a desert mirage by heating up the air around the interferometer and watching the fringes appear and disappear. Although a mirage doesn't involve lasers, you can easily see how a rise in temperature causes a change in the refractive index of air. A "real" mirage looks like a shimmering oasis that awaits a weary traveler, but in reality, it is air set in motion by the heat. The different densities of the air cause unusual refraction effects.

Fringe Counter

Some experiments move the fringes too quickly and all you see is a blur. A counter circuit can be used to count the number of light-to-dark or dark-to-light transitions of the shifting fringe pattern, even if the fringes move several thousand times per second. See FIG. 9-13 and parts list in TABLE 9-2.

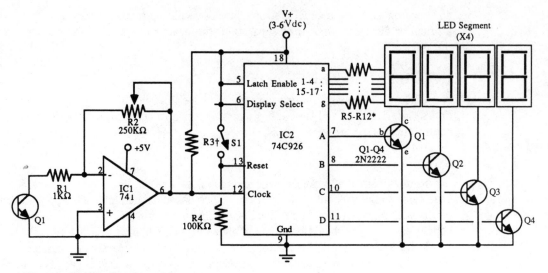

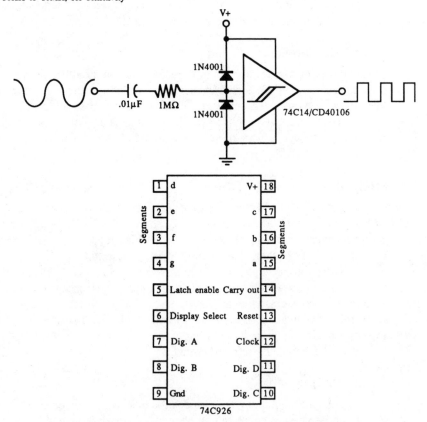

* Not required if +V is 4V or less
† Optional, 10KΩ to 10MΩ, for sensitivity

FIG. 9-13. *Circuit schematic for counting fringes. (A) Complete circuit using the National Semiconductor 74C926 all-in-one counter chip; (B) Adding a Schmitt trigger to provide a clean square wave input for the counter chip (insert it between pin 6 of IC1 and pin 12 of IC2; (C) Pinout diagram for the 74C926.*

Table 9-2. Fringe Counter Circuit Parts List

Full Counter

IC1	741 op amp
IC2	National Semiconductor 74C926 integrated four-digit counter IC
R1	1 kilohm resistor
R2	250 Kilohm potentiometer
R3	10 kilohm to 10 megohm resistor
R4	100 kilohm resistor
R5-R12	330 ohm resistor (not required if supply voltage is under 5 Vdc)
Q1-Q4	2N2222 transistor
Q5	Infrared phototransistor
LED1-4	Common-cathode seven-segment LED display
S1	SPST switch

Optional Wave-Shaping Electronics

1	0.01 μF disc capacitor
1	1 megohm resistor
2	1N4001 diode
1	74C14 or 40106 Schmitt trigger IC

All resistors are 5 to 10 percent tolerance, ¼ watt. All capacitors are 10 to 20 percent tolerance, rated 35 volts or more.

Remove the ground-glass viewing screen on the interferometer. Place the phototransistor behind a simple focusing lens (such as a 20 to 40 mm focal length biconvex lens), at least two or three feet from the interferometer. At this distance, the fringe pattern should be fairly large and the lens and transistor should be able to discriminate separate circular fringes.

Connect the counter to the output of the amplifier and reset it to 0000. Move the sled and watch the counter. It should read some number. Note that the accuracy won't be 100 percent, but the counter should be able to read at least 90 to 95 fringe changes out of 100 (accuracy drops dramatically if the interferometer is exposed to vibrations). You can improve the count accuracy by turning out all room lights.

OTHER TYPES OF INTERFEROMETERS

The Michelson/Twyman-Green apparatus is only one of many types of interferometers developed over the last 75 years or so. A variety of interferometer types are shown in FIG. 9-14. These interferometer designs using corner cubes do not reflect the the beam back into the laser cavity. This back-to-the-source reflection can perturb the laser wavelength, making fringe counts meaningless. Note that more accurate fringe counts can also be obtained using thick plate beam splitters or cube beam splitters, where unwanted reflections and satellite beams are either non-existent or can be masked off

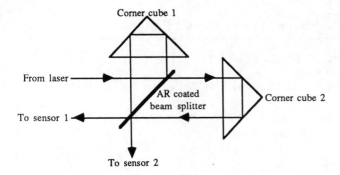

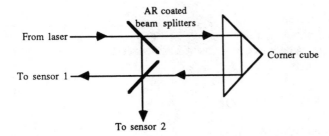

FIG. 9-14. *Corner cubes can be used to make interferometers that don't cause reflected light to re-enter the laser. Three different approaches are shown here.*

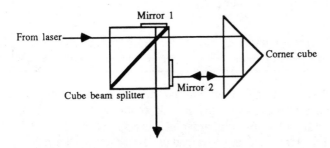

using black tape. (Full details on both plate and cube beam splitters, as well as corner-cube (porro) prisms, are in Chapter 3, "Introduction to Optics.")

Another type of interferometer is shown in FIG. 9-15 (a parts list for this and the remaining experiments in this chapter is in TABLE 9-3). This is called a Lloyd's Mirror interferometer and consists of a double-concave lens, a double-convex lens, and a

Table 9-3. Lloyd's Mirror Interferometer Parts List

1	Laser
1	Double-concave lens (10 to 20 mm in diameter)
1	Double-convex lens (20 to 30 mm in diameter)
1	Microscope slide
1	Viewing screen

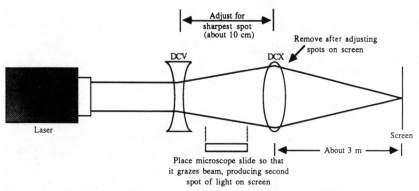

FIG. 9-15. *A Lloyd's Mirror interferometer consists of a laser, two lenses, and microscope slide (or other piece of flat glass). Arrange the components as shown and watch the interference fringes appear at the screen.*

microscope slide. By placing the components as shown in the figure, it's possible to calculate the wavelength of the light using the formula for double-slit diffraction (see the previous chapter for more details). Note that the double-convex converging lens is removed from the light path in order to see the fringes on the viewing screen.

One interesting interference effect can be used to dazzle an audience during a light show. Simply shine a laser beam onto a front-surface mirror and position the mirror so that the beam strikes a wall or ceiling. Dip a cotton swab in rubbing alcohol and spread the alcohol over the mirror. As the alcohol dries, you see constantly moving lightforms swirling on the wall or ceiling. Tilt the mirror at an angle and some of the alcohol will run down the mirror producing more effects.

Other effects of interference can be demonstrated using a microscope slide. Hold the slide up to a slightly expanded laser beam, as shown in FIG. 9-16. Some of the light is internally reflected inside the slide until it finally exits and strikes the screen. Interference fringes appear on the screen because of the many reflections of light inside the glass.

Another experiment shows the effects of interference caused by heat expansion. Spread the beam slightly with a double-concave lens and shine it through a microscope slide. Touch a hot soldering iron to the glass and watch the the fringes appear around the point of contact with the iron.

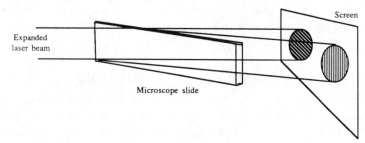

FIG. 9-16. *An expanded beam passed through a microscope slide shows interference fringes when the beam spots are projected on a screen.*

10

Introduction to
Semiconductor Lasers

The typical 5 mW helium-neon gas laser measures almost two inches in diameter by 10 to 15 inches in length. Imagine stuffing it all into a size no larger than the dot in the letter *i*! Such is the semiconductor laser. A close relative to the ordinary light-emitting diode, the semiconductor laser is made in mass quantities from wafers of gallium arsenide or similar crystals.

In quantity, low- to medium-power semiconductor lasers cost from $5 to $35. Such lasers are used in consumer products such as compact audio disc players and laser disc players, as well as bar-code readers and fiberoptics data links. With the proliferation of these and other devices, the cost of laser semiconductors (or laser diodes) is expected to drop even more.

This chapter presents an overview of the diode laser: how it's made, the various types that are available, and how to use them in your experiments. The low cost of semiconductor lasers— typically $10 on the surplus market—make them ideal for school or hobbyist projects where a tight budget doesn't allow for more expensive gas lasers.

THE INSIDES OF A SEMICONDUCTOR LASER

The basic configuration of the diode laser (sometimes called an *injection laser*) is shown in FIG. 10-1. The laser is composed of a pn junction, similar to that found in transistors and LEDs. A chunk of this material is cut from a larger silicon wafer, and the ends are cleaved precisely to make the diode chip. Wires are bonded to the top and bottom. When current is applied, light is produced inside the junction. As it stands, the device is an LED—the light is not coherent.

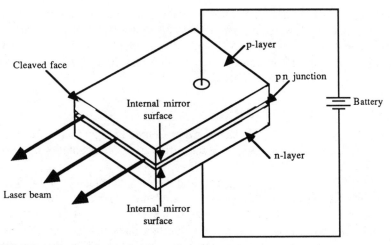

FIG. 10-1. *Design of a semiconductor laser chip showing cleaved face and pn junction.*

An increase in current causes an increase in light output. The cleaved faces act as partially reflective mirrors that bounce the emitted light back and forth within the junction. Once amplified, the light exits the chip. This light is temporally and spatially coherent, but because of the design of the diode chip, is not very directional. The beam of most laser diodes is elliptical, with a spread of about 10 to 35 degrees.

The first laser diodes, created in 1962 shortly after the introduction of the ruby and helium-neon lasers, were composed of a single material forming one junction—a *homojunction*. These could be powered only in short pulses because the heat produced within the junction would literally cause the diode to explode. Continuous output could only be achieved by dipping the diode in a cryogenic fluid such as liquid nitrogen (with a temperature of −196 degrees C, or −320 degrees F).

As manufacturing techniques improved, additional layers were added in varying thicknesses to produce a *heterojunction diode*. The simplest heterojunction semiconductor lasers have a gallium arsenide (GaAs) junction topped off by layers of aluminum gallium arsenide (AlGaAs). These can produce from 3 to 10 watts of optical output when driven by a current of approximately 10 amps. At such high outputs, the diode must be operated in pulsed mode.

Typical specifications for *single heterostructure* (sh) lasers call for a pulse duration of less than 200 nanoseconds. Most drive circuits operate the diode laser conservatively with pulse durations under 75 or 100 nanoseconds. Output wavelength is generally between 780 nm and 904 nm.

A *double heterostructure* (dh) laser diode is usually made by sandwiching a GaAs junction between two AlGaAs layers. This helps confine the light generated within the chip and allows the diode to operate continuously (called *continuous wave*, or *cw*) at room temperature. The wavelength can be altered by varying the amount of aluminum in the AlGaAs material. The output wavelength can be between 680 nm and 900 nm, with 780 nm being most common.

Power output of a double heterostructure laser is considerably less than with a single heterostructure diode. Most dh lasers produce 3 to 5 mW of light, although some high-

145

output varieties can generate up to 500 mW yet can still be operated at room temperature (indeed, some high-cost cw lasers can produce up to 2.6 watts of optical power, but these are rare and very expensive). High-output laser diodes come in T0-3 transistor-type cases and are mounted on suitable heat sinks. A typical application for high-output lasers is long-haul (long distance) fiberoptic data links.

POWERING A DIODE LASER

Drive circuits for both sh and dh lasers are presented in Chapter 11, "Laser Power Supplies." But it's worthwhile here to discuss the drive requirements necessary for operating diode lasers.

Single heterostructure lasers are typically driven by applying a high-voltage, short-duration pulse. The duration of the pulse is controlled by an RC network, as shown in the basic schematic in FIG. 10-2, and the pulse is delivered by a power transistor. Care must be exercised to ensure that the pulse duration does not exceed the maximum specified by the manufacturer. Longer pulses cause the laser to overheat, annihilating itself in a violent puff of smoke.

Double heterostructure semiconductor lasers can be operated either in pulsed or cw mode. In pulsed mode, the diode is driven by short, high-energy spikes, as with an sh laser. Power output may be on the order of several watts, but because the pulses are short in duration, the average power is considerably less. In cw mode, a low-voltage constant current is applied to the laser outputs in a steady stream of light. Cw lasers and drive circuits are used in compact disc players where the light emitted by the laser is even more coherent than the beam from the revered He-Ne tube.

Forward-drive current for most cw lasers is in the neighborhood of 60 to 80 mA. That's 50 to 200 percent higher than the forward current used to power light-emitting diodes. If a cw laser is provided less current, it can still emit light, but it won't be laser light. The device lases only when the threshold current is exceeded—typically a minimum of 50 to 60 mA. Conversely, if the laser is provided too much current, it generates excessive heat and is soon destroyed.

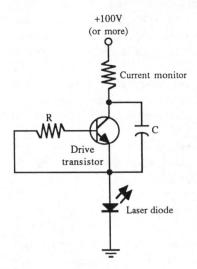

FIG. 10-2. *One way to drive a single heterostructure laser diode. The transistor is driven in avalanche mode, producing short-duration pulses of current.*

Monitoring Power Output

All laser diodes are susceptible to changes in temperature. As the temperature of a semiconductor laser increases, the device becomes less efficient and its light output falls. If the temperature decreases, the laser becomes far more efficient. With the increase in output power, there is a risk of damaging the laser, so most cw drive circuits incorporate a feedback loop to monitor the temperature or output power of the device and adjust its operating current accordingly.

Sensing temperature change requires an elaborate thermal sensing device and complicated constant-current reference source. An easier approach is to monitor the light output of the laser. When the output increases, current is decreased. Conversely, when the output decreases, current is increased.

To facilitate the feedback system, the majority of cw laser diodes now incorporate a built-in photodiode monitor. This photodiode is positioned at the opposite end of the diode chip, as shown in FIG. 10-3, and samples a small portion of the output power. The photodiode is connected to a relatively simple comparator or op amp circuit. As the power output of the laser varies, the current (and voltage) of the photodiode monitor changes. The feedback circuit tracks these changes and adjusts the voltage (or current) supplied to the laser. The feedback circuit can be designed around discrete parts or a custom-made IC. Actual driving circuits using both designs are presented in the following chapter. There is also a schematic for driving a cw laser in pulsed mode.

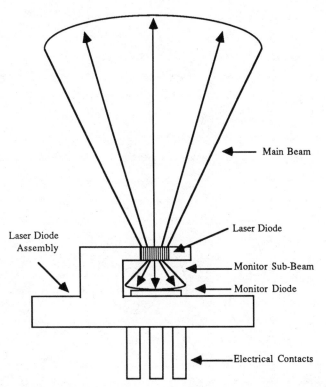

FIG. 10-3. *The orientation of the laser and monitor photodiode chips in a typical double-heterostructure semiconductor laser.*

Connecting the Laser to the Drive Circuit

The laser and photodiode are almost always ganged together, using one of two approaches. Either the anode of the laser is connected to the cathode of the photodiode, or the cathodes are grouped together. That leaves three terminals for connecting the diode to the control circuits. Schematic diagrams for the two approaches are illustrated in FIG. 10-4. A sample terminal layout for the popular Sharp laser diodes (as used in bar code readers and compact disc players) is shown in FIG. 10-5.

There is a danger of damaging a laser diode by improperly connecting it to the drive circuit. Connecting a 60 to 80 mA current source to the photodiode will probably burn it out and can destroy the entire laser. Moral: follow the hook-up diagram carefully. If no diagram came with the laser diode you received, write to the seller or manufacturer and ask for a copy of the specifications sheet or application note.

HANDLING AND SAFETY PRECAUTIONS

While the latest semiconductor lasers are hearty, well-made beasts, they do require certain handling precautions. And, even though they are small, they still emit laser light that can be potentially dangerous to your eyes. Keep these points in mind:

★ Always make sure the terminals of a laser diode are connected properly to the drive circuit (I've covered this already but it's most crucial).

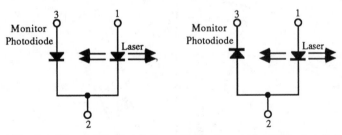

FIG. 10-4. *Two ways of internally connecting laser and monitor photodiode.*

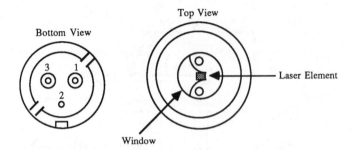

FIG. 10-5. *Package outline and terminal configuration for the Sharp LT020 laser diode.*

★ Never apply more than the maximum forward current (as specified by the manufacturer), or the laser will burn up. Use the pulser drive (see Chapter 11) if you are not using the laser with a monitor photodiode feedback circuit.

★ Handle laser diodes with the same care you extend to CMOS devices. Wear an anti-static wrist strap while handling the laser, and keep the device in a protective, anti-static bag until ready for use.

★ Use only a grounded soldering pencil when attaching wires to the laser diode terminals. Limit soldering duration to less than 5 seconds per terminal.

★ Never connect the probes of a volt-ohmmeter across the terminals of a laser diode (the current from the internal battery of the meter can damage the laser).

★ Use only batteries or *well-filtered* ac power supplies. Laser diodes are susceptible to voltage transients and can be ruined when powered by poorly filtered line-operated supplies.

★ Take care not to short the terminals of the laser during operation.

★ Avoid looking into the window of the laser while it is operating, even if you can't see any light coming out. This is especially important if you have added focusing or collimating optics.

★ Mount the laser diode on a suitable heatsink, preferably larger than 1 inch square. Use silicone heat transfer paste to assure a good thermal contact between the laser and the heatsink. You can buy heatsinks ready made or construct your own. Some ideas for heatsinks appear in the next section.

★ Insulate the connections between the laser diode and the drive to minimize the chance of short circuits. Use shielded three-conductor wire to reduce induction from nearby high-frequency sources.

★ Laser diodes are subject to the same CDRH regulations as any other laser in its power class. Apply the proper warning stickers and advise others not to stare directly into the laser when it is on.

★ Unless otherwise specified by the manufacturer, clean the output window of the laser diode with a cotton swab dipped in ethanol. Alternatively, you can use optics-grade lens cleaning fluid.

MOUNTING AND HEATSINKS

Most laser diodes lack any means by which to mount them in a suitable enclosure. Their compact size does not allow for mounting holes. However, with a bit of ingenuity, you can construct mounts that secure the laser in place as well as provide the recommended heatsinking. One approach is to clip the laser in place using a fuse holder, as shown in FIG. 10-6. You might have to bend the holder out a bit to accommodate the laser. Mount the clip on a small piece of aluminum or a TO-220 heatsink. Use silicone paste at the junction of all-metal pieces; this assists in proper heat transfer.

Another method, detailed in FIG. 10-7, is to drill a hole the same diameter as the laser in an aluminum heatsink. Use copper retaining clips (available at the hobby store) to secure the laser in place. Once again, apply silicone paste to aid in heat transfer.

Some lasers are available on the surplus market, like that shown in FIG. 10-8, and are already attached to a heatsink and mount. The mount doubles as a rail for collimating and beam-shaping optics. You can use the laser with or without these optics, of course, or substitute with your own.

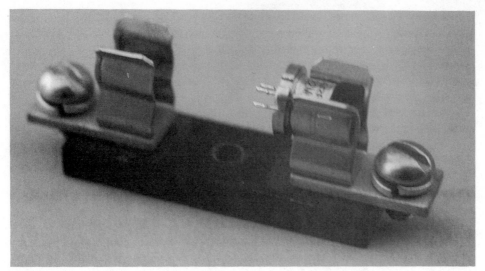

FIG. 10-6. *A fuse clip can be used as a simple heatsink for a semiconductor laser.*

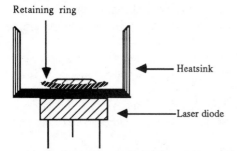

Retaining ring

Heatsink

Laser diode

FIG. 10-7. *Use a flexible copper retaining ring to hold a diode to the heatsink. Use silicone heatsink paste to aid in proper heat transfer.*

FIG. 10-8. *A commercially made "sled" with laser diode (left side) and beam-shaping optics installed.*

150

SOURCES FOR LASER DIODES

Laser diodes are seldom sold at the neighborhood electronics store, and as of this writing, Radio Shack does not carry the device as a replacement or experimenter's item. That leaves buying your laser diodes directly from the manufacturer, through an authorized manufacturer's representative, or through surplus. Buying direct from the manufacturer or rep assures you of receiving prime, new goods, but the cost can be high. Average cost for a new 3 to 5 mW laser cw diode is about $30. Names and addresses of manufacturers are in Appendix A. You can locate local representatives by writing to the manufacturer, or look in the Yellow Pages under ''Electronics—Wholesale and Retail.''

The same or similar device on the surplus market is about $10 to $15, depending on the power output. Several of the surplus mail-order dealers listed in Appendix A offer sh and dh laser diodes; write them for a current catalog. Many also provide kits and ready-made drive/power supply circuits. Be aware that, at this time, most surplus laser diodes are take-outs, meaning that they were used in some product that was later retired and scrapped. While buying used He-Ne tubes can be a chancy affair, the risk of buying pre-owned laser diodes is minimal. Like all solid-state electronics, the life span of a laser diode is extremely long—in excess of 5,000 to 10,000 hours of continuous use.

BUILD A POCKET LASER DIODE

You can build a complete laser in a box about the size of a pack of cigarettes. FIGURE 10-9 shows the basic layout; TABLE 10-1 provides the parts list. You can use just about any of the drive circuits presented in Chapter 11 to power the laser. In all cases, you can mount the components on a universal solder PCB and fit the whole thing in a 3¼-by-2⅛-by-1⅛-inch plastic experimenter's box. Drill holes for the switch, power jack, and lens tube. The lens hole should be ¾-inch in diameter.

Saw off a solderless RG59U coaxial connector and mount the laser inside (check the connector style first to be sure the diode fits snugly). Use all-purpose adhesive to secure a 10-mm double-concave lens (with a focal length of about 15 mm) inside one end of a 1-inch length of a ⁷⁄₁₆-inch (I.D.) brass tube. The tube is available at most hobby stores. Fit the laser diode in the lens tube and mount the tube in the enclosure. Use all-purpose adhesive to keep it in place. You can adjust the spacing between the laser and lens later.

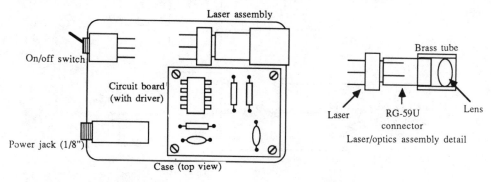

FIG. 10-9. *Basic layout for the pocket laser showing ON/OFF switch, power jack, laser assembly, and driver board.*

Table 10-1. Pocket Laser Parts List

1	Laser diode; Sharp LT020, LT022, or equivalent
1	RG59U solderless video connector
1	1-inch length 7/16-inch (I.D.) brass tube
1	10 mm diameter, 15 mm focal length double-convex lens
1	SPST switch (DPDT switch for dual-ended supply)
1	1/8-inch jack (2- or 3-conductor, depending on supply)
1	Driver board (see Chapter 12)
1	3¼-by-2⅛-by-1⅛-inch plastic project box

Wire the components as shown in FIG. 10-10. Install the switch, power jack, and drive board in the box. Temporarily apply power to the circuit board and dim the lights. Point the lens toward a lightly colored wall at a distance of no more than a few inches. Adjust the distance between laser and lens by sliding the RG59U connector in or out of the brass tube until the spot on the wall is bright and well-defined. You will see rings in the beam; this is normal. A grainy speckle in the spot means that the diode is emitting laser light. If you don't see the speckle, the laser might not be driven with enough current.

When everything looks ok, dab a small drop of all-purpose adhesive on the RG59U connector and brass tube to keep the laser from coming loose. Don't apply too much glue, because you might need to readjust the laser later on. Close up the box and fit a set of four "AA" batteries in a battery holder. Place the battery holder in a box measuring at least 2½ by 2½ by 1 inch (see FIG. 10-11). Use three-conductor shielded microphone cable as the power cord, as shown, and solder a 1/8-inch stereo plug on the end. To use the battery pack, simply plug it into the power jack on the laser box.

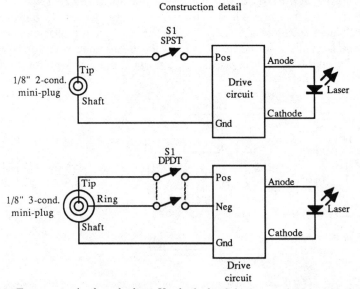

FIG. 10-10. *Two ways to wire the pocket laser. Use the dual-ended power supply if the driver board requires it.*

Table 10-2. Pocket Laser Power Pack Parts List

1	4-cell "AA" battery holder
1	⅛-inch plug (2- or 3-conductor, depending on supply)
1	2½-by-2½-by-1-inch (minimum) project box Batteries

Note that the 6-volt battery pack is meant for use with the pulsed drive circuit described in Chapter 12. Other drive schemes call for a 12-volt supply or for a split ±5-volt supply. Use the appropriate type of batteries, connected in parallel and/or in series, to provide the required voltage level. You might need to add voltage regulators (small TO-92 case) or zener diodes to maintain or regulate the supply voltages.

You might want to combine the laser/drive components in the same box as the batteries. You can fit everything in a project box measuring 6¼ by 3¾ by 2 inches, or even less if you are careful how you mount the components. Make sure that the batteries are placed in a convenient location so that they can be easily changed when they wear out.

USING THE POCKET LASER

The light from the pocket laser is largely invisible unless you happen to own see-in-the-dark infrared glasses or an IR viewing card (a card coated with a chemical that reacts to infrared radiation). The faint red glow of the laser is discernible only in darkness and when the lens is focused on a nearby wall. That makes applications such as laser pointers out of the question. But the pocket laser is far from useless. As you'll learn in future chapters, you can use this basic configuration to create a collimated free-air laser light communicator.

You can also use the pocket laser as the head-end for a fiberoptic data link or as a means to experiment with interferometry. Although the beam is difficult (if not impossible) to see without some sort of IR viewing device or infrared viewing card, you

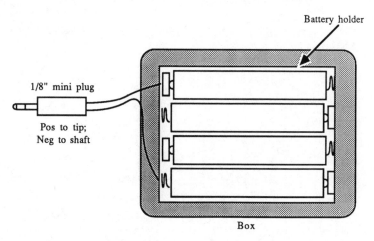

FIG. 10-11. *Install the batteries in a project box and terminate the battery holder leads with a ⅛-inch mini plug (two- or three-conductor as needed).*

can detect the interferometric fringes with an ordinary photodetector. Connect the photodetector to an audio amplifier, as shown in Chapter 9, and you can *hear* the fringes move. Connected to a counter, you can even count the number of fringes that go by.

An advanced project might be to use the laser to make near-infrared holograms. Although most films are already sensitive to near-infrared radiation, you can obtain better results if you use an emulsion specifically formulated for the 780 to 880 nm range of most laser diodes. Be sure that the film has very high resolution, or the hologram won't turn out.

11

Laser Power Supplies

Imagine a world without electricity. Without the motive force of electricity and more importantly a way to harness it, we would be without 90 percent of our creature comforts. Everything from the family car to the kitchen food processor operates on electrical power, and without juice, these things would come to a grinding halt.

The same is true of lasers and their support systems. Without power, your laser is useless and no more worthwhile than a rock paperweight. In this chapter, you'll learn how to construct universal power supplies, including:

★ High-voltage power supplies for operating a helium-neon laser; both 117 volt ac and 12 volt dc versions.
★ Regulated power supplies for diode lasers.

Low-voltage power supplies for operating electronic equipment are in the next chapter. There you'll find designs for single- and dual-voltage regulated supplies, including an all-purpose adjustable version and battery pack regulators.

ABOUT HELIUM-NEON POWER SUPPLIES

A helium-neon laser tube must be connected to a high-voltage power supply or it won't work. You have two options to provide the required juice: buy a ready-made laser power supply or build your own. Commercially made power supplies for helium-neon lasers are available from a variety of sources, and if you are just starting out, this is the best route to go. As detailed in Chapter 5, you need to be sure that the supply is rated for the tube you are using. Some tubes require more operating current than others and might not work properly with a power supply that can't deliver the milliamps.

He-Ne laser power supplies you build yourself are not overly complicated and they don't need lots of parts. But the parts they do require can be hard to find. Specifically, the laser power supply must use high-voltage diodes and capacitors—the higher the rating, the better. The 1N4007 diode is rated at 1 kV, the minimum you can use. Such diodes are bound to burn out when running a laser that consumes more than 5 milliamps, so 3- to 10-kV diodes are preferred. High-voltage capacitors of the typical values used in laser power supplies—.1 to 0.001 μF, are even harder to find. Most high-voltage capacitors have very low values, usually in the tens of picofarads.

Perhaps the most troublesome component is the transformer. The *ideal* laser power supply transformer is specially made to conform to the specifications required by the job, but a number of ready-made step-up switching type transformers can effectively be used. The hard part is finding them. The typical transformer for use in a dc-operated helium-neon laser steps up 12 volts to between 300 and 1,000 volts. High-voltage transformers designed for use with photocopiers can also be used. These transform 117 Vac to 1,000 to 4,000 Vac. Most laser tubes require between 1,200 and 3,000 volts.

A local surplus or electronics outlet can carry suitable high-voltage diodes, capacitors, and transformers, but you might have better luck trying surplus mail-order outlets. See Appendix A for a list of mail-order surplus dealers. Ask for their latest catalog, and if you don't see the items you want, write or call. Some outlets carry stock that is not included in the general distribution catalog.

Appendix A also lists several sources for laser components. These include Meredith Instruments, Information Unlimited, and General Science & Engineering. These mail-order companies are prime sources of laser power supply components, and you should obtain their catalogs before beginning any serious laser project.

Many also offer power supply kits, with all the parts conveniently pre-packaged for you. In fact, one of the dc power supplies discussed below is available (at the time of this writing) in kit form from General Science & Engineering. In the event that the kit is no longer available, you can still construct the power supply using the schematic and parts list, provided in this chapter.

Before building any of the laser power supplies described in this chapter, read the following very carefully:

★ Any laser power supply delivers high voltages that, under certain circumstances, can injure or kill you. Use extreme caution when building, testing, and using these power supplies.

★ Do not attempt to build your own power supply unless you have at least some knowledge of electronics and electronic construction.

★ Although the power supply projects are not difficult to construct, they should be considered suitable only for intermediate to advanced hobbyists.

★ Power supplies and laser tubes retain current even after electricity has been removed. Be sure to short out the output of the power supply before touching the laser or high-voltage leads.

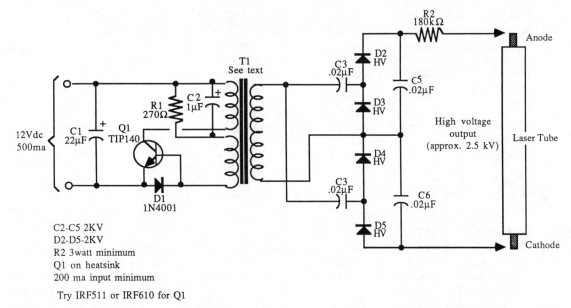

C2-C5 2KV
D2-D5-2KV
R2 3watt minimum
Q1 on heatsink
200 ma input minimum

Try IRF511 or IRF610 for Q1

FIG. 11-1. *An easy approach to building a high-voltage He-Ne power supply. Be wary of the high voltages present at the secondary of T1.*

BASIC HE-NE 12-VOLT POWER SUPPLY

The schematic in FIG. 11-1 shows a basic, no-frills power supply suitable for use with helium-neon tubes rated at 0.5 to 1 milliwatt. TABLE 11-1 contains the parts list. The circuit is shown more as a lesson in high-voltage power supply design than a full-fledged project. You will probably want to supplement the supply with additional features, such as a 10-kV trigger transformer or current feedback circuit. A number of books provide details on advanced high-voltage power supplies; see Appendix B for a selected list.

Table 11-1. Basic Dc He-Ne Power Supply Parts List

R1	270 ohm resistor
R2	180 kilohm resistor, 3 to 5 watt
C1	22 μF electrolytic capacitor
C2	1 μF electrolytic capacitor
C3-C6	0.02 μF capacitor, 1 kV or more
D1	1N4001 diode
D2-D5	High-voltage diode (3 kV or more)
Q1	TIP 140 power transistor
T1	High-voltage dc-to-dc converter transformer; see text for specifications.

All resistors are 5 to 10 percent tolerance, ¼ watt, unless otherwise indicated. All capacitors are 10 to 20 percent tolerance, rated 35 volts or more, unless otherwise indicated.

At the heart of the power supply is a dc switching transformer, T1. This oscillation transformer is designed for use as a dc-to-dc converter and is available from Meredith Instruments. It has the following characteristics:

★ Input voltage (primary): 6 volts
★ Output voltage (secondary): 330 volts
★ Ferrite core size: EE19
★ Maximum power output: 7 watts
★ Oscillation frequency: 15 kHz
★ Winding ratio (Ns/Np): 57.4

Operation of the Power Supply

Here's how the power supply works: Q1, R1, and C1 form a Hartley-type astable multivibrator (a free-running oscillator) that switches the incoming 12 volts dc between the two secondary windings of the transformer. T1 is a high-turns-ratio transformer that steps up the incoming voltage to about 1,000 volts. Capacitors C2 through C5 as well as diodes D2 through D5 form a voltage multiplier that increases the voltage to about 2,500 volts at approximately 3 to 4 mA.

Resistor R2 is an important component. All laser tubes are current-sensitive and try to consume as much current as the power supply will deliver. The resistor limits the current to a safe level; without it, the tube might burn out. Resistor R2 is chosen for a typical 1 mW laser tube. If your laser sputters or doesn't fire, the resist or value might be too high or too low, causing the tube to be unstable. Later you will see how to test the power supply to discover how much current the tube is drawing and to adjust R2 to deliver just the minimum to keep the tube lasing.

Note that the battery power requirement is rather steep. The power supply consumes about 350 mA of current, so you should use only heavy-duty batteries. Although the supply will work on "C" alkaline cells, you'll have better luck with "D" cells. The best results are obtained when using high-output lead-acid or gelled electrolyte batteries. A pair of 6-volt, 4 AH batteries will power the laser for several hours before needing a recharge.

The power supply works best when the input voltage is as close to 12 volts as possible. Because most batteries deliver a range of voltages during their discharge period, you might want to add the regulator circuit provided in FIG. 11-2 (parts list in TABLE 11-2). The schematic uses a positive 12-volt regulator that requires about 1 volt as "overhead."

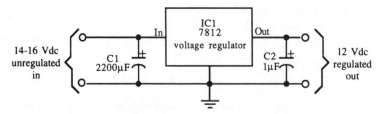

FIG. 11-2. *Use a 7812 voltage regulator with a 14- to 16-volt supply to regulate the voltage to the high-voltage power supply presented in FIG. 11-1.*

Table 11-2. 12 Vdc Battery Regulator Parts List

IC1	7812 +12 Vdc voltage regulator
C1	2200 μF electrolytic capacitor
C2	1 μF electrolytic capacitor

All capacitors are 10 to 20 percent tolerance, rated 35 volts or more.

When fed by the typical lead-acid or gelled electrolyte battery—which have an average output of about 13.8 volts—approximately 12 volts reaches the power supply.

Components R1 and C1 determine the frequency rate of the circuit. By adjusting R1, you change the frequency and therefore the output voltage of T1. If your supply is having trouble igniting and running your laser tube, try a slightly higher or lower value for R1.

Building the Circuit

The basic power supply should be constructed on a printed circuit board. Component placement is not crucial, but you should allow as much room as possible for the high-voltage components. Keep the anode lead as short as possible (2 to 4 inches) and place the ballast resistor close to the anode terminal on the laser tube.

Testing the Current Output

Resistor R2, the ballast resistor, determines the amount of current delivered to the tube. Although you can calculate the exact value of the resistor using design formulas, you need to know the parameters of the particular tube you are using. Dial the meter to read dc milliamps. Turn on the power supply and watch the meter. The current should not exceed 6 or 7 mA (it probably won't with the basic power supply described earlier, anyway). If the current is too high, you should immediately remove the power. Short the leads of the power supply to remove any remaining current, and replace R2 with

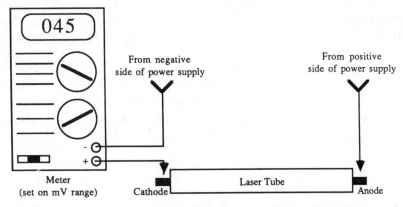

FIG. 11-3. *Connect a volt-ohmmeter (set to read milliamps) as shown to determine the amount of current consumed by the laser tube.*

159

a higher value resistor. Be sure to use a resistor rated to at least 3 to 5 watts. Re-apply power and take a new reading.

Most likely, the laser will sputter or not turn on at all. The usual cause is a ballast resistor that is either too high or too low; either way, the sputtering is caused by unstable operation and can usually be corrected by selecting another ballast resistor. The tube will not ignite or lase if the current is less than about 3.5 mA. If the tube stays on without sputtering and the current output is between about 3.5 and 6 mA, you have selected the proper ballast resistor.

If the ballast resistor is too low, excessive current will flow through the tube, damaging it or at the least severely shortening its life. Besides doing harm to the tube, the power supply consumes excessive current, prematurely draining battery power. You will realize the longest battery life by careful selection of the ballast resistor.

Note that some sputtering is caused by arcing of the anode and cathode leads. Be sure the leads are securely attached to the power supply and the laser. You can often see the result of arcing by turning off the lights and looking carefully for a tell-tale blue glow around the anode and cathode terminals. The glow is a corona caused by the ionization of air by high-voltage discharge.

PULSE-MODULATED DC-OPERATED HE-NE SUPPLY

The basic helium-neon laser power supply is good for low-output tubes, but it doesn't deliver sufficient current for higher power and hard-to-start tubes. The advanced power supply shown in FIG. 11-4 can be used with 1- to 5-mW tubes, depending on the transformer you use. The parts list for this supply is included in TABLE 11-3.

About the Circuit

The advanced laser power supply uses an LM555 timer IC as a pulse width modulator (PWM). Two potentiometers, R12 and R13, adjust the width of the output pulses from the 555, and therefore change the currents used to trigger and operate the tube.

Capacitor C5 and resistors R8, R9, R12, and R13 determine the pulse width of the 555. Initially, the R12 and R13 are dialed to their center positions and relay R1 is de-energized, effectively removing R8 and R13 from the circuit. When 12 volts is applied to the circuit, the 555 pulses and triggers Q1, C1, and R2. This in turn drives transformer T1. This transformer steps up the 12 volts to approximately 1,000 volts. Capacitors C7 through C10 and diodes D4 through D19 form a four-stage cascaded voltage multiplier that increases the output to about 3,500 volts.

If the tube doesn't fire, adjust R12 to increase the duty cycle of the 555 pulses. When the tube ignites, sensing resistors R3 through R6 trigger Q2, which closes relay R1. That brings R8 and R13 into the circuit. Adjusting R13 controls the duty cycle of the 555 while the tube is operating. Shortening the duty cycle of the pulses decreases the current delivered to the tube; lengthening the duty cycle increases the current.

Building the Circuit

The PWM power supply can be built on a perforated board or printed circuit board. When using a perforated board, be sure that lead lengths are kept to a minimum and

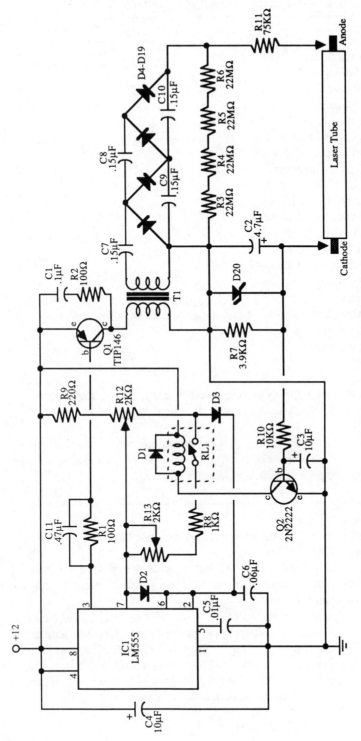

FIG. 11-4. *The circuit schematic for the pulse-width-modulated power supply.*

Table 11-3. Pulse Width Modulated Power Supply

IC1	555 timer IC
R1-R2	100 ohm resistor
R3-R6	22 megohm resistor
R7	3.9 kilohm resistor
R8	1 kilohm resistor
R9	220 ohm resistor
R10	10 kilohm resistor
R11	75 kilohm resistor, 3-5 watt
R12,R13	2 kilohm potentiometer
C1	0.1 μF
C2	4.7 μF electrolytic capacitor
C3,C4	10 μF electrolytic capacitor
C5	0.01 μF disc capacitor
C6	0.06 μF disc capacitor
C7-C10	0.15 μF capacitor, 3 kV or more
C11	0.47 μF capacitor
D1-D3	1N914 diode
D4-D19	High-voltage diodes (3 kV or more; four diodes in series for each diode symbol in schematic)
Q1	TIP146 (on heatsink)
Q2	2N2222
RL1	12-volt SPST relay
T1	High-voltage step-up transformer; 9-volt primary, 375-volt secondary
Misc.	Heatsink for Q1, high-dielectric wire for connecting tube to supply

All resistors are 5 to 10 percent tolerance, ¼ watt, unless otherwise indicated. All capacitors are 10 to 20 percent tolerance, rated 35 volts or more, unless otherwise indicated. Kit is available from General Science and Engineering.

that the high-voltage capacitors and diodes are not placed too close together. To prevent arcing, place the diodes at 45-degree angles.

The leads for the anode and cathode should be 6 inches or shorter. Reduce the chance of arcing by wrapping high-voltage dielectric tape around the leads. Or, slip a length of neoprene aquarium tubing over the wires.

Construct clips for the laser terminals as detailed in Chapter 6, ''Build a He-Ne Laser Experimenter's System.'' You can also form heavy-duty steel or copper wire and bend it in a clip shape. Make the clip slightly smaller than the diameter of the laser tube terminals. When made properly, the wire should clip securely around the tube. Wrap a length of high-voltage dielectric tape around the clip and terminal to hold them in place. Be sure that you don't cover the mirrors on either end of the laser.

Using the Power Supply

Operating the power supply is straightforward. Once the tube is secured, rotate potentiometers R12 and R13 to their center positions. Apply power and watch the tube. Slowly rotate R12 until the tube triggers. You will hear the relay click in. If it chatters and the tube sputters, keep turning R12. If the tube still won't ignite, rotate R13 slightly.

Once the tube lights and stays on, rotate R13 so that the tube begins to sputter and the relay clatters. This marks the threshold of the tube. Advance R13 just a little until the tube turns back on and remains steady. Every tube, even those of the same size and with the same output, have slightly different current requirements, so you will need to readjust R12 and R13 for every tube you own.

Resistor R11 is the ballast, limiting current to the tube. The schematic shows a 80-kilohm resistor, but you can experiment with other values to find one that works best with your tube. If the laser doesn't trigger or run after adjusting R12 and R13, try reducing the value of the ballast resistor. Use a voltmeter, as explained in the previous section, to monitor the output current to ensure against passing excessive current through the tube.

AC-OPERATED HE-NE POWER SUPPLY

The schematic in FIG. 11-5 shows a basic ac-operated helium-neon power supply (parts list in TABLE 11-4). A high-voltage transformer converts the 117 Vac line current to 1,000 volts or more. The voltage multiplier increases the working voltage while rectifying the ac. The circuit shows a transformer with a 1,000 volt secondary and the voltage multiplier section used in the basic dc-operated power supply presented earlier in this chapter. The output voltage is about 4 to 5 kV (rectified and unloaded). Some tubes require extra

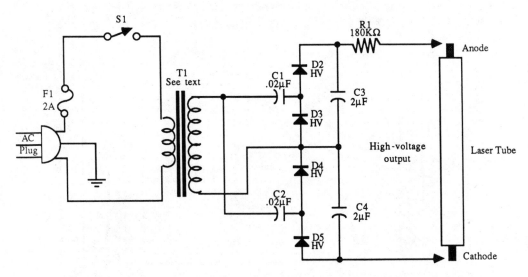

FIG. 11-5. *Minimum configuration for an ac-operated high-voltage He-Ne power supply.*

Table 11-4. Basic Ac He-Ne Power Supply Parts List

R1	180 kilohm resistor, 3 to 5 watt
C1-C4	2 μF capacitor, 1 kV or more
D1-D4	High-voltage diode (3 kV or more)
T1	High-voltage step-up transformer; see text for specifications.
S1	SPST switch
F1	Fuse (2 amp) and holder ac plug

voltage to start and might need a transformer with 2,000- to 2,500-volt secondary. Note that the circuit is basic and lacks current sensing or high-voltage start capabilities.

ENCLOSING THE HE-NE POWER SUPPLIES

Laser power supplies should never be used without placing them in protective, insulating enclosures. After you have built and tested your power supply, tuck it safely in a plastic enclosure. If you plan on using the supply to power a variety of tubes, mount heavy-duty (25-amp) banana jacks to provide easy access to the anode and cathode leads. Keep the jacks separated by *at least* 1 inch and apply high-voltage putty around all terminals to prevent arcing.

A functional schematic for a completely self-contained, rechargeable battery pack/power supply is shown in FIG. 11-6. The parts list for the battery pack/power supply is provided in TABLE 11-5. Note the addition of the key switch, power-on indicator, and battery-charging terminal. The key switch prevents unauthorized used of the power supply and acts as the main ON/OFF switch. For maximum security, you should get the

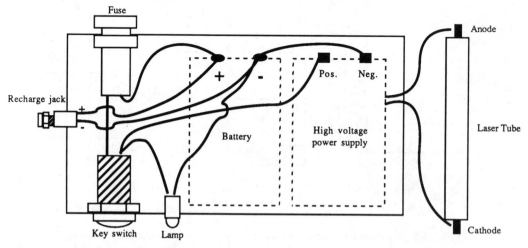

FIG. 11-6. *Hookup diagram for building an all-in-one battery-operated helium-neon laser power supply. The battery can be recharged in the enclosure without removing it.*

Table 11-5. Battery Pack/He-Ne Supply Parts List

B1	12 Vdc battery
PS1	Modular 12 Vdc He-Ne high-voltage power supply
F1	Fuse holder (fuse: 2 amps)
S1	SPST key switch
LA1	LED indicator (with built-in dropping resistor)
J1	¼-inch, 2-conductor phone jack
Misc.	Project box, grommet (for output leads), high-dielectric wires for connection to laser

kind of switch where the key can be removed only when it is in the OFF position (CDRH requirements call for a key that cannot be removed in the ON position).

The power-on indicator is simply an LED with a current-dropping resistor. The battery-charging terminal provides a means to recharge the batteries without removing them from the enclosure. Note that you can operate the laser while the battery recharger is connected, but in most cases, the power supply will consume too much current and the batteries will not be recharged.

A fuse is added to provide protection against an accidental short circuit in the battery compartment. Lead-acid and gelled electrolyte batteries can easily burn plastic and even metal when their terminals are shorted. The fuse helps prevent accidental damage and fire.

ABOUT LASER DIODE POWER SUPPLIES

As discussed in Chapter 10, laser diodes come in two basic forms: single- and double-heterostructure. The single-heterostructure (or sh) diodes are regarded as the "older" variety and can only be operated in pulsed mode (unless you cool them with a cryogenic fluid, such as liquid nitrogen). Sh laser diodes are capable of multi-watt operation, but only when the pulses are 200 nanoseconds or shorter. Therefore, sh diode supplies may be built around some type of astable multivibrator.

Double-heterostructure (dh) laser diodes can be operated in pulsed or CW modes. Like any diode, excessive currents can destroy the laser, so you must take precautions to operate the unit within its design parameters. Dh lasers are capable of multi-watt operation when used in pulsed mode, but most are designed for CW operation and emit 1 to 10 mW of light energy. Remember that although you can often see a red glow from a diode laser, this light represents only a fraction of the total radiation from the diode. The bulk of the radiation is in the near-infrared spectrum and is largely invisible to your eyes.

PULSED SINGLE-HETEROSTRUCTURE INJECTION DIODE SUPPLY

A common method for powering an sh injection diode is shown in FIG. 11-7 (see the parts list in TABLE 11-6). The power supply provides pulses of about 10 to 20 amps at a short duration of around 50 ns. The supply provides sufficient drive current to exceed the threshold of the laser (typically about 7 or 8 amps), with some room to spare. The

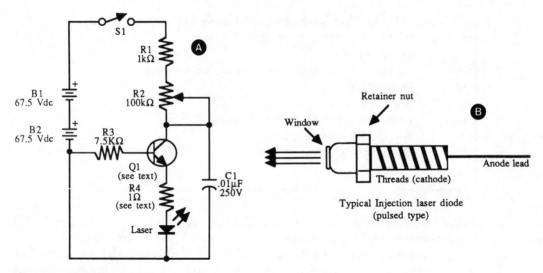

FIG. 11-7. *(A) High-current drive circuit for a single heterostructure laser diode. (B) Power leads for the typical sh laser diode, showing single lead for the anode.*

Table 11-6. Single Heterostructure Laser Pulsed Driver Parts List

R1	1 kilohm resistor
R2	100 kilohm potentiometer
R3	7.5 kilohm resistor
R4	1 ohm resistor, carbon composition, 5 watts
C1	0.01 μF capacitor, 250 V or higher
Q1	2N2222 or equivalent; see text
B1,B2	67.5 Vdc batteries
Misc.	Single heterostructure laser diode, heatsink

All resistors are 5 to 10 percent tolerance, ¼ watt, unless otherwise indicated.

laser might still glow at currents less than threshold, but the light won't be stimulated emission. In other words, the device will not emit laser light but behave like an expensive LED.

The sh laser diode circuit uses a common npn transistor operated in avalanche mode. The batteries are 67.5-volt type (NEDA 217, Eveready number 416) used in older tube-type equipment. You'll have better luck finding the required batteries at an electronic store specializing in communications or ham gear. The price can be steep—up to $10 each depending on the source—so make sure they are fresh before you sign the check.

Quality control in low-cost plastic npn transistors is not great, so not all transistors will work in the circuit. The schematic calls for a 2N2222, but you might need to experiment with several until you find one that oscillates in the circuit. Construct the circuit using component leads that are as short as possible and test the transistor by substituting the laser with a short piece of copper wire (magnet wire works well). Use an oscilloscope across current-monitor resistor R4 (1 ohm, carbon composition) and watch for the pulses from the transistor. Avoid the use of a logic probe, as most are not designed for circuits exceeding 18 volts.

After you have determined that the transistor is oscillating (adjust R2 as needed), substitute the laser, being careful to observe polarity. Most sh lasers use the case as the cathode and the single lead as the anode. Yours *might be different*, so be sure to check the specifications or information sheet that came with the unit. The diode operates at a wavelength of about 904 nm, which is beyond that of normal human vision, so don't expect the same bright red beam that's emitted by a helium-neon laser. You can test the operation of the laser by using one of the infrared sensors described in the previous chapters.

PULSED DOUBLE-HETEROSTRUCTURE INJECTION DIODE SUPPLY

The popularity of compact audio discs, as well as many forms of laser bar-code scanning, have made double-heterostructure laser diodes plentiful in the surplus market. A number of sources (many of which are listed in Appendix A) offer dh laser diodes for prices ranging from $5 to $15. Depending on the power output of the laser, new units are even affordable. A typical 5 mW laser diode lists for about $25 to $30 in low quantities. Sharp is a major manufacturer of dh laser diodes; write them for literature and a price list.

As discussed in the previous chapter, one of the most attractive features of dh laser diodes is that they work with low voltage power supplies. A dh laser can easily be run off a single 9-volt transistor battery. However, dh laser diodes are sensitive to temperature. They become more efficient at lower temperatures, and their power output increases. Unless the temperature is very low (such as when the diode is immersed in liquid nitrogen, as described in Chapter 22), the increase in power output can damage the laser. That's why most dh lasers are equipped with a monitor photodiode. The current output of the monitor photodiode is used in a closed-loop feedback circuit to keep the power output of the laser constant.

Although dh lasers are designed for CW operation, they can also be used in pulse mode. An astable multivibrator, such as a 555 timer, can be used to pulse the laser. A circuit is shown in FIG. 11-8, with a parts list in TABLE 11-7. Because the laser is pulsed, the forward current can exceed the maximum allowed for CW operation (generally 60 to 80 mA). However, care must be taken to keep the pulses short. Pulses longer than about a 50 percent duty cycle (half on, half off) can cause damage to the laser. Duty cycle is not a critical consideration when the current is maintained under 80 mA. The circuit shown in the figure lets you alter the frequency of the astable multivibrator (and therefore the duty cycle).

A closed-loop feedback system constantly watches over the output of the monitor photodiode and maintains the proper current to the laser diode. One such circuit is shown in FIG. 11-9. This circuit is designed around the IR3C02 chip, which is a special-purpose IC manufactured by Sharp. See TABLE 11-8 for a list of required parts. This IC is made for use with their extensive line of dh lasers, and while hard to find, it is relatively inexpensive (obtain the chip through Sharp's parts service or from a distributor dealing with Sharp components).

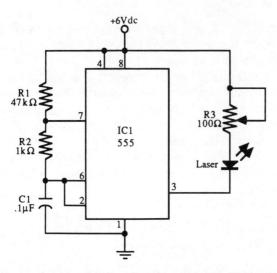

FIG. 11-8. *A double heterostructure laser can be connected to a 555 timer IC for pulse operation. With the components shown, pulse rate is about 300 Hz; pulse width is about 3 milliseconds.*

Table 11-7. Pulsed Double Heterostructure Laser Power Drive Parts List

IC1	555 timer IC
R1	47 kilohm resistor
R2	1 kilohm resistor
R3	100 kilohm potentiometer
C1	0.1 μF disc capacitor
Misc.	Double heterostructure laser diode, heatsink

All resistors are 5 to 10 percent tolerance, ¼ watt. All capacitors are 10 to 20 percent tolerance, rated 35 volts or more.

Another method using discrete components is shown in FIG. 11-10 (parts list in TABLE 11-9). Here, an op amp, acting as high-gain comparator, checks the current from the monitor photodiode. As the current increases, the output of the op amp decreases, and output of the laser drops. The gain of the circuit—the ratio between the incoming and outgoing current—is determined by the settings of R1, R4, and R5.

The circuit in the schematic was adapted from an application note for an RCA C86002E laser diode and uses a CA3130 CMOS op amp. You can readily modify the circuit if you use another op amp or laser diode. Both output transistors are available through most larger electronics outlets, but if you have trouble locating them, you might have luck substituting them with a single TIP120 Darlington power transistor.

168

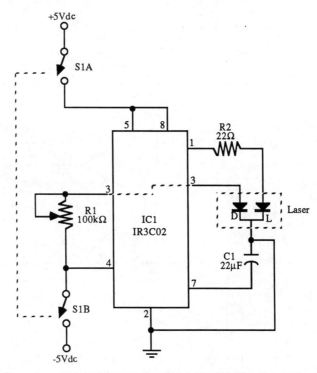

FIG. 11-9. *Basic schematic for the Sharp IR3C02 laser diode driver IC.*

Table 11-8. Sharp IC Laser Drive Parts List

IC1	Sharp IR3C02 laser diode drive IC
R1	100 kilohm resistor
R2	22 ohm resistor
C1	22 μF electrolytic capacitor
S1	DPDT switch
Misc.	Double heterostructure laser diode (such as Sharp LT020), heatsink

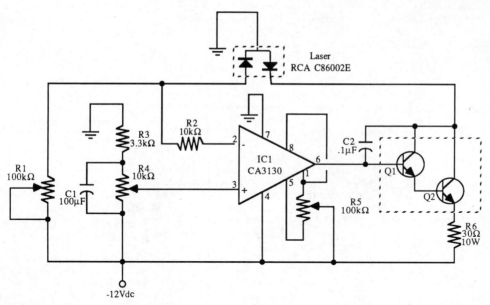

FIG. 11-10. *One way to automatically adjust drive current using a discrete op amp. Use the transistors specified or replace with a suitable Darlington power transistor (such as TIP 120).*

Table 11-9. Op Amp Laser Drive Parts List

IC1	RCA CA 3130 operational amplifier
R1,R5	100 kilohm potentiometer
R2	10 kilohm resistor
R3	3.3 kilohm resistor
R4	10 kilohm potentiometer
R6	30 ohm, 10 watt resistor
C1	100 μF electrolytic capacitor
C2	0.1 μF disc capacitor
Q1	2N2101 transistor
Q2	2N3585 transistor
Laser	RCA C86002 (or equivalent laser diode)

All resistors are 5 to 10 percent tolerance, ¼ watt, unless otherwise indicated. All capacitors are 10 to 20 percent tolerance, rated 35 volts or more.

12

Build an Experimenter's Power Supply

Many laser projects require a steady supply of low-voltage dc, typically between 5 and 12 volts. You may use one or more batteries to supply the juice, but if you plan on doing lots of laser experiments, you'll find that batteries are both inconvenient and anti-productive. Just when you get a circuit perfected, the battery goes dead and must be recharged.

A stand-alone power supply that operates on your 117 Vac house current can supply your laser system designs with regulated dc power without the need to install, replace, or recharge batteries. You can buy a ready-made power supply (they are common in the surplus market) or make your own.

Several power supply designs follow that you can use to provide operating juice to your laser circuits. The designs show you how to construct a:

* 5-volt dc regulated power supply
* 12-volt dc regulated power supply
* Quad ± 5- and ± 12-volt regulated power supply
* Adjustable (3 to 20 volts dc) regulated power supply.

Note that the power supplies presented within this chapter are similar with the exception of different values for capacitors, diode bridges, and other components. You may use the schematics to create power supplies of different voltage levels. The multi-voltage supply is designed to provide the four voltages common in laser support systems: $+5$ volts, $+12$ volts, -5 volts, and -12 volts. These voltages are used by motors, solenoids, and ICs.

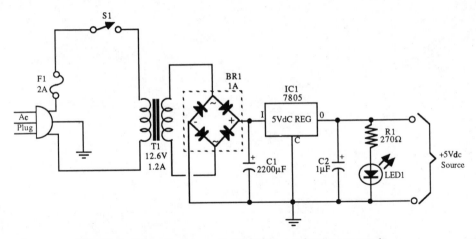

FIG. 12-1. *Schematic diagram for the 5 Vdc regulated power supply.*

SINGLE-VOLTAGE POWER SUPPLY

Refer to FIGS. 12-1 and 12-2 for schematics of the single-voltage power supplies. FIGURE 12-1 shows the circuit for a +5-volt supply; FIG. 12-2 shows the circuit for a +12-volt supply. There are few differences between them, so the following discussion applies to both. For the sake of simplicity, we'll refer just to the +5-volt circuit. Parts lists for the two supplies are provided in TABLES 12-1 and 12-2.

For safety, the power supply must be enclosed in a plastic or metal chassis (plastic is better as there is less chance of a short circuit). Use a perforated board to secure the components and solder them together using 18- or 16-gauge insulated wire. Do not use point-to-point wiring where the components are not secured to a board.

Alternatively, you can make your own circuit board using an etching kit. Before constructing the board, collect all the parts and design the board to fit the specific parts you have. There is little size standardization when it comes to power supply components and large value electrolytic capacitors, so pre-sizing is a must.

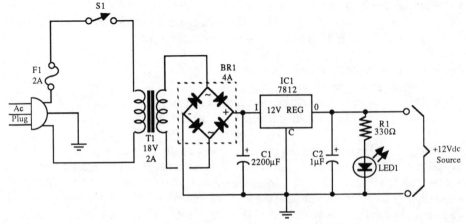

FIG. 12-2. *Schematic diagram for the 12 Vdc regulated power supply.*

Table 12-1. 5 Vdc Power Supply Parts List

IC1	7805 +5 Vdc voltage regulator
R1	270 ohm resistor
C1	2200 μF electrolytic capacitor
C2	1 μF electrolytic capacitor
BR1	Bridge rectifier, 1 amp
LED1	Light-emitting diode
T1	12.6-volt, 1.2-amp transformer
S1	SPST switch
F1	Fuse (2-amp)
Misc.	Ac plug, cord, fuse holder, cabinet

All resistors are 5 to 10 percent tolerance, ¼ watt. All capacitors are 10 to 20 percent tolerance, rated 35 volts or more.

To explain the circuit in FIG.12-1, note the incoming ac routed to the primary terminals on a 12.6-volt transformer. The "hot" side of the ac is connected through a fuse and a single-pole single-throw (SPST) toggle switch. With the switch in the OFF (open) position, the transformer receives no power so the supply is off.

The 117 Vac is stepped down to about 12.6 volts. The transformer specified here is rated at 2 amps, sufficient for the task at hand. Remember that the power supply is limited to delivering the capacity of the transformer (and later, the voltage regulator). A bridge rectifier, BR1, converts the ac to dc (shown schematically in the dotted box). You can also construct the rectifier using discrete diodes (connect them as shown within the box).

When using the bridge rectifier, be sure to connect the leads to the proper terminals. The two terminals marked with a " ~ " connect to the transformer. The " + " and " − "

Table 12-2. 12 Vdc Power Supply Parts List

IC1	7812 +12 Vdc voltage regulator
R1	330 ohm resistor
C1	2200 μF electrolytic capacitor
C2	1 μF electrolytic capacitor
BR1	Bridge rectifier, 4-amp
LED1	Light-emitting diode
T1	18-volt, 2-amp transformer
S1	SPST switch
F1	Fuse (2-amp)
Misc.	Ac plug, cord, fuse holder, cabinet

All resistors are 5 to 10 percent tolerance, ¼ watt. All capacitors are 10 to 20 percent tolerance, rated 35 volts or more.

terminals are the output and must connect as shown in the schematic. A 5-volt, 1-amp regulator, a 7805, is used to maintain the voltage output at a steady 5 volts.

Note that the transformer supplies a great deal more voltage than is necessary. This is for two reasons. First, lower-voltage 6.3- or 9-volt transformers are available, but most do not deliver more than 0.5 amp. It is far easier to find 12- or 15-volt transformers that deliver sufficient power. Second, the regulator requires a few extra volts as "overhead" to operate properly. The 12.6-volt transformer specified here delivers the minimum voltage requirement, and then some.

Capacitors C1 and C2 filter the ripple inherent in the rectified dc at the outputs of the bridge rectifier. With the capacitors installed as shown (note the polarity), the ripple at the output of the power supply is negligible. LED1 and R1 form a simple indicator. The LED glows when the power supply is on. Remember the 270-ohm resistor; the LED will burn up without it.

The output terminals are insulated binding posts. Don't leave the output wires bare, or they could accidentally touch one another and short the supply. Solder the output wires to the lug on the binding posts, and attach the posts to the front of the power supply chassis. The posts accept bare wires, alligator clips, or even banana plugs.

Differences in the 12-Volt Version

The 5- and 12-volt versions of the power supply are basically the same, but with a few important changes. Refer again to FIG. 12-2. First, the transformer is rated for 18 volts at 2 amps. The 18-volt output is more than enough for the overhead required by the 12-volt regulator and is commonly available. You may use a transformer rated at between 15 and 25 volts.

The regulator, a 7812, is the same as the 7805 except that it puts out a regulated +12 volts instead of +5 volts. Use the T series regulator (TO-220 case) for low-current applications and the K series (TO-3) for higher capacity applications. Lastly, R1 is increased to 330 ohms.

MULTIPLE-VOLTAGE POWER SUPPLY

The multi-voltage power supply is like four power supplies in one. Rather than using four bulky transformers, however, this circuit uses just one, tapping the voltage at the proper locations to operate the +5, +12, −5, and −12 regulators.

The circuit, as shown in FIG. 12-3, is composed of two halves. One half of the supply provides +12 and −12 volts; the other half provides +5 and −5 volts. Each side is connected to a common transformer, fuse, switch, and wall plug. See TABLE 12-3 for a parts list.

The basic difference between the multi-voltage supply and the single-voltage supplies described earlier in this chapter is the addition of negative power regulators. Circuit ground is the center tap of the transformer. Make two boards, one for each section. That is, one board will be the ±5-volt regulators and the other board will contain the ±12-volt regulators. The supply provides approximately 1 amp for each of the outputs.

Use nylon binding posts for the five outputs (ground, +5, +12, −5, −12). Clearly label each post so you don't mix them up when using the supply. Check for proper operation with your volt-ohmmeter.

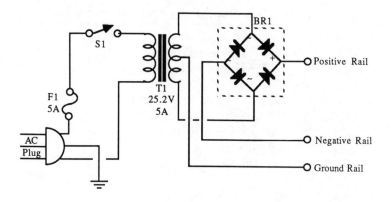

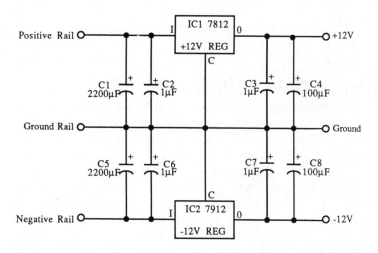

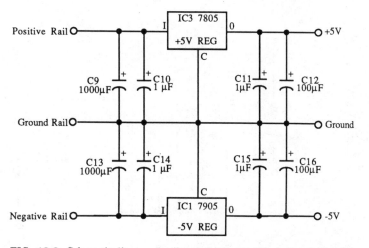

FIG. 12-3. *Schematic diagram for the quad power supply (±5 and 12 volts).*

175

Table 12-3. Quad Power Supply Parts List

IC1	7812 +12 Vdc voltage regulator
IC2	7912 −12 Vdc voltage regulator
IC3	7805 +5 Vdc voltage regulator
IC4	7905 −5 Vdc voltage regulator
C1,C5	2200 μF electrolytic capacitor
C2,C3,	1 μF electrolyic capacitor
C6,C7,C10,C11,C14,C15	
C4,C8,	100 μF electrolytic capacitor
C12,C16	
C9,C13	1000 μF electrolytic capacitor
C1,C5	2200 μF electrolytic capacitor

All capacitors are 10 to 20 percent tolerance, rated 35 volts or more.

ADJUSTABLE-VOLTAGE POWER SUPPLY

The adjustable power supply uses an LM317 adjustable voltage regulator. With the addition of a few components, you can select any voltage between 1.5 to 37 volts. By using a potentiometer, you can select the voltage you want by turning a knob.

The circuit shown in FIG. 12-4 is a no-frills application of the LM317, but it has everything you need to build a well-regulated, continuously adjustable, positive-voltage power supply. See TABLE 12-4 for the parts list. The regulator is rated at over 3 amps so you must mount it on a heavy-duty heatsink. Although you don't need to forcibly cool the regulator and heatsink, it's a good idea to mount them on the outside of the power supply cabinet, for example on the top or back.

Remember that the case of the regulator is the output, so be sure to provide electrical insulation from the heatsink, or a short circuit could result. Use a TO-3 transistor mounting

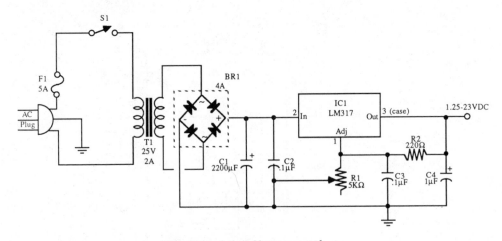

FIG. 12-4. *Adjustable power supply.*

Table 12-4. Adjustable Power Supply Parts List

IC1	LM317 adjustable positive voltage regulator
R1	5 kilohm potentiometer
R2	220 ohm resistor
C1	2200 μF electrolytic capacitor
C2,C3	0.1 μF disc capacitor
C4	1 μF electrolytic capacitor
BR1	Bridge rectifier, 4-amp
T1	25-volt, 2-amp (or more) transformer
S1	SPST switch
F1	5-amp fuse
Misc.	Ac plug, cord, fuse holder, cabinet

All resistors are 5 to 10 percent tolerance, ¼ watt. All capacitors are 10 to 20 percent tolerance, rated 35 volts or more.

and insulator kit. It has all the hardware and insulating washers you need. Apply silicone grease to the bottom of the regulator to aid in heat transfer.

INSPECTION AND TESTING

All of the dc power supplies should be inspected and tested before use. Be particularly wary of wires or components that could short out. Visually check your wiring and check for problems with a volt meter. When all looks satisfactory, apply power and watch for signs of problems. If any arcing or burning occurs, immediately unplug the supply and check everything again. When all appears to be operating smoothly, check the output of the power supply to ensure that it is providing the proper voltage.

BATTERY PACK REGULATORS

Voltage regulators can also be used with battery packs for portable equipment. A 5-volt regulator can be used with a single 6-volt battery to provide a steady supply of 5 volts. The schematic in FIG. 12-5 shows how to connect the parts. Refer to TABLE 12-5 for a parts list. Alternatively, use a 12-volt regulator. The battery should put out a nominal 13 volts to accommodate for the 1- to 1.2-volt drop across the regulator. Most lead-acid and gelled electrolyte batteries put out 13.8 volts when fully charged. See TABLE 12-6 for a chart of voltage values for various types of batteries.

BATTERY RECHARGERS

With a rechargeable battery, you can use it once, zap new life into it, use it again, and repeat the process several hundred—even thousands—of times before wearing it out. The higher initial cost of rechargeable batteries more than pays for itself after the third or fourth recharging.

Rechargeable batteries can't be revived simply by connecting them to a dc power supply. The dc supply delivers too much current and tries to charge the battery too

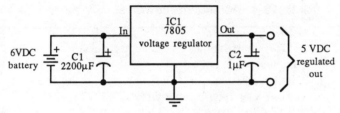

FIG. 12-5. *Battery pack regulator.*

quickly. If you are recharging gelled electrolyte or lead-acid batteries, you might be able to get away with using an ac power adapter, the kind designed for video games, portable tape recorders, and other battery-operated equipment (the output *must* be dc). By design, these adapters limit their maximum current to between 250 to 600 mA. A 300 mA recharger can be effectively used on batteries with capacities of 2.5 AH to 5 AH. A 400 mA or 500 mA ac adapter can be used on batteries with capacities of 3.5 AH to 6.5 AH.

However, one problem is that you must be careful the battery doesn't stay on charge much longer than 12 to 16 hours. Leaving it on for a day or two can ruin the battery. This is especially true of lead-acid batteries. The circuit shown in FIG. 12-6 minimizes the danger of overcharging.

Table 12-5. 5 Vdc Battery Voltage Regulator

IC1	7805 +5 Vdc voltage regulator
C1	2200 μF electrolytic capacitor
C2	1 μF electrolytic capacitor

All capacitors are 10 to 20 percent tolerance, rated 35 volts or more.

Table 12-6. Battery Voltage Levels

Battery	Newly Charged	Nominal	Discharged
Alkaline Ni-cad	1.4 volts	1.2 volts	1.1 volts
Power/1 cell*	2.3 volts	2.0 volts	1.6 volts
Power/multi	6.5 volts	6.0 volts	4.8 volts
Power/multi	13.8 volts	12.0 volts	9.6 volts

*Gelled electrolyte and lead-acid battery; single cell, 6 volt-(three cells in series), 12-volts (six cells in series).

178

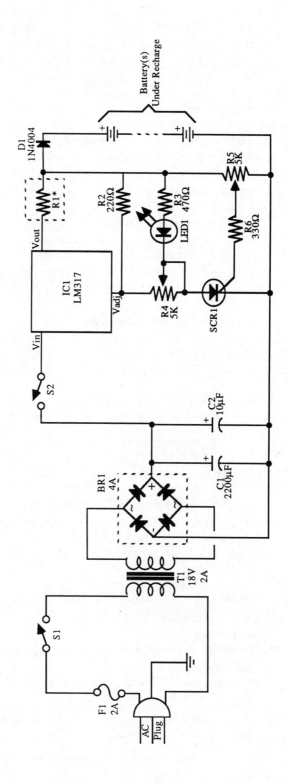

FIG. 12-6. *Circuit diagram for the battery charger. See page 180 for values of R1 and pg 182 for settings for R4 and R5.*

Build the Universal Battery Recharger

The universal battery recharger shown in FIG. 12-6 is built around the LM317 adjustable voltage regulator IC. As indicated in TABLE 12-7, this IC comes in a TO-3 transistor case and should be used with a heatsink to provide for cool operation. The heatsink is absolutely necessary when recharging batteries at 500 mA or higher.

The circuit works by monitoring the voltage level at the battery. During recharging, the circuit supplies a constant-current output; the voltage level gradually rises as the battery charges. When the battery nears full charge, the circuit removes the constant-current source and maintains a regulated voltage to complete or maintain charging. By switching to constant-voltage output, the battery can be left on charge for periods longer than recommended by the manufacturer.

Before you build the circuit, you should consider the kind of batteries you want recharged. You'll have to consider whether you will be recharging 6-volt or 12-volt batteries (or both) and the maximum current output that can be safely delivered to the battery (use the 10 percent rule or follow the manufacturer's recommendations).

Resistor R1 determines the current flow to the battery. Its value can be found by using this formula:

$$R1 = 1.25/Icc$$

where Icc is the desired charging current in mA. For example, to recharge a battery at 500 mA (0.5 amp), the calculation for R1 is 1.25/0.5 or 2.5 ohms. TABLE 12-8 lists

Table 12-7. Universal Battery Charger Parts List

IC1	LM317 adjustable positive voltage regulator
R1	See text; Table 12-8
R2	220 ohm resistor
R3	470 ohm resistor
R4,R5	5 kilohm, 10-turn precision potentiometers
R6	330 ohm resistor
C1	2200 μF electrolytic capacitor
C2	10 μF electrolytic capacitor
D1	1N4004 diode
BR1	Bridge rectifier, 4-amp
SCR1	200-volt silicon controlled rectifier (1 amp or more)
LED1	Light-emitting diode
S1,S2	SPST switch
T1	18-volt, 2-amp transformer
F1	2-amp fuse
Misc.	Ac plug, cord, fuse holder, cabinet, heatsink for LM317, binding posts for battery under charge

All resistors are 5 to 10 percent tolerance, ¼ watt, unless otherwise indicated. All capacitors are 10 to 20 percent tolerance, rated 35 volts or more.

Milliamperes	Ohms
50	25.00
100	12.50
200	6.25
400	3.13
500	2.50

Table 12-8. Common Currents and Resistor Values

common currents for recharging and the calculated values of R1. For currents under 400 mA, you can use a 1-watt resistor. With currents between 400 mA and 1 amp, use a 2-watt resistor.

If the resistor you need isn't a standard value, choose the closest one to it as long as the value is within 10 percent. If not, use two standard-value resistors, in parallel or in series, to equal R1. If you'd like to make the charger selectable, wire a handful of resistors to a one-pole multi-position rotary switch, as shown in FIG. 12-7. Dial in the current setting you want.

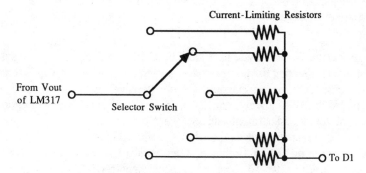

FIG. 12-7. *Rotary switch for selectable change currents.*

The output terminals can be banana jacks, alligator clips, or any other hardware you desire. You might want to use banana jacks and construct cables that can stretch between the jacks and the batteries or systems you want to recharge. For example, you can connect the charger to a 12-volt He-Ne laser battery pack. The pack is outfitted with a common ¼-inch phone plug for easy connection to the laser. To recharge the battery, you simply remove the cable attaching it to the laser and replace it with the one from the recharger.

Building The Circuit. For best results, build the circuit on a printed circuit board. Alternatively, you can wire the circuit on perforated board. Wiring is not critical, but you should exercise the usual care, especially in the incoming ac section. Be sure that you provide a fuse for your recharger.

Calibrating The Circuit. After the circuit is built, it must be calibrated before use. First set R4, the voltage adjust. This potentiometer sets the end-of-charge voltage. Then set the trip point, which is adjusted by R5. Follow these steps.

1. Before attaching a battery to the terminals and turning the circuit on, set variable resistors R4 and R5 to their mid ranges. With the recharger off, use a volt-ohmmeter to calibrate R4, referring to TABLE 12-9. Adjust R4 until the ohmmeter displays the proper resistance for the current setting you've chosen for the charger.
2. Connect a 4.7k, 5-watt resistor across the output terminals of the charger (this approximates a battery load). Apply power to the circuit. Measure the output across the resistor. For 12-volt operation with gelled electrolyte cells and lead-acid batteries, the output should be approximately 13.8 volts; for 6-volt operation, the output should be approximately 6.9 volts. If you don't get a reading or if it is low, adjust R5. If you still don't get a reading or if it is considerably off the described mark, turn R4 a couple of times in either direction.
3. Connect the volt-ohmmeter between ground and the wiper of R5, the trip-point potentiometer. Turn R5 until the meter reads zero. Turn the charger off.
4. Remove the 4.7k resistor, and in its place connect a partially discharged battery to the output terminals (be sure to use a *dis*charged battery), observing the correct polarity. Turn the charger on and watch the LED. It should not light.
5. Connect the volt-ohmmeter across the battery terminals and measure the output voltage. Monitor the voltage until the desired output is reached (see step 2, above).
6. When you reach the desired output, adjust R5 so that the LED glows. At this point, the constant-current source is removed from the output, and the battery float charges at the set voltage.

Application Notes. If you have both 6- and 12-volt batteries to charge, you might find yourself readjusting the potentiometers each time. A better way is to construct two battery rechargers (the components are inexpensive) and use one at 6 volts and the other at 12 volts. Alternatively, you can wire up a selector switch that chooses between two sets of voltage adjustment and trip-point pots.

At least one manufacturer of the LM317, National Semiconductor, provides extensive application notes on this and other voltage regulators. Refer to the *National Linear Databook Volume 1 (1987)* if you need to recharge batteries with unusual supply voltages and currents.

Table 12-9. Values for R4

R1	6-volt (in ohms)	12-volt (in ohms)
25.00	1578	2950
12.50	1497	2799
6.25	1457	2724
3.13	1437	2686
2.50	1433	2679

Depending on your battery and the tolerances of the components you use, you might need to experiment with the values of two other resistors. If the output voltage cannot be adjusted to the point you want (either high or low), increase or decrease the value of R2. If the LED never glows, or glows constantly, adjust the value of R6. Be careful not to go under about 200 ohms for R6, or the SCR could be damaged.

When recharging a battery, you know it has reached full charge when the LED goes on. To be on the safe side, turn the charger off and wait five to 10 seconds for the SCR to unlatch. Reapply power. If the LED remains lit, the battery is charged. If the LED goes out again, keep the battery on charge a little longer.

BATTERY MONITORS

A battery monitor simply provides an aural or visual indicator that a battery is either delivering too much or too little voltage. FIGURE 12-8 shows a schematic for a simple "window comparator" battery monitor (see TABLE 12-10 for a parts list). It is designed to be used with 12-volt batteries, but you can substitute one or more of the zener diodes for use with other voltages.

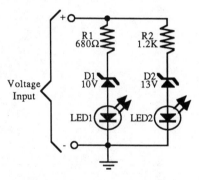

FIG. 12-8. *A simple battery condition indicator. Choose the zener diodes to provide a "window" for over/under voltage indication.*

Table 12-10. Battery Monitor Dual LED Parts List

R1	680 kilohm resistor
R2	1.2 kilohm resistor
D1	10 volt zener
D2	13 volt zener
LED1,2	Light emitting diodes

All resistors are 5-10 percent tolerance, ¼-watt.

In normal operation, LED1 glows when the voltage from the battery is at least 10 volts. It is also desirable to know if the battery is delivering too much voltage, so a second zener diode is used. If LED2 is on, the circuit is receiving too much power, and it could be damaged. More likely, however, the battery level will drop, and LED1 will grow dim or flicker off completely. If LED1 is not lit or is dim, the battery needs to be recharged.

13

Free-Air Laser Light Communications

Because light is at such a high frequency in the electromagnetic spectrum, it is an even better medium for communications than radio waves. Lasers are perfect instruments for communications links because they emit a powerful, slender beam that is least affected by interference and is nearly impossible to intercept.

This chapter explains the basics of laser light communications using both helium-neon and semiconductor lasers. You'll discover the different ways light can be modulated and cajoled into carrying an analog signal from a microphone or FM radio. The following chapter details advanced projects in laser light communications.

LIGHT AS A MODULATION MEDIUM

Higher frequencies in the radio spectrum provide greater bandwidth. The bandwidth is the space between the upper and lower frequencies that define an information channel. Bandwidth is small for low-frequency applications such as AM radio broadcasts, which span a range 540 kHz to 1600 kHz. That's little more than 1 MHz of bandwidth, so if there are 20 stations on the dial, that's only 50 kHz per deejay.

Television broadcasts, including both VHF and UHF channels, span a range from 54 MHz to 890 MHz, with each channel taking up 6 MHz. Note that the 6 MHz bandwidth of the TV channel provides more than 100 times more room for information than the AM radio band. That way, television can pack more data into the transmission.

Microwave links, which operate in the gigahertz (billions of cycles per second) region, are used by communications and telephone companies to beam thousands of phone calls in one transmission. Many calls are compacted into the single microwave channel because the bandwidth required for one phone conversation is small compared to the overall bandwidth provided by the microwave link.

Visible light and near-infrared radiation has a frequency of between about 430 to 750 terahertz (THz)—or 430 to 750 trillion cycles per second. Thanks to the immense bandwidth of the spectrum at these high frequencies, one light beam can simultaneously carry all the phone calls made in the United States, or almost 100 million TV channels. Of course, what to put on those channels is another thing!

Alas, all of this is theoretical. Transmitters and receivers don't yet exist that can pack data into the entire light spectrum; the current state of the art cannot place intelligent information at frequencies higher than about 25 or 35 gigahertz (billion cycles per second). It might take a while for technology to advance to a point where the full potential of light beam communications can be realized.

Even with these limitations, light transmission offers additional advantages over conventional techniques. Light is not as susceptible to interference from other transmissions, and when squeezed into the arrow-thin beam of a laser, is highly directional. It is difficult to intercept a light beam transmission without the intended receiver knowing about it. And, unlike radio gear, experimenting with even high-power light links does not require approval from the Federal Communications Commission. Businesses, universities, and individuals can test lightwave communications systems without the worry of upsetting every television set, radio, and CB in the neighborhood (however, CDRH regulations must be followed).

On the down side, light is greatly affected by weather conditions, and unlike low frequencies such as AM radio, it does not readily bounce off objects. Radar (low-band microwave) pierces through most any weather and bounces off just about everything.

EXPERIMENTING WITH A VISIBLE LED TRANSMITTER

It's easy to see how laser lightwave communication links work by first experimenting with a system designed around the common and affordable visible light-emitting diode. The LED provides a visual indication that the system is working and allows you to see the effects of collimating and focusing optics.

The LED communications link, like any other, consists of a transmitter and receiver. An LED is used as the transmitting component and a phototransistor is used as the receiving component. To facilitate testing, a radio or cassette player is used as the transmission source. You listen to the reception at the receiver using headphones. In Chapter 14, "Advanced Projects in Laser Communication," you'll learn how to transmit computer and remote-control data through the air via a laser beam.

Just about any LED will work in the circuit shown in FIG. 13-1, but if you want to operate the link over long distances (more than 5 or 10 feet), you should use a high-output LED, such as the kind described in Chapter 4, "Experimenting with Light and Optics." After you test the visible LED, you can exchange it with one or more high-output infrared LEDs to extend the working distance. However, you enjoy the greatest range using an infrared or visible laser (we'll get to that shortly).

Building the Transmitter

The transmitter, with parts indicated in TABLE 13-1, is designed around a 555 timer IC. The 555 generates a modulation frequency upon which the information you want

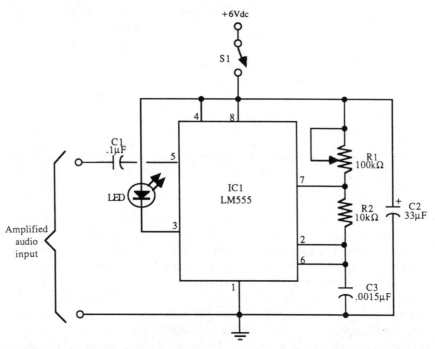

FIG. 13-1. *Schematic diagram for the pulse frequency modulated LED transmitter. Adjust frequency by rotating R1. With components shown, frequency range is between 8 and 48 kHz.*

to send is placed. The output frequency of the 555 changes as the audio signal presented to the input changes. This modulation technique is commonly referred to as pulse frequency modulation, or PFM, and is shown diagrammatically in FIG. 13-2. The signal can be received using a simple amplifier, as shown later in this section, but for best response, a receiver designed to "tune in" to the PFM signal is desired. Advanced receivers are discussed later.

Table 13-1. Transmitter Circuit Parts List

IC1	LM555 timer IC
R1	100 kilohm potentiometer
R2	10 kilohm resistor
C1	0.1 μF disc capacitor
C2	33 μF electrolytic capacitor
C3	0.0015 μF mica or Hi-Q disc capacitor
LED1	Light-emitting diode (see text)
S1	SPST switch

All resistors are 5 to 10 percent tolerance, ¼ watt. All capacitors are 10 to 20 percent tolerance, rated 35 volts or more, unless otherwise indicated.

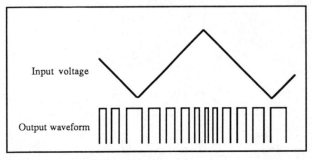

FIG. 13-2. *Comparison of input voltage and width of the output waveform.*

Construct the transmitter in a small project box. Power comes from a single 9-volt transistor battery. The switch lets you turn the circuit on and off and the potentiometer allows you to vary the relative power delivered to the LED. In actuality, adjusting the pot changes the modulation frequency, which in turn changes the pulse width, which in turn changes the current delivered to the LED. Got that?!

In any case, the entire range is beyond human hearing and above the audio signals in frequency that you will be transmitting. You can readily increase the modulation frequency to the upper limit of the components used in the transmitter and receiver, but lowering them into the 20-to 20,000-Hz region of the audio spectrum causes an annoying buzz. Any sourcebook on using the LM555 timer IC will show you how to calculate output frequency for astable operation.

Mounting details are provided in FIGS. 13-3 and 13-4; parts are shown in TABLE 13-2. Solder an LED to the terminals of a ⅛-inch phone plug jack, and mount the jack in the base of a ¾-inch PVC end plug, as shown in FIG. 13-3. If the plug is rounded on the end, file it flat with a grinder or file. Lightly countersink the hole so that the shaft of the phone jack is flush to the outside of the plug. Countersinking also helps the shaft of the jack to poke all the way through the thick-walled PVC fitting.

The transmitter and LED connect via an ⅛-inch plug that is mounted so that it extrudes through the project box, as detailed in FIG. 13-4. Use a ⁵⁄₁₆-inch 18 nut to hold the plug in place. The ⅛-inch mini plug used in the prototype is threaded for ⁵⁄₁₆-inch 18 threads, but not all plugs are the same. Check yours first.

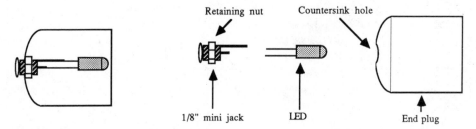

FIG. 13-3. *How to mount the LED in a PVC end plug. The same approach is used for the receiver phototransistor.*

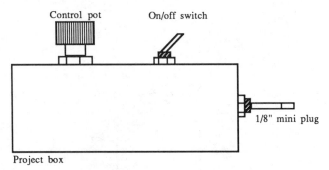

FIG. 13-4. *The project box, shown with ⅛-inch mini plug for connecting to the LED.*

Table 13-2. Plug and Box Transmitter Parts List

1	¾-inch schedule 40 PVC end plug
1	⅛-inch miniature phone jack
1	⅛-inch miniature phone plug
1ea.	Project box, knob for potentiometer, 6 Vdc battery holder (4 "AA").

Attach the transmitter into the LED by plugging it in. Install a 9-volt battery and turn the transmitter on. The LED should glow. You won't be able to test the transmitter circuit until you build the receiver.

Building the Receiver

The receiver, shown in FIG. 13-5, is designed around the common LM741 op amp and an LM386 audio amplifier. See TABLE 13-3 for a parts list. Power is supplied via two 9-volt batteries (to provide the 741 with a dual-ended supply). A switch turns the circuit on and off (interrupting both positive and negative battery connections) and a potentiometer acts like a volume/gain control.

You can listen to the amplified sounds through headphones or a speaker, or you can connect the output of the receiver to a larger amplifier. A good, handy outboard amplifier to use is the pocket amp available at Radio Shack. The pocket amp accepts an external input and has its own built-in speaker.

Construct the receiver in a plastic project box. The one used for the prototype measured 2¾ by 4⅛ by 1⁹⁄₁₆ inches and was more than large enough to accommodate the circuit, batteries, switch, potentiometer, and output jack.

The receiving phototransistor is built into a PVC end plug in the same manner as the transmitter LED, described above. Mount the phototransistor as shown in FIG. 13-3, being sure to note the orientation of the transistor leads and jack terminals. Although the circuit will work if you connect the phototransistor backwards, sensitivity will be greatly reduced.

Connect the receiver to the phototransistor, install two batteries, plug in a set of headphones, and turn the power switch on (but don't put the headphones on just yet). Adjust the potentiometer midway through its travel and point the phototransistor at an

189

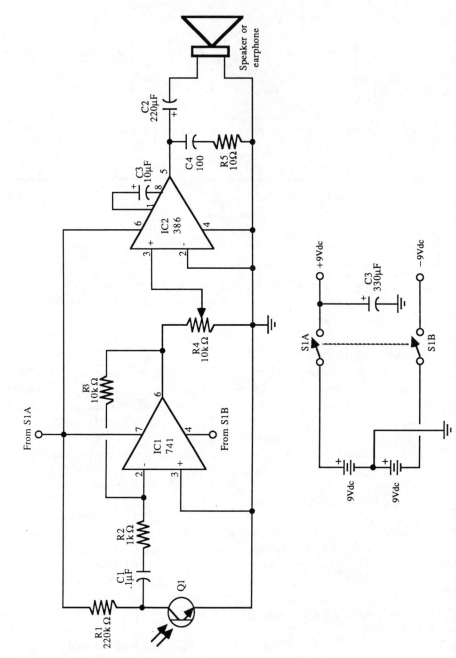

FIG. 13-5. *The universal laser light detector. The output of the LM386 audio amplifier can be connected to a small 8-ohm speaker or earphone. Two 9-volt batteries provide power. Decrease R1 to lower sensitivity; increase R3 to increase gain of the op amp (avoid very high gain or the op amp might oscillate).*

Table 13-3. Universal Receiver Parts List

IC1	LM741 operational amplifier IC
IC2	LM386 audio amplifier IC
R1	220 kilohm resistor
R2	1 kilohm resistor
R3	10 kilohm resistor
R4	10 kilohm potentiometer
R5	10 ohm resistor
C1	0.1 μF disc capacitor
C2	220 μF electrolytic capacitor
C3	10 μF electrolytic capacitor
C4	100 μF electrolytic capacitor
Q1	Infrared phototransistor
S1	DPDT switch

All resistors are 5 to 10 percent tolerance, ¼ watt. All capacitors are 10 to 20 percent tolerance, rated 35 volts or more.

incandescent lamp. You should hear a buzzing sound through the headphones (the buzzing is the lamp fluctuating under the 60-cycle current).

If you don't hear the buzz, adjust the volume control until the sound comes in. Should you still not hear any sound, double check your wiring and the batteries. Even with the phototransistor not plugged in you should hear background hiss. No hiss might mean that the circuit is not getting power or the headphone jack is not properly wired.

The receiver can be used with the LED lightwave link as well as all the other communications projects in this chapter (as well as most of those in the remainder of this book). Its wide application makes it an ideal all-purpose universal laser beam receiver. When I refer to the "universal receiver," this is the one I'm talking about.

Using the Lightwave Link

Once the receiver checks out, you can test the transmitter. Switch off the lights or move to a darkened part of the room. Turn on the transmitter source (radio, tape player) and aim the transmitter LED at the receiver phototransistor. Adjust the controls on the receiver and transmitter until you hear sound. You might hear considerable background hiss and noise, caused by other nearby light sources. If you use the communications link outdoors in sunlight, the infrared radiation from the sun might swamp (overload) the phototransistor, and the sound could be drastically reduced or cut off completely. The transmitter and receiver works best in subdued light.

Test the sensitivity and range of the communications link by moving the receiver away from the transmitter. Depending on the output of the LED, the range will be limited to about 5 feet before reception drops out.

Extending the Range of the Link

Most all phototransistors are most sensitive to infrared light. The peak spectral sensitivity depends on the makeup of the transistor, but it is generally between about 780 and 950 nm in the near-infrared portion of the spectrum. A red LED has a peak spectral output of about 650 nm, considerably under the sensitivity of the phototransistor. A solar cell offers a wider spectral response and can provide greater range. The best type of solar cell to use is the kind encased in plastic like the phototransistor (many have a built-in lens). Connect the cell in the circuit as shown in FIG. 13-6.

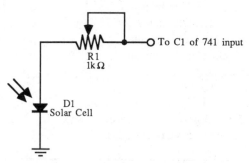

FIG. 13-6. *How to connect a solar cell to the input of the universal laser light detector. The cell provides better sensitivity in the visible light range than an infrared phototransistor.*

The solar cell is sensitive to a wide range of colors. The light spectrum above or below the red radiation from the LED isn't needed for reception, so block it with a red filter. Test the effectiveness of the filter by temporarily taping it to the front of the solar cell.

You can also use the filter with the phototransistor to help limit the incoming radiation to the red wavelengths. Even though the phototransistor is designed to be most sensitive to near-infrared radiation, it can still detect light at other wavelengths, especially red. One or two layers of red acetate placed over the phototransistor can increase the range in moderate light conditions by several feet.

The best way to increase the working distance of the communications link is to add lenses to the LED and/or the phototransistor. The PVC end plug makes it fairly easy to add lenses to both transmitter and receiver components. Mount a simple double-convex or plano-convex lens in the end plug. The PVC rings hold the lens in place and let you easily adjust the distance between the lens and phototransistor. If the lens has a focal length of more than 10 to 15 mm, attach a coupling to the end plug and stuff the lens in the coupling. Again, use PVC rings to hold the lens in place.

Be sure that you position the lens at the proper focal point with respect to the *junction* of the LED or phototransistor. If you don't know the focal length of the lens you are using, test it following the instructions provided earlier in this book.

You can see the effect of the lens on the transmitted light by pointing the LED against a lightly colored wall. At close range and with the lens properly adjusted, you should actually see the junction of the LED projected on the wall (assuming you are not using an LED with a diffused case). You might also see a faint halo around the junction; this is normal and is caused by light emitted from the sides of the LED.

With the lens(es) attached, try the lightwave link again and test its effective range. With extended range comes increased directionality, so you must carefully aim the transmitter element at the receiver. A simple focusing lens on both receiver and transmitter should extend the working distance to a hundred feet or more. Test the real effectiveness of the system at night outside. The darkness will also help you better aim the transmitter. At 100 feet, the light from the LED will be dim, but you should be able to spot it if you know where to look for it.

Note that the plastic case of the LED and phototransistor acts as a kind of lens, and that can alter the effective focal length of the system. Experiment with the position of the lens until the system is working at peak performance.

ACOUSTIC MODULATION

In 1880, Alexander Graham Bell, with his assistant Sumner Tainter, demonstrated the first *photophone,* a mechanical contraption using sunlight or collimated artificial light to transmit and receive voice signals over long distances. Its operation was simple. The system used a lightweight membrane similar to reflective Mylar as a voice diaphragm. A bright beam of light, typically from the sun, was pointed at the diaphragm, which vibrated when a person talked into it. The vibration then caused the light to fluctuate in syncopation with the sound. A receiver, located some distance away, demodulated the fluctuating light levels and turned the beam back into the talker's voice.

Bell had great hopes for the photophone, and in fact had predicted that it would be a bigger hit than the telephone. But the problems of poor range in inclement weather doomed the photophone as just another scientific curiosity. Had Bell used a laser with his photophone, he would have been able to greatly increase the range of the device. Of course, clouds, fog, and heavy rain would have still reduced the working distance of the laser photophone, limiting it to a clear-weather communications device.

You can easily duplicate Bell's photophone, adding the laser as a high-tech improvement. The process of transmitting low-frequency audio signals via a photophone-like device is more accurately termed *acousto-modulation*. You can use a stretched membrane as the acoustic vibrating element or adapt a surplus speaker as a "light switch."

Stretched Membrane Modulator

Thin reflective Mylar is a fairly common find among the mail order surplus outfits, as well as local army/navy surplus shops. Reflective Mylar, or a reasonable facsimile made with generic acetate, is used to produce parachutes for radiosonde equipment, the thermal layer on camping blankets, high-tech jewelry, radar jammer streamers, and lots more. Price is reasonable. A small 2-by-2-foot square sheet of reflective (or "aluminized") acetate costs about a dollar on the surplus market and a little more when you buy it from a commercial dealer. One small square is all you need.

Refer to TABLE 13-4 for a parts list for the stretched membrane modulator. Secure the Mylar sheet inside a 4- to 6-inch diameter embroidery hoop as shown in FIG. 13-7. The hoop allows you to open the two halves, insert the material, and pull it tight as the two halves are tightened together. The idea is to pull the Mylar as taut as possible.

Mount the hoop on a wooden or plastic base. Set up a speaker behind the hoop and direct a laser beam at the Mylar. Activate the speaker using a radio, tape player, or

193

Table 13-4. Mylar Hoop Modulator Parts List

1	4- to 6-inch diameter circular wood or plastic embroidery hoop
1	4- to 6-inch diameter full-range speaker
2	3-by-3-inch wood block (for base and speaker mount); ½-inch plywood or pine
2	1-by-½-inch corner angle bracket
4	$8/32$ by 1½-inch bolts, nuts, flash washers
1	¼-inch 20 nut and washer (for tripod)
1	Portable camera tripod

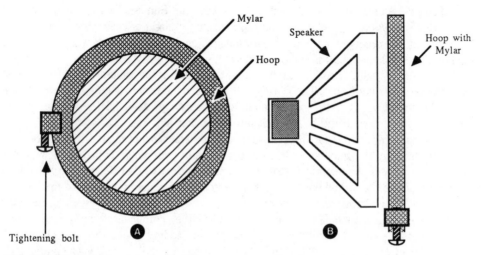

FIG. 13-7. *Basic arrangement for the Mylar speaker modulator. (A) Front view with Mylar stretched in embroidery hoop; (B) Side view with speaker behind the hoop.*

amplified microphone. As the speaker vibrates, it oscillates the Mylar and thus modulates the laser beam.

The modulation is not electronic or electromechanical, but *positional* or *geometrical*. You can see the effect of the modulation by positioning the laser so the beam strikes a distant target. When the Mylar vibrates, the beam is displaced at the target, making squiggles and odd shapes. Position a receiving element such as a solar cell at the target and you can register the movement by sensing the varying intensity of the beam (actually, the intensity falls off as the beam moves off-axis to the center of the cell).

The universal laser receiver, described earlier in this chapter, can be readily used to capture and demodulate the signal transmitted over the beam. The solar cell is connected to the receiver as shown in FIG. 13-6.

The ideal size for the solar cell depends on the divergence of the beam and the distance between the laser and receiver. Beam divergence with most helium-neon lasers is only about one milliradian (less on high quality tubes). Placing the target 50 meters away produces a spot of about 50 mm across (about 2 inches). That means you can use a silicon solar cell that's 2 inches in diameter and capture all or most of the beam. Modulation that causes the beam to wander off-axis to the cell generates a change of voltage.

You can readily calculate the approximate beam spread at any distance by multiplying the divergence in radians by the distance in meters. For example:

$$\text{Divergence (radians)} = 0.001$$
$$\text{Distance} = 200$$

0.001 times 200 equals 0.2, or 200 millimeters.

Another example: What is the spread at 1 km using a laser with a divergence of 1.2 mrad? Answer: 1.2 meters. That's a small amount considering that the beam travels over half a mile. You can reduce beam divergence by adding collimating optics to the output of the laser. Chapter 8 provides details on building laser collimating optics.

Speaker Cone Modulator

An interesting effect used in many light shows is created by mounting a mirror in front of a speaker (the mirror can also be mounted directly on the speaker). A laser beam, reflected off the mirror, bounces around on the wall or screen in time to the music (various mirror/speaker mounting techniques are discussed more fully in Chapter 19).

You can use the same technique to transmit audio information over the air. Simply place a receiving element at the spot where the beam lands. For best results, keep the amplitude of the speaker at a low level so the beam doesn't deflect more than a few degrees. Use a 2- or 3-inch diameter silicon solar cell as the receiver element. You can use the speaker to transmit music from a radio or tape player, or rig up the speaker to an amplifier and microphone and broadcast your own voice.

Even with the speaker turned down low, wide deflection of the beam becomes a problem when transmitting over long distances. The beam covers a larger area at the target as the distance between the receiver and transmitter is increased. It is generally impractical to enlarge the sensing area by more than 4 or 5 inches in diameter, so another approach is recommended. This idea comes from Roger Sontag at General Science and Engineering. Instead of bouncing the light off of a mirror, cut an edge off the cone of a speaker and use it as a "shutter". As the speaker cone vibrates, it alternately passes and cuts off the laser beam. The system requires careful alignment, but the deflection of the beam at the receiver is minimal.

The best speakers to use are those that measure 4 to 6 inches in diameter and have a deep taper. Avoid using a speaker where the cone lies flat in the frame.

Mount the speaker in a swivel mount so that you can adjust its height and angle. Place a helium-neon or cw diode laser to one side of the speaker so that the beam skims across the top of the cut portion of the cone. Energize the speaker with a fairly powerful amplifier (but don't exceed the wattage rating of the speaker), and watch for the cone to move in and out in response to the sound. Now look at the target and watch it flicker as the speaker moves. If the cone doesn't block the beam, or blocks the beam entirely, readjust the position of the speaker as needed.

You can use the universal laser beam receiver described earlier in this chapter to capture the signal on the modulated beam. The intensity of the beam could swamp the phototransistor, so place a set of polarizers in front and vary their rotation to reduce the beam intensity to a usable level.

ELECTRONIC MODULATION OF HELIUM-NEON LASERS

Agreeably, using a sheet of plastic or a dissected speaker does not represent a high-tech approach to laser modulation. Although it might appear otherwise, it's fairly easy to modulate the beam of a helium-neon laser, and using only a handful of parts at that. Two approaches are provided here: both have an effective bandwidth of around 0 Hz to 3 kHz, making them suitable for most voice and some music transmission schemes.

Transformer

A transformer placed in line with the high-voltage power supply and cathode of the tube can be used to vary the current supplied to the tube. This causes the intensity of the beam to vary. This is amplitude modulation, the same technique used in AM radio broadcasts.

Although you can use a number of transformers as the modulating element, Dennis Meredith of Meredith Instruments suggests you use a public address power output transformer. It's ideal for the job because of its high turns ratio—the ratio of wire loops in the primary and secondary. You wire the transformer in reverse to the typical application: the speaker terminals from a hi-fi or amplifier connect to the "output" of the transformer and the laser connects to the "input."

PA transformers are available from almost any electronics parts store, including Radio Shack, who offers a good one for under $5. PA transformers are rated by their voltage, usually either 35 or 70 volts. Get the higher voltage rating. There are several terminals on the transformer. Connect the speaker terminals to the common and 8-ohm terminals; connect the laser cathode, as shown in FIG. 13-8, to the common and one of the wattage terminals (parts list in TABLE 13-5). Experiment with the wattage terminal that yields the most modulation. The prototype seemed to work best using the 5-watt terminal.

The cathode passes some current, so touching its leads can cause a shock. Isolate the transformer and wires in a small project box, like the one shown in FIG. 13-9. Five-way binding posts (fancy banana jacks) are used for the cathode connections; the audio input is an 1/8-inch miniature phone jack.

Transistor

Who wants to lug around a bulky and heavy transformer when you can provide modulation to the He-Ne tube using a simple silicon transistor? This next mini-project provides a seed that you can use to design and build an all-electronic analog or digital

1	70-volt PA transformer
1	1/8-inch miniature jack
2	5-way binding posts (25-amp)
1	Project box

Table 13-5.
Transformer Modulator Parts List

phototransistor can also increase sensitivity and reduce background noise. Infrared filters are available at most photographic stores; surplus is another good source. Many IR filters may appear dark red or purple or even completely black. You might not be able to see through the filter, but it is practically transparent to near-infrared radiation.

OTHER MODULATION AND DEMODULATION TECHNIQUES

Amplitude modulation is susceptible to interference from changing light levels. This can lead to noise and poor system response. The pulse frequency modulation technique used with the visible LED and diode laser system rejects noise and is not as sensitive to changes in the intensity of the source beam.

The universal laser light receiver can be used to capture the signal transmitted over an AM or PFM modulated beam. A better approach is the circuit shown in FIG. 13-13. It uses a LM565 phase-locked loop (PLL) adjusted so that its center frequency matches the center frequency of the transmitter—about 40 kHz.

An audio signal impressed upon the 555 in the transmitter changes the center frequency. This change is detected by the PLL as an error signal. The amount of error

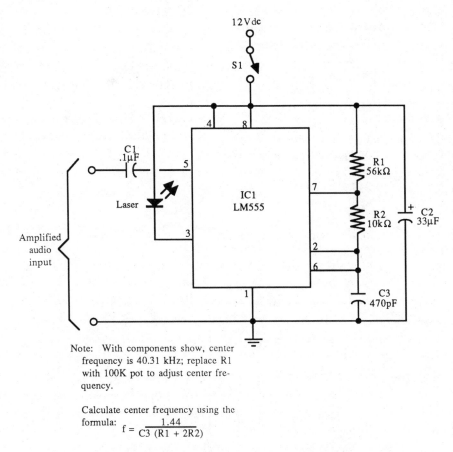

Note: With components show, center frequency is 40.31 kHz; replace R1 with 100K pot to adjust center frequency.

Calculate center frequency using the formula:
$$f = \frac{1.44}{C3\ (R1 + 2R2)}$$

FIG. 13-14. *Circuit diagram for the laser diode transmitter.*

signal is proportional to the frequency of the original audio signal. Therefore, tapping the error signal pin on the PLL chip and then amplifying it retrieves the audio that was transmitted over the beam.

The circuits for the transmitter and receiver appear in FIGS. 13-13 and 13-14 (see TABLES 13-7 and 13-8 for a list of required parts). The transmitter is virtually the same as the one presented earlier in the chapter but with no provision for adjusting the center frequency. The receiver uses the 565 PLL (other PLLs can be used) and a trim pot to adjust the circuit for the exact center frequency of the transmitter. You can connect the transmitter as discussed above, or amplify it and apply the signal to the laser diode.

With the circuits complete and set up, aim the laser at the phototransistor and provide an audio signal for transmission. Adjust the trim pot on the receiver until you hear the audio carried over the light beam. The components used in both receiver and transmitter can drift, so you might need to touch up the trim pot control to re-align the center frequency.

IC1	LM741 op amp IC
IC2	LM565 PLL IC
IC3	LM386 audio amplifier IC
R1	220 kilohm resistor
R2	1 kilohm resistor
R3	10 kilohm resistor
R4	6.8 kilohm resistor
R5	10 kilohm potentiometer
R6	10 ohm resistor
C1	0.1 μF disc capacitor
C2	0.001 μF silvered mica capacitor
C3	0.001 μF disc capacitor
C4	0.047 μFdisc capacitor
C5,C6	10 μF electrolytic capacitor
C7	220 μF electrolytic capacitor
C8	100 μF electrolytic capacitor
Q1	Infrared phototransistor
S1	DPDT switch

Table 13-7. Pulse Frequency Modulator Receiver Parts List

IC1	LM555 timer IC
R1	56 kilohm resistor
R2	10 kilohm resistor
C1	0.1 μF disc capacitor
C2	33 μF electrolytic capacitor
C3	470 pF silvered mica capacitor
S1	SPST switch
	Laser

Table 13-8. Pulse Frequency Modulator Transmitter Parts List

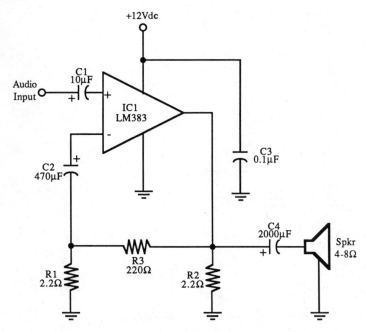

FIG. 13-15. *An 8-watt audio amplifier, designed around the LM383 integrated amp. The IC must be installed on a suitable heatsink.*

Table 13-9. Eight-Watt Audio Amplifier

IC1	LM383 audio amplifier
R1,R2	2.2 ohm resistor
C1	10 μF electrolytic capacitor
C2	470 μF electrolytic capacitor
C3	0.1 μF disc capacitor
C4	2000 μF electrolytic capacitor
SPKR	8-ohm speaker

Table 13-10. Sixteen-Watt Audio Amplifier

IC1,IC2	LM383 audio amplifier
R1,R3	220 ohm resistor
R2,R4	2.2 ohm resistor
R5	1 megohm resistor
R6	100 kilohm potentiometer
C1,C7	10 μF electrolytic capacitor
C2,C5	470 μF electrolytic capacitor
C3,C4,C6	0.2 μF disc capacitor
SPKR	8-ohm speaker

203

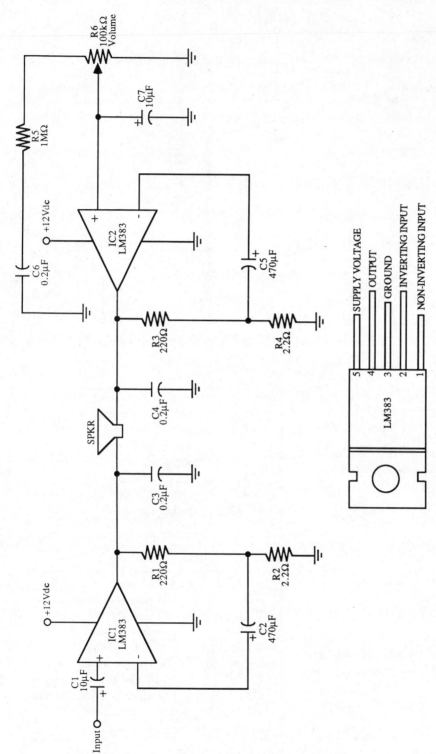

FIG. 13-16. *A 16-watt audio amplifier, designed around two LM383 integrated amps. The ICs must be installed on a suitable heatsink.*

AUDIO AMPLIFIER CIRCUITS

The universal receiver has a built-in LM386 integrated amp. The sound output is minimal, but the chip is easy to get, it's cheap, and it can be wired up quickly. It's perfect for experimenting with sound projects.

If you need more sound output or must amplify the audio input for the transformer or transistor modulator, try the circuit in FIG. 13-15 (parts list in TABLE 13-9). You can use it instead of the LM386 in the receiver or in addition to it. The circuit is designed around an LM383 8-watt amplifier IC. The IC comes mounted in a TO-220-style transistor package, and you should use it with a suitable heatsink. FIGURE 13-16 shows a higher output 16-watt version using two LM383's (parts list in TABLE 13-10). Note that the LM383 IC is functionally identical to the TDA2002 power audio amplifier.

14

Advanced Projects in Laser Communication

The last chapter presented a number of basic free-air laser light communications projects. You learned how to modulate a He-Ne laser beam using a transformer, transistor, and even a piece of Mylar foil stretched in a needlepoint hoop. You also learned various ways to electronically modulate laser diodes and recover the transmitted audio signal. This chapter presents advanced projects in free-air laser-beam communication. Covered are methods of remotely controlling devices and equipment via light and how to link two computers by a laser beam.

TONE CONTROL

Everyone is familiar with Touch-Tone dialing: pick up the phone and push the buttons. You hear a series of almost meaningless tones, but to the equipment in the telephone central office, those tones are decoded and used to dial the exact phone you want out of the millions in the world. You can use the same technique as a remote control for actuating any of a number of devices, such as motors, alarms, lights, doors, you name it. The tone signals are sent from transmitter to receiver via a laser light beam. With the right setup, you can remotely control devices up to several miles away, and without worry of interference or FCC regulations.

The Touch-Tone (or more simply "tone control") system supports up to 16 channels. Each channel is actuated by a pair of tones. Tone selection depends on the buttons pressed on the keypad. Dividing a common 16-key keypad into a matrix of 4 by 4, as illustrated in FIG. 14-1, shows how the tones are distributed. For example, pressing the number 5 key actuates the 770 Hz and 1336 Hz tones. Pressing the number 9 key actuates both the 852 Hz and 1477 Hz tones.

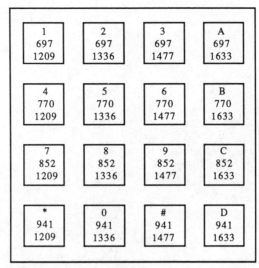

FIG. 14-1. *Keypad tones for tone dialing matrix.*

The dual tones help prevent accidental triggering, but they also present a somewhat difficult decoding dilemma. The first tone dialing circuits used tuned components that were expensive and difficult to maintain. As tone dialing caught on, custom-made ICs were developed that dispensed with the tuned circuits. Until recently, these ICs have been rather costly and required some sophisticated interface electronics. Now, Touch-Tone decoding ICs cost under $15 and operate with only two or three common components.

Most telephones use a matrix of 12 keys, not 16, so the last column of buttons isn't used. The circuits in the phone may or may not be able to reproduce the tones for the last column, but most off-the-shelf Touch-Tone dialing ICs are capable of the full 16-channel operation.

Making Your Own Controller

Dialing other phones is not the goal of this project, but using a phone dialer as a remote controller is. You have several alternatives for making the controller.

★ Salvage the innards of a discarded phone (must have tone dialing).
★ Use a portable, hand-held tone dialing adapter.
★ Build the controller from scratch using a keypad and dialing chip.

The last approach allows you full access to all 16 combinations of tones. The other two approaches limit you to the 12 keys on a standard telephone—digits 0 through 9 as well as the # and * symbols.

Salvaging the keypad and circuits from a phone requires some detective work on your part, unless you happen to receive a schematic (not likely), but the advantage is you get the entire controller as one module. The biggest disadvantage is that the dialing circuits may require odd operating voltages.

FIG. 14-2. *A battery-operated portable pocket tone dialer (shown with hookup leads attached).*

The portable tone-dialing adapter, such as the one in FIG. 14-2, is an easier approach, with the added benefit of an easy-to-carry (fits in your pocket) module that runs on battery power. The adapter is meant for use with rotary or pulse dial phones when you need to access services that respond to Touch Tones (long-distance services, computer ordering, etc.). You place the adapter against the mouthpiece of the phone and press the buttons.

FIGURE 14-3 shows how easy it is to modify the dialer for use as a tone source for the PFM laser diode modulator. Alternatively, you can connect the output of the dialer to an audio amp and process it through the transformer or transistor He-Ne laser modulator described in the previous chapter. The connections to the dialer's speaker can remain in place, thereby providing you with audible feedback that the controller is working and sending out tones. See TABLE 14-1 for the parts list.

1	Pocket Touch-Tone dialer
1	⅛-inch miniature plug
Misc.	Wire

Table 14-1. Pocket Dialer Parts List

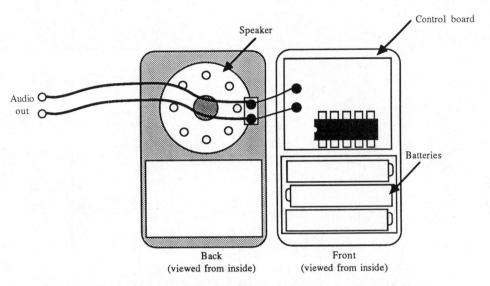

FIG. 14-3. *Wiring diagram for adding external hookup leads to the pocket dialer.*

The Tone Receiver

A number of electronics outlets, such as Radio Shack, sell an all-in-one tone receiver chip. This chip deciphers the dual tones and provides a binary weighted decoded output (8, 4, 2, 1). Add to the basic circuit a 4-of-16 demultiplexer, as shown in FIG. 14-4, and you can control up to 16 different devices. When used with a telephone keypad or dialing adapter, you can decode the first 12 digits. Use the universal laser light detector (presented in the last chapter) to capture the beam and amplify it for the receiver. See the parts list for the tone receiver in TABLE 14-2.

INFRARED PUSH BUTTON REMOTE CONTROL

The Motorola MC14457 and MC14458 chips form the heart of a useful remote control receiver-transmitter pair. The chips are available through Motorola distributors as well as several mail-order outlets and retail stores (including, at the time of this writing, Digikey and Circuit Specialists; see Appendix A for addresses). Price is under $15 for the pair.

Table 14-2. Tone Dialer Receiver Parts List

IC1	SSI202 tone decoder IC
IC2	74154 IC
RI	10 megohm resistor
R2-R5	1 kilohm resistor
LED1-4	Light-emitting diode
XTAL1	3.57 MHz (colorburst) crystal

All resistors are 5 to 10 percent tolerance, ¼ watt. All capacitors are 10 to 20 tolerance.

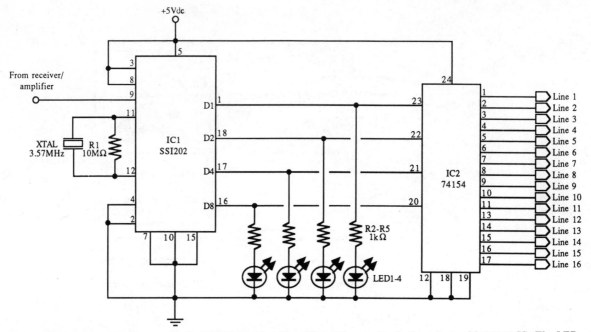

FIG. 14-4. *A schematic diagram for the SSI202 tone-decoding chip and how to decode the outputs with 74154 IC. The LEDs provide a visual indication of the binary output of the tone-decoding chip.*

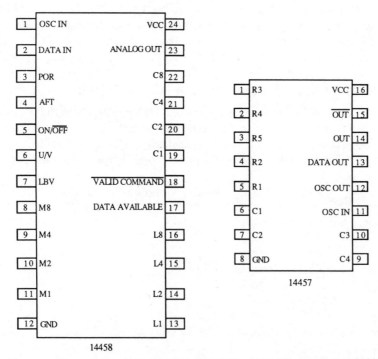

FIG. 14-5. *Pinout diagrams for the Motorola 14457 and 14458 remote control chips.*

The 14457 is the transmitter and can be used with up to 32 push-button switches (we'll be using 12). Pushing a switch commands the chip to send a binary serial code through the laser (preferably a diode laser). The 14458 chip receives the signal. Decoded output pins on the 14458 receiver chip can be connected directly to a controlled device, such as a relay or LED, a counter, or a computer port.

FIGURE 14-5 shows the pinout diagrams for the two chips. The 14457 comes in a small, 16-pin DIP package, and with all the other components added in, takes up a space of less than 2 inches square. The chip uses CMOS technology to conserve battery power, and when no key is pressed, the entire thing shuts down. Battery power is used only when a key is depressed.

Building the Transmitter

The basic hookup diagram for the 14457 transmitter circuit is shown in FIG. 14-6; the parts list is provided in TABLE 14-3. Note the oscillator and tank circuit connected to pins 11 and 12. The oscillator required by the 14457 (and a matching one for the 14458 receiver) is a hard-to-find ceramic resonator, not a standard crystal.

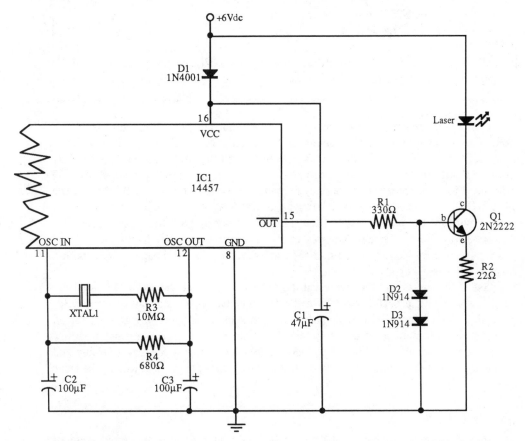

FIG. 14-6. *Hookup diagram for the timing and output portion of the 14457 transmitter IC. XTAL1 is a 300-650 kHz ceramic resonator, as discussed in the text.*

211

Table 14-3. Remote Control Transmitter Parts List

Transmitter

IC1	Motorola MC14457 remote control transmitter IC
R1	330 ohm resistor
R2	22 ohm resistor
R3	10 megohm resistor
R4	680 ohm resistor
C1	47 μF electrolytic capacitor
C2,C3	0.001 μF disc capacitor
Q1	2N2222 transistor
D1	IN4001 diode
D2,D3	IN914 diode
XTAL1	300 to 600 kHZ ceramic resonator
Misc.	Laser diode

Keypad

Q1-Q4	2N2222 transistor
Misc.	Matrix keypad or switches

All resistors are 5 to 10 percent tolerance, ¼ watt. All capacitors are 10 to 20 percent tolerance, rated 35 volts or more.

The ceramic resonator works just like a crystal but comes in frequencies under 1 MHz. The exact value of the resonator isn't critical, I found, as long as it is within a range of about 300 to 650 kHz and the resonators for both chips are identical. I successfully used a 525 kHz resonator in the prototype circuit. A common ceramic resonator is 455 kHz, used as an intermediate frequency in many receivers. The inverted output of the 14457 ($\overline{OUT}$) is applied to the modulating input of the laser (see the previous chapter).

So much for the output stage of the transmitter; what about the input stage? FIGURE 14-7 shows how to connect a series of 12 push-button switches to the column and row inputs of the 14457. You can use separate switches, but a cheaper way is with a surplus telephone keypad.

Just about any wiring technique can be used to construct the 14457 transmitter, but because you'll probably want to make the unit handheld, stay away from wire-wrapping. The stems of wire-wrapping posts and sockets are too long and will fatten the controller considerably. Use a set of four "AA" batteries or a 9-volt transistor battery to power the transmitter. You might need additional batteries for the amplifier and modulator, depending on the type used. And, you must also provide power to the laser itself.

Building the Receiver

Short-haul communications links do not require amplification on the input stage of the receiver. But if you plan on using the receiver some distance from the transmitter

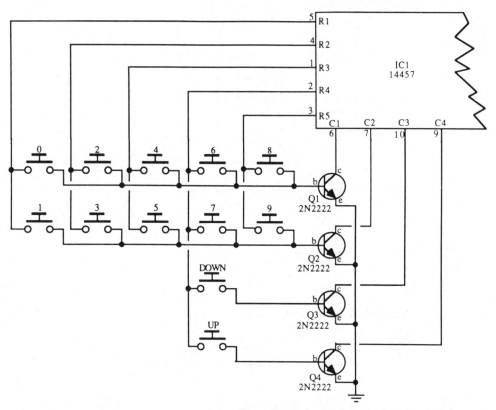

FIG. 14-7. *Keypad connection for adding 12 push buttons to the 14457 transmitter. You may easily add additional rows of buttons by connecting them to Q3 and Q4 in the manner shown for Q1 and Q2.*

(500 meters or more), an extra amplification stage is recommended. You may use the universal amplifier found in Chapter 13.

The basic wiring diagram for the 14458 receiver chip is shown in FIG. 14-8 (parts list in TABLE 14-4). Once again, the ceramic resonator is used as a timing reference for the IC. One inverter from a 4069 is used to provide an active element and driver for the oscillator circuit.

All that's required now is to connect the devices to be controlled to the output lines, shown in FIG. 14-9. Note the various sets of outputs and the $\overline{VC}$ function pin. The $\overline{VC}$ line goes HIGH when all but the number keys are pressed. The pin is used in some advanced decoding schemes as a function bit. The chart in TABLE 14-5 shows what happens when the 12 keys are pressed (the chip can accommodate another 20 push buttons; see the manufacturers data sheet for more information).

For most routine applications, you need only to connect the controlled device to pins 19 through 22 (labeled C1, C2, C4, and C8). These are binary weighted, and by connecting a 4028 one-of-ten decoder IC to the receiver, you can individually control up to 10 control devices or functions.

The receiver is wired to accept a single keypress on the transmitter as a complete command. The receiver can also be made to wait until *two* keys are pressed (this is used because in television and VCR applications, you are able to dial in multi-digit

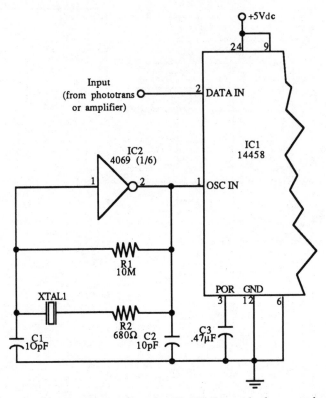

FIG. 14-8. *The timing portion of the 14458 receiver IC. XTAL must be the same value as the resonator used with the 14457 transmitter. Connect a phototransistor or amplifier to pin 2 of the chip (see Chapter 13 for details on a suitable amplifier).*

Table 14-4. Remote Control Receiver Parts List

Receiver

IC1	Motorola MC14458 remote-control receiver IC
IC2	4069 CMOS hex inverter gate
R1	10 megohm resistor
C1,C2	10 pF disc capacitor
C3	0.47 μF disc capacitor
XTAL	300 to 600 kHz ceramic resonator

All resistors are 5 to 10 percent tolerance, ¼ watt. All capacitors are 10 to 20 percent tolerance, rated 35 volts or more.

Output Enhancements

IC2	4028 CMOS decoder IC
or	
IC2	4001 NOR gate IC
IC3	4029 CMOS counter IC

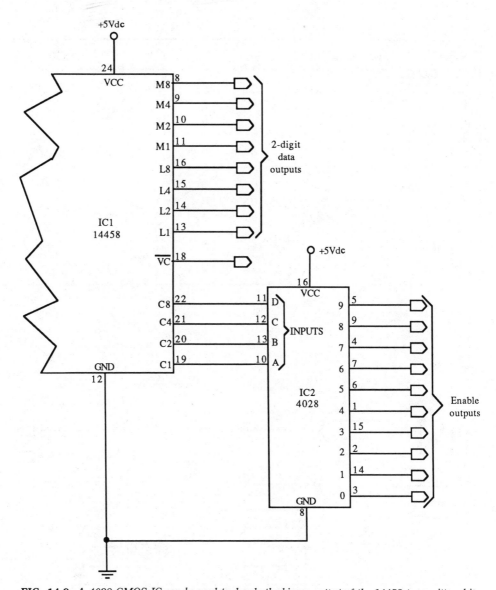

FIG. 14-9. *A 4028 CMOS IC can be used to decode the binary output of the 14458 transmitter chip.*

channels). The two-digit data outputs are used when the chip is this two-digit mode. To change from one- to two-digit mode, disconnect the power leads to pins 9 and 6.

FIGURE 14-10 shows how you can use the UP and DOWN functions with a counter to provide variable step control. Note the DOWN ENABLE and UP ENABLE lines. When pressing the UP or DOWN buttons on the transmitter, the UP and DOWN codes are sent continually rather than just one per push, as with the other buttons.

The enable outputs from the 4028 can only handle 10 functions, so the UP and DOWN functions are integrated with the number functions. That is, when the DOWN button is pressed, the output at the 4028 chip is the same as if you pressed the number 6 button

215

Table 14-5. Key Decoding

Key	Row	Column	FB	C8	C4	C2	C1	$\overline{VC}$ Pulse
0	1	1	0	0	0	0	0	
1	1	2	0	0	0	0	1	
2	2	1	0	0	0	1	0	
3	2	2	0	0	0	1	1	
4	3	1	0	0	1	0	0	
5	3	2	0	0	1	0	1	
6	4	1	0	0	1	1	0	
7	4	2	0	0	1	1	1	
8	5	1	0	1	0	0	0	
9	5	2	0	1	0	0	1	
DOWN	4	3	1	0	1	1	0	X
UP	4	4	1	0	1	1	1	X

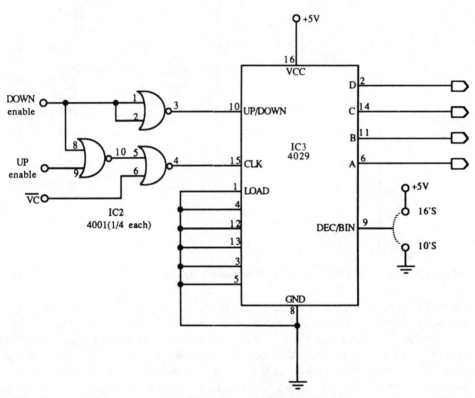

FIG. 14-10. *Decoding for the UP and DOWN functions. The BCD output of the 4029 can be interfaced to a display or computer. The 4029 chip can be made to count by 16s or 10s by connecting pin 9 to ground or V+ as shown.*

(binary code 0110). However, the $\overline{VC}$ pin is toggled HIGH, which can be used in further decoding. Similarly, when the UP button is pressed, the output of the 4028 chip is the same as if you pressed the number 7 button (binary code 0111).

To enable you to count the number of UP and DOWN pulses, connect the UP ENABLE input of the circuit shown in the figure to the number 7 output of the 4028 and the DOWN ENABLE input to the number 6 output. The $\overline{VC}$ pin acts to gate the circuit so that the counter doesn't count when numbers 7 and 6 are pressed.

DATA TRANSMISSION

The UART (Universal Asynchronous Receive/Transmit) chip converts parallel to serial and serial to parallel. It's much more involved than a shift register that simply converts parallel data to pure serial form, or vice versa.

The UART allows you to send data to devices like printers, plotters, and modems and yet be assured that all the information you are sending is getting there intact. Built into the chip are provisions for sending and receiving at the same time, for adding parity

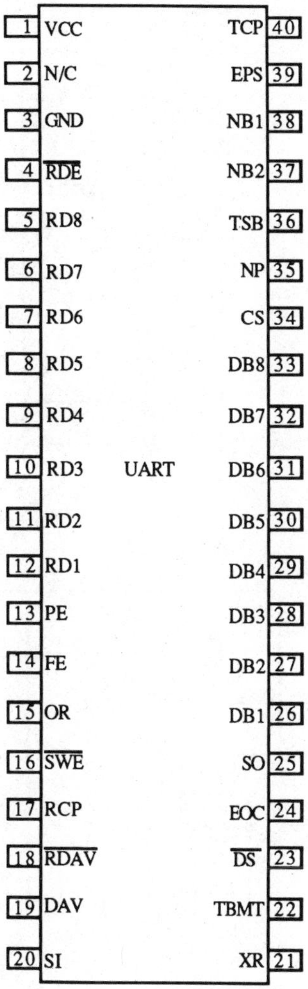

FIG. 14-11. *Pinout diagram for the AY3-1015 (or equivalent) UART chip.*

Pin	Signal		Signal	Pin
1	VCC		TCP	40
2	N/C		EPS	39
3	GND		NB1	38
4	RDE		NB2	37
5	RD8		TSB	36
6	RD7		NP	35
7	RD6		CS	34
8	RD5		DB8	33
9	RD4		DB7	32
10	RD3	UART	DB6	31
11	RD2		DB5	30
12	RD1		DB4	29
13	PE		DB3	28
14	FE		DB2	27
15	OR		DB1	26
16	SWE		SO	25
17	RCP		EOC	24
18	RDAV		$\overline{DS}$	23
19	DAV		TBMT	22
20	SI		XR	21

217

bits and stop bits to the serial data train, and more. A pinout diagram for the IC is shown in FIG. 14-11. Note that a number of other UARTs will work as well and that these chips might even have the same pinouts. The functions of the pins are listed in any UART spec sheet.

For all their sophistication, however, UART chips are surprisingly inexpensive—under $5 or $6. They require accurate timing, however, which means the addition of a crystal and a baud-rate generator (the generator can be replaced by other circuits, but in the long run, the generator is a better choice). With all the components added, a UART system costs about $15.

You can arrange the UARTs in a number of ways. For example, a computer such as the IBM PC has its own UART built into it. You can connect it to a laser modulator and send serial data through the light beam. You can either receive the data and process it through the serial port on the remote computer, or convert it to parallel form with a receiver UART.

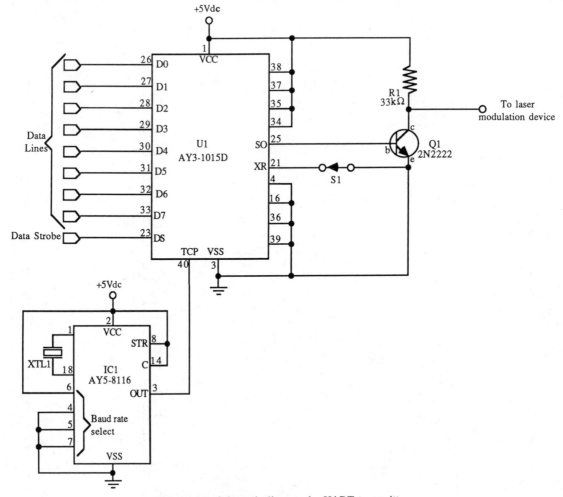

FIG. 14-12. *Schematic diagram for UART transmitter.*

218

Table 14-6. Transmitter UART Parts List

IC1	AY3-1015D UART IC
IC2	AY5-8116 baud-rate generator
R1	33 kilohm resistor
Q1	2N2222 transistor
S1	SPST switch (momentary, normally closed)
XTL1	3.57 MHz (colorburst) crystal

All resistors are 5 to 10 percent tolerance, ¼ watt.

The Commodore 64 computer lacks a UART device and is ideal for the UART link used in this project. The link connects to the Commodore by way of the computer's User Port. The User Port is a bidirectional parallel port, though this project uses it as an output device only—the UART connected to the Commodore 64 is a transmit-only device. You can convert to transceiver operation by rewiring it.

The circuit in FIG. 14-12 shows how to connect the transmitter UART to the User Port of the "host" Commodore 64. The circuit, with a parts list provided in TABLE 14-6, can be connected to the laser diode PFM modulator (introduced in the last chapter). Alternatively, you can connect the transmitter UART to an audio amplifier and transformer or transistor modulator to operate a helium-neon laser tube. The receiver UART shown in FIG. 14-13 can be used as a stand-alone remote-control device. Or, it can be connected to another computer, such as a Commodore 64, for the purpose of receiving signals from the host machine. See TABLE 14-7 for a list of parts for the receiver UART.

The output of the receiver UART is an 8-bit binary code. This code can be used to remotely control up to 256 functions. One way to operate up to 16 devices, such as solenoids, alarms, or motors, is shown in FIG. 14-14. The UART is connected to a 4028 1-of-10 decoder. The first 10 digits (00000000 through 00001010) are decoded and applied to relays or opto-isolators where they can be used to drive any of a number of output devices.

How The UART Works

In the schematic in FIG. 14-12, 8-bit parallel data from the Commodore 64 (or other computer) is routed to the data lines on the UART. When the computer is ready to send the byte, it pulses the STROBE line high (the line might be called DATA READY or something similar). The UART converts the data to serial format and sends it through the serial output (SO) pin. The speed of the data leaving the output is determined by the baud-rate generator. The COM8116 dual baud-rate generator sets the speed of the transmission and reception, and it is hooked up here to be rather slow—about 300 baud. This means that the UART sends serial data at the rate of roughly 300 bits per second (equivalent to 30 bytes per second).

The receiving UART is connected almost in reverse to the transmitting UART. The receiver uses a baud-rate generator that is operating at the same frequency as the

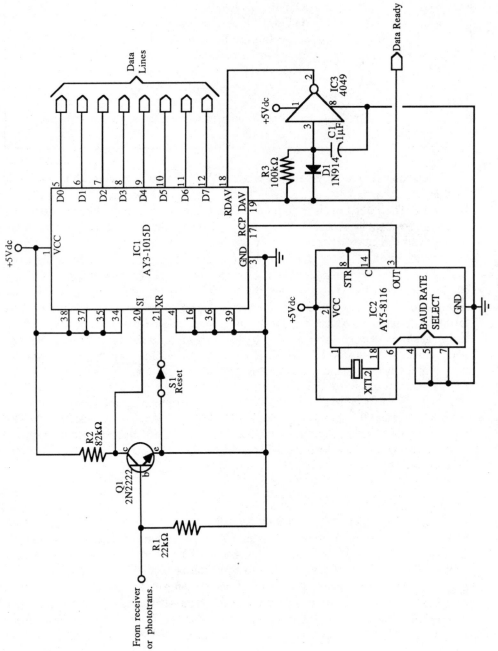

FIG. 14-13. *Schematic diagram for UART receiver.*

Table 14-7. Receiver UART Parts List

IC1	AY3-1015D UART IC
IC2	AY5-8116 baud-rate generator
IC3	4049 CMOS hex inverter IC
R1	22 kilohm resistor
R2	82 kilohm resistor
R3	100 kilohm resistor
C1	1 μF tantalum capacitor
Q1	2N2222 transistor
D1	1N914 diode
S1	SPST switch (momentary, normally closed)
XTL1	3.57 MHz (colorburst) crystal

All resistors are 5 to 10 percent tolerance, $\frac{1}{4}$ watt. All capacitors are 10 to 20 percent tolerance, rated 35 volts or more.

transmitter. The receiver is equipped with an IR photodetector. If you could see infrared light, you'd see the laser diode flash on and off very rapidly as the data passed. The ON and OFF periods are equal to 0's and 1's, or "spaces" and "marks" as they are called in serial communications.

The amplified output is applied to the serial data pin on the UART. When an entire word is received, the UART places it on the parallel data output pins and pulses the

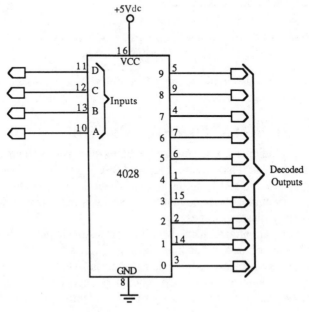

FIG. 14-14. *Four lines from the UART receiver can be decoded into 10 lines using a 4028 IC.*

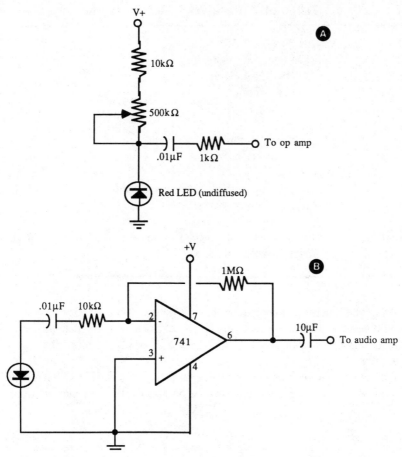

FIG. 14-15. *Ways to connect an undiffused red LED to an op amp. (A) Adjustable sensitivity; (B) Preset sensitivity.*

DATA AVAILABLE pin. In this circuit, a short time delay is used to automatically reset the UART so it processes the next word.

ALTERNATE HE-NE LASER PHOTOSENSOR

Conventional photodiodes are engineered to be most sensitive to light wavelengths in the near-infrared region. That automatically reduces their effectiveness in receiving a modulated signal over the red beam of a helium-neon laser. This loss in sensitivity is not a major concern in most laser communications links, but it can present problems if the laser-to-receiver distance is very long.

More importantly, an infrared-sensitive phototransistor or photodiode is susceptible to swamping by the infrared radiation of the sun. Even with red filters, it's difficult to "tune out" the sun while receiving the modulated signal over the laser beam. A better approach is to use a red light-emitting diode. Though not specifically designed to detect light, an LED can be easily adapted for the purpose. Inventor and magazine columnist

Forrest Mims III has written extensively on this topic; check out back issues of *Popular Electronics* and *Modern Electronics* for details.

Sensitivity in the red region of the visible spectrum is accentuated, because the LED is designed to emit red light. You obtain best results when using clear, non-diffused red LEDs. Sample hookup diagrams appear in FIG. 14-15. In my tests, the white light and infrared content of a nearby desk lamp changed the output of the circuit by a few tenths of a volt. But shining a red He-Ne laser at the LED caused the output to swing to about a volt or two of the supply voltage.

15

Lasers and Fiberoptics

Most everyone has encountered fiberoptics at one time or another. Many telephone calls—both local and long distance— are now carried at least partway by light shuttling through a strand of plastic or glass. Fiberoptics are now used on some of the higher end audio systems as a means to prevent digital signals from interfering with analog signals. And fiberoptic sculptures, in vogue in the late 1960s but coming back in style today, look like high-tech flowers that seem to burst out in brightly colored lights.

An optical fiber is to light what PVC pipe is to water. Though the fiber is a solid, it channels light from one end to the other. Even if the fiber is bent, the light will follow the path in whatever course it takes. Because light acts as the information carrier, a strand of optical fiber no bigger than a human hair can carry the same information as about 900 copper wires. This is one reason why fiberoptics is used increasingly in telephone communications.

Laser light exhibits unique behavior when transmitted through optical fiber. This chapter discusses how to work with optical fibers, ways to interface fibers to a laser, and many interesting applications of laser/fiberoptic links.

HOW FIBEROPTICS WORK

The idea of optical fibers is over 100 years old. British physicist John Tyndall once demonstrated how a bright beam of light was internally reflected through a stream of water flowing out of a tank. Serious research into light transmission through solid material started in 1934 when Bell Labs was issued a patent for the *light pipe*.

In the 1950's, the American Optical Corporation developed glass fibers that transmitted light over short distances (a few yards). The technology of fiberoptics really took off in about 1970 when scientists at Corning Class Works developed long-distance optical fibers.

All optical fibers are composed of two basic materials, as illustrated in FIG. 15-1: the core and the cladding. The *core* is a dense glass or plastic material where the light actually passes through as it travels the length of the fiber. The *cladding* is a less dense sheath, also of plastic or glass, that serves as a refracting medium. An optical fiber may or may not have an outer *jacket*, a plastic or rubber insulation for protection.

Optical fibers transmit light by *total internal reflection (TIR)*. Imagine a ray of light entering the end of an optical fiber strand. If the fiber is perfectly straight, the light will pass through the medium just as it passes through a plate of glass. But if the fiber is bent slightly, the light will eventually strike the outside edge of the fiber.

If the angle of incidence is great (greater than the critical angle), the light will be reflected internally and will continue its path through the fiber. But if the bend is large and the angle of incidence is small (less than the critical angle), the light will pass through the fiber and be lost. The basic operation of fiberoptics is shown in FIG. 15-2.

Note the *cone of acceptance*; the cone represents the degree to which the incoming light can be off-axis and still make it through the fiber. The cone of acceptance (usually 30 degrees) of an optical fiber determines how far the light source can be from the optical axis and still manage to make it into the fiber. Though the cone of acceptance might seem generous, fiberoptics perform best when the light source (and detector) are aligned to the optical axis.

Optical fibers are made by pulling a strand of glass or plastic through a small orifice. The process is repeated until the strand is just a few hundred (or less) micrometers in diameter. Although single strands are sometimes used in special applications, most optical fibers consist of many strands bundled and fused together. There might be hundreds or even thousands of strands in one fused optical fiber bundle. Separate fused bundles can also be clustered to produce fibers that measure 1/16 of an inch or more in diameter.

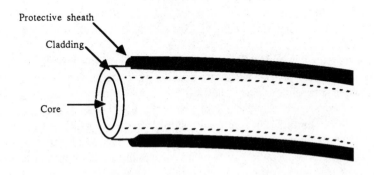

FIG. 15-1. *Design of the typical optical fiber.*

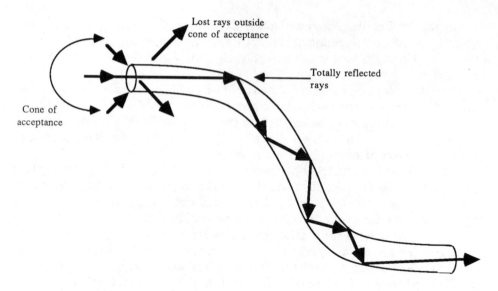

FIG. 15-2. *Operation of an optical fiber, with cone of acceptance.*

TYPES OF OPTICAL FIBERS

The classic optical fiber is made of *glass*, also called *silica*. Glass fibers tend to be expensive and are more brittle than stranded copper wire. But they are excellent conductors of light, especially light in the infrared region between 850 and 1300 nm. Less expensive optical fibers are made of *plastic*. Though light loss through plastic fibers is greater than with glass fibers, they are more durable. Plastic fibers are best used in communications experiments with near-infrared light sources—in the 780 to 950 nm range. This nicely corresponds to the output wavelength and sensitivity of commonplace infrared emitters and detectors.

Optical fiber bundles can be coherent or incoherent. The terms don't directly relate to laser light or its properties but to the arrangement of the individual strands in the bundle. If the strands are arranged so that the fibers can transmit a pictorial image from one end to the other, it is said to be coherent. The vast majority of optical fibers are incoherent, where an image or special pattern of light is lost when it reaches the other end of the fiber.

The cladding used in optical fibers can be one of two types—step-index and graded-index. *Step-index* fibers provide a discrete boundary between more dense and less dense regions between core and cladding. They are the easiest to manufacture, but their design causes a loss of coherency when laser light passes through the fiber. That means coherent light in, largely incoherent light out. The loss of coherency, which is due to light rays traveling on slightly different paths through the fiber, reduces the efficiency of the laser beam. Still, it offers some practical benefits, as you'll see later in this chapter.

There is no discrete refractive boundary in *graded-index* fibers. The core and cladding media slowly blend, like an exotic tropical drink. The grading acts to refract light evenly, at any angle of incidence. This preserves coherency and improves the efficiency of the fiber. As you might have guessed, graded-index optical fibers are the most expensive of the bunch. Unless you have a specific project in mind (like a 10-mile fiber link), graded-index fibers are not needed, even when experimenting with lasers.

WORKING WITH FIBEROPTICS

Where do you buy optical fibers? While you could go directly to the source, such as Dow Corning, American Optical, or Dolan-Jenner, a cheaper and easier way is through electronic and surplus mail order. Radio Shack sells a 5-meter length of jacketed plastic fiber; Edmund Scientific offers a number of different types and diameters of optical fibers. You can buy most any length you need, from a sampler containing a few feet of several types to a spool containing thousands of feet of one continuous fiber.

A number of surplus outfits, such as Jerryco and C&H Sales, offer optical fibers from time to time (stocks change, so write for the latest catalog before ordering). You might not have much choice over the type of fiber you buy, but the cost will be more than reasonable.

Although it's possible to use an optical fiber by itself, serious experimentation requires the use of fiber couplings and connectors. These are mechanical splices used to connect two fibers together or to link a fiber to a light emitter or detector. Be aware that good fiberoptic connectors are expensive. Look for the inexpensive plastic types meant for non-military applications. You can also make your own home-built couplings. Details follow later in this chapter.

Optical fibers can be cut with wire cutters, nippers, or even a knife. But care must be exercised to avoid injury from shards of glass that can fly out when the fiber is cut (plastic fibers don't shatter when cut). Wear heavy cotton gloves and eye protection when working with optical fibers. Avoid working with fibers around any kind of food serving or preparation areas, because tiny bits of glass can inadvertently and invisibly settle on food, plates, etc.

One good way to cut glass fiber is to gently nick it with a sharp knife or razor, then snap it in two. Position your thumb and index finger of both hands as close to the nick as possible, then break the fiber with a swift downward motion (snapping upwards increases the chance of glass shards flying off toward you).

Whether snapped apart or cut, the end of the fiber should be prepared before splicing it to another fiber or connecting it to a light emitter or detector. The ends of the cut fiber can be polished using extra fine grit aluminum oxide wet/dry sandpaper (330 grit or higher). Wet the sandpaper and gently grind the end of the fiber on it. You can obtain good results by laying the sandpaper flat on a table and holding the fiber in your hands. Rub in a circular motion and take care to keep the fiber perpendicular to the surface of the sandpaper. If the fiber is small, mount it in a pin vise.

Inspect the end of the fiber with a high-powered magnifying glass (a record player stylus magnifier works well). Shine a light through the opposite end of the fiber. The magnified end of the fiber should be bright and round. Recut the fiber if the ends look crescent shaped or have nicks in them.

FIBEROPTIC CONNECTORS

Commercially made fiberoptic connectors are pricy, even the plastic AMP Optimate Dry Non-Polish (DNP) variety. A number of mail-order firms such as Digi-Key and Jameco

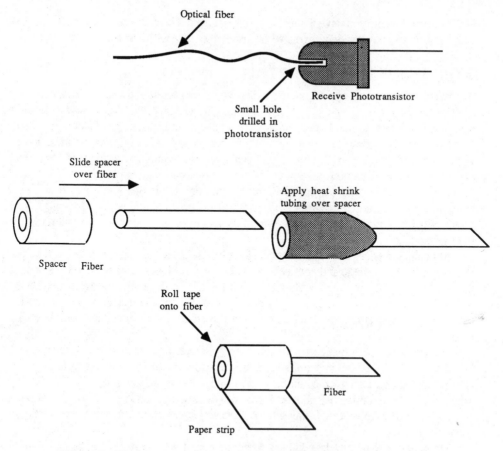

FIG. 15-3. *Ways to connect fiberoptics to phototransistors, LEDs, and laser diodes.*

offer splices, for joining two fibers, and connectors, for attaching the fiber to emitters and detectors. Depending on the manufacturer and model, the connectors are made to work with either the round- or flat-style phototransistors and emitters.

You can make your own connectors and splices for home-brew laser experiments. FIGURE 15-3 shows several approaches. An easy way to splice fibers is to use small heatshrink tubing. Cut a piece of the tubing to about ½ inch. After properly cutting (and polishing) the ends of the fiber, insert them into the tubing and heat lightly to shrink. Best results are obtained when the tubing is thick-walled.

Optical fibers can be directly connected to photodiodes by drilling a hole in the casing, inserting the fiber, and bonding the assembly with epoxy. Be sure that you don't drill into the semiconductor chip itself. Keep the drill motor at a fairly slow speed to avoid melting the plastic casing. Work slowly.

A strand of optical fiber can be held in place using a pin vise (remove the outer jacket, if any, and tighten the chuck around the fiber) or by using solderless insulated spade tongues. These tongues are designed for terminating copper wire but can be successfully used to anchor almost any size of optical fiber to a bulkhead. The laser, be it He-Ne or semiconductor, can then be aimed directly into the cone of acceptance of the fiber.

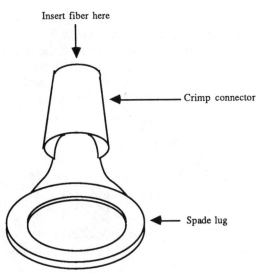

Insert fiber here

Crimp connector

Spade lug

FIG. 15-4. *A solderless crimp lug can be used to secure the end of an optical fiber to a circuit board or bulkhead.*

Spade tongues, as shown in FIG. 15-4, are available in a variety of sizes to accommodate different wire gauges. Use #6 (22 to 18 gauge) for small optical fibers and #8 (16 to 14 gauge) for larger fibers. Secure the fiber in the spade tongue by crimping with a crimp tool. Do not exert too much pressure or you will deform the fiber. If the fiber is loose after crimping, dab on a little epoxy to keep everything in place.

The "FLCS" package, shown in FIG. 15-5, is a low-cost fiberoptic connector available from a variety of sources including Radio Shack, Circuit Specialists, and many Motorola

FIG. 15-5. *The popular plastic "FLCS" optical fiber connector.*

semiconductor representatives. It can be easily adapted for use with laser diodes by cutting off the back portion. This exposes the optical fiber.

After removing the emitter diode, file or grind off the back end of the connector, as shown in FIG. 15-6. You can also drill out the back of the connector with a ⁵⁄₃₂-inch bit. Mount the connector and laser on a circuit board or perf board, and be careful to align the laser so that its beam directly enters the end of the fiber.

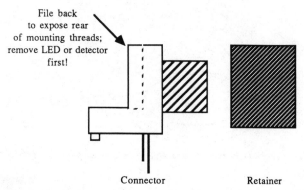

FIG. 15-6. *How to modify an "FLCS" connector to open the back portion. With the back open, you can shine a laser beam inside the connector and into the optical fiber.*

BUILD A LASER DATA LINK

A number of educational fiberoptic kits are available (see Appendix A for sources) and at reasonable cost. For example, the Edu-Link (available through Advanced Fiberoptics Corp., Circuit Specialists, Edmund, and others) contains pre-etched and drilled PCBs for a small transmitter and receiver along with a short length of jacketed plastic or glass fiber. The emitter LED and photodetector are housed in plastic connectors, and the circuits provide input and output pins for sending and receiving digital data.

You can use the Edu-Link or the circuits shown in FIGS. 15-7 and 15-8 to build your own fiberoptic transmitter and receiver. Parts lists for the two circuits are provided in TABLES 15-1 and 15-2. Note that the Edu-Link, as well as other fiberoptic data communications kits, use a resistor to limit current to the emitter LED. The value of the resistor is typically 100 to 220 ohms. At 100 ohms and a 5 Vdc supply, current to the emitter LED is 35 milliamps (assuming a 1.5-volt drop through the LED). That isn't enough to operate a dual-heterostructure laser diode, so the resistor must be exchanged for a lower value. A 56 ohm resistor will provide about 62 mA current to the laser diode; a 47-ohm resistor will deliver about 74 mA current. A feedback mechanism for controlling the output of the laser diode is not strictly required because the device is used in pulsed mode.

If you are using the Edu-Link system and have not yet assembled the boards, construct the transmitter and receiver circuits, but don't mount the emitter LED. Connect a Sharp LT020, LT022, or equivalent low-power dh laser diode to the transmitter circuit. Dismantle the emitter and modify the connector for use with the laser diode as detailed in the previous section.

230

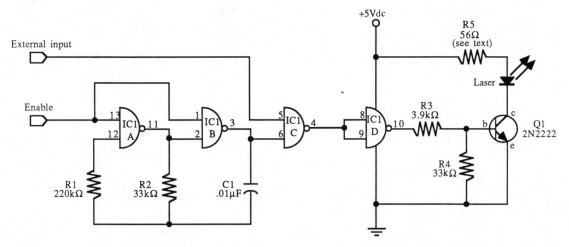

FIG. 15-7. *Schematic for experimenter's data transmitter using optical fibers and a laser diode. Transmission frequency of the free-running oscillator is approximately 3 kHz.*

The transmitter circuit of FIG. 15-7 includes a built-in oscillator. Trigger it by placing the ENABLE and DATA IN lines HIGH. Next, connect a logic probe to the DATA OUT pin of the receiver. Apply power to the two circuits and watch for the pulses at the DATA OUT pin. If pulses are not present, double-check your work and make sure that the fiberoptic connection between the transmitter and receiver is aligned and secure and that the output laser diode is properly aligned to the fiber. Remove power and disconnect the logic probe.

In normal operation, place the ENABLE pin LOW and connect any serial data to the DATA IN pin. Be sure the incoming signal is TTL compatible or does not exceed the supply voltage of the transmitter. You can use the circuit to transmit and receive ASCII computer

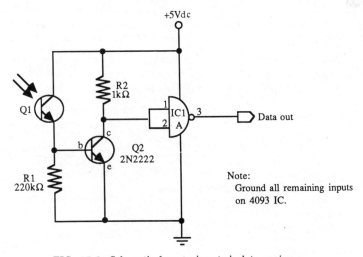

FIG. 15-8. *Schematic for experimenter's data receiver.*

Table 15-1. Optical Fiber Transmitter Parts List

IC1	4093 CMOS NAND gate IC
R1	220 kilohm resistor
R2	33 kilohm resistor
R3	3.9 kilohm resistor
R4	33 kilohm resistor
R5	56 ohm (nominal) resistor
C1	0.01 μF disc capacitor
	Laser diode
Q1	2N2222 transistor

All resistors are 5 to 10 percent tolerance, ¼ watt. All capacitors are 10 to 20 percent tolerance, rated 35 volts or more.

data, remote-control codes, or other binary information. A number of circuits in Chapter 14, "Advanced Projects in Laser Communication," show how to transmit computer data by laser over the air.

A number of these projects can easily be adapted for use with the fiberoptic transmitter and receiver. For example, you can connect the output pin of the Motorola 14557 remote control transmitter chip to the DATA IN pin of the Edu-kit transmitter circuit. The matching Motorola 14558 receiver chip connects to the DATA OUT terminal on the Edu-kit receiver circuit.

The CMOS chips in the transmitter and receiver can be operated with supplies up to 18 volts, but care must be taken to adjust the value of R5, the resistor that limits current to the laser diode. An increase in supply current must be matched with an increase in resistance or the laser diode could be damaged.

Use Ohm's law to compute the required value ($R = E/I$). Assume a 1.2- to 1.5-volt drop through the diode.

Example:

Supply voltage—12 volts.
Voltage drop through laser diode—1.5 volts.
Working voltage—10.5 volts (12 − 1.5).
Desired current (60 to 80 mA)—70 mA nominal..
Resistor to use—150 ohms (10.5 / 0.070 = 150).

Table 15-2. Optical Fiber Receiver Parts List

IC1	4093 CMOS NAND gate IC
R1	220 kilohm resistor
R2	1 kilohm resistor
Q1	Infrared phototransistor
Q2	2N2222 transistor

All resistors are 5 to 10 percent tolerance, ¼ watt.

MORE EXPERIMENTS WITH LASERS AND FIBEROPTICS

Besides data transmission, fiberoptics can be used to:

★ Transmit analog data (modulate a diode or He-Ne laser and pass it through the fiber).

★ Detect vibration and motion.

★ Route laser light to remote locations.

★ Separate a laser beam into several shafts of light.

This is only a partial listing; there are literally dozens of useful and practical applications for lasers and fiberoptics. Some hands-on projects follow.

Vibration and Movement Detection

A fiberoptic strand doesn't make the best medium for transmitting laser-light analog data. Why? The fiber itself can contribute to noise. As mentioned earlier, when a beam of coherent laser light is passed through a conventional step-index optical fiber, the rays travel different paths, and the light that exits is largely incoherent. You can see the effects of this interference by shining a helium-neon laser through an optical fiber. Point the exit beam at a white piece of paper and you'll see a great deal of speckle. The speckle is the constructive and destructive interference, created inside the fiber, as the laser light rays travel from one end to the other.

This interference—which is most prominent in low-cost plastic optical fibers—is normally an undesirable side effect. However, it can be put to good use as a vibration and motion detection system. Connecting a phototransistor to the exit-end of the fiber lets you monitor the light output. Movement of the fiber causes a change in the way the light is reflected inside, and this changes the coherency (or incoherency, depending on how you look at it) of the beam. A simple audio amplifier connected to the phototransistor allows you to hear the movement.

FIGURE 15-9 shows a setup you can use to test the effects of fiberoptic vibration and motion. A parts list for the system is provided in TABLE 15-3. More advanced projects using this technique are in Chapter 16, "Experiments in Laser Seismology." The noise is sometimes a hiss and sometimes a "thrum." Depending on the length of the fiber and type of motion, you might also hear low- or high-pitched squeals. These squeals can change pitch as the fiber or phototransistor is slowly moved.

The squeals are caused by the Doppler effect of *optical heterodyning*, a process whereby two rays of light at slightly different frequencies meet. The two basic (or fundamental) frequencies mix together, creating two additional frequencies. One is the sum of the two fundamental frequencies and the other is the difference.

How are the different frequencies of light created in the first place? Remember that the speed of light slows down as it passes through a refractive medium. There is a strict relationship between the wavelength, speed, and frequency of light. Because the frequency of light can't be altered (at least by ordinary refraction), that means the wavelength must be shifted in direct proportion to the change in the speed of light.

The change in wavelength caused by refraction is rather small (on the order of a few hundred hertz) and is dependent on the original light wavelength and the refractive index

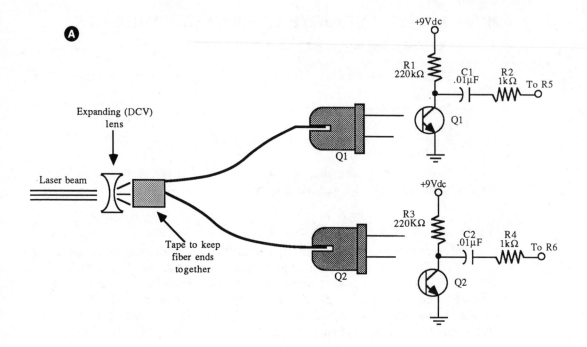

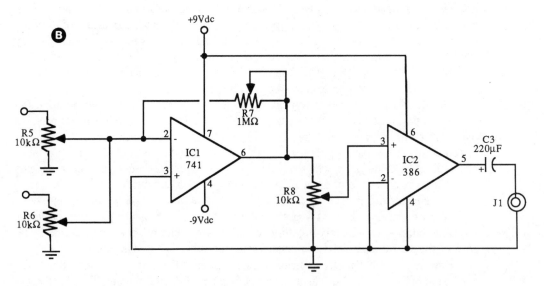

FIG. 15-9. *Hookup diagram for connecting a pair of phototransistors to op amp and audio amplifier. (A) Physical connection (showing beam expansion into both fibers) and phototransistor hookup; (B) Op amp and audio amplifier.*

Table 15-3. Vibration Detection System Parts List

IC1	LM741 operationsl amplifier IC
IC2	LM386 audio amplifier IC
R1,R3	220 kilohm resistor
R2,R4	1 kilohm resistor
R5,R6,R8	10 kilohm potentiometer
R7	1 megohm potentiometer
C1,C2	0.01 μF disc capacitor
C3	220 μF electrolytic capacitor
Q1,Q2	Infrared phototransistors
J1	Audio output jack (¼- or ⅛-inch)
Misc.	DCV expansion lens (approx. 6 to 10 mm diameter)

All resistors are 5 to 10 percent tolerance, ¼ watt. All capacitors are 10 to 20 percent tolerance, rated 35 volts or more.

of the medium. The change in wavelengths causes a beat frequency—a sum and difference. The sum frequency is extremely high and can't be heard, but the difference frequency might only be 200 to 500 Hz and can be readily detected with an ordinary phototransistor and audio amplifier.

Optical heterodyning is most conspicuous when only two coherent rays of light meet. In an optical fiber, dozens and even hundreds of rays of internally reflected laser light might meet at the phototransistor, and the result can sound more like cacophonous noise than a distinct tone. A Michelson interferometer (see Chapter 9) reveals optical heterodyning much more readily.

You might also notice a varying tone when sampling the beam directly from the laser. Even though lasers are highly monochromatic, they can still emit several frequencies of light, with each frequency spaced only fractions of a nanometer apart. As these frequencies meet on the surface of the photodetector, they cause heterodyning or beat frequencies. When the difference frequency is 20 kHz or less, you can hear them. You can precisely measure the difference frequencies using an oscilloscope. The tones heard when sampling the beam directly from a laser are most prominent with short tubes and when they are first turned on.

Separating Beam With Optical Bundles

By grouping together one end of two or more fused bundles, you can separate the beam of a laser into many individual sub-beams. The laser light enters the common end (where all the bundles are tied together), and exits the opposite end of each individual fiber. Some optical fibers come pre-made with four or more grouped strands (used most often in automotive dashboard application) or you can make your own.

The pencil-thin beam of the typical He-Ne laser is too narrow to enter all the fibers at once, so the beam must be expanded. Place a bi-concave or plano-concave lens in front of the entrance to the bundles. Adjust the distance between the lens to the bundle until the beam is spread enough to enter all the fibers.

Split bundles can be used to experiment with optical heterodyning as well as to split the beam of one laser into several components. Each beam can be used in a separate optical fiber system. For example, you must use a four-fiber split bundle to provide illumination for four-fiberoptic intrusion detection systems. Each sub-system is placed in a quadrant around the protected area and has its own phototransistor, making it easier to locate the area of disturbance.

Interfacing Fiberoptics to a Computer

Many of the advanced applications of lasers and fiberoptics require interface to a computer. In Chapter 16, "Experiments in Laser Seismology," you'll learn how to connect a phototransistor to a computer via an analog-to-digital converter.

16

Experiments in Laser Seismology

At first you hear it. It sounds like the low rumbling of a woofer speaker, yet the sounds are coming from nowhere and everywhere. Then the windows join the strange session of music-making and begin to rattle, followed by the eerie creaks of the wooden beams in the house.

Then you feel it, a swaying and pumping motion like a carnival ride gone haywire. Within a few seconds, you realize it's not a large truck passing outside or the heavy thump of a jet breaking the sound barrier, but an earthquake.

Earthquakes are among the most frightening natural phenomenon, feared most because of their stealthy suddenness. Even though geologists and seismologists have been measuring earthquakes for decades with sophisticated instruments on land, in the air, and even in space, predicting tremors is an inexact science. The best seismologists can do is warn that a "big earthquake is due soon"—expect it anytime between now and the next century.

Fortunately, massive earthquakes on land are rare. Many of the largest earthquakes occur out at sea, and while they can cause enormous tidal waves (such as the Japanese tsunami), earthquakes at sea seldom topple buildings or swallow up people. Earthquakes on the West Coast in Southern California are almost a dime a dozen, and most of them so faint that they cannot be felt.

Earthquakes can happen anywhere, and even tremors that occur hundreds of miles away can be detected with the proper instruments. Most seismographs use complex and massive electromagnetic sensors to detect earthquakes, both near and far, but you can readily build your own compact seismograph using a laser and a coil of fiberoptics.

When constructed properly, a laser/fiberoptic seismograph can be just as sensitive as an electromagnetic seismograph costing several thousand dollars.

This chapter covers construction details of a laser/fiberoptic seismograph as well as useful information on how to attach the seismograph to a personal computer.

THE RICHTER SCALE

A number of scales are used to quantify the magnitude of earthquakes. The best known and most used is the Richter scale, named after Charles F. Richter, a pioneer in seismology research. The Richter scale is a logarithmic measuring system ranging from 1 to 10 (and theoretically from 10 to 100), where each increase of 1 represents a ten-fold increase in earthquake magnitude. However, the actual energy released by the earth during the quake can be anywhere between 30 and 60 times for each increase of one digit. Each numeral is further broken down into units of 10, so earthquakes are often cited as 4.2 or 5.6 on the Richter scale.

To give you an idea of how the Richter scale works (and why it can cause confusion), consider the difference between a 3.0 and 4.0 earthquake. Both are difficult to detect without instruments (although some people say they can feel the swaying motion of a 4.0 earthquake). But because the magnitude is increased logarithmically by a factor of 10 per numeral and the actual energy released could be 60 times as great, the difference between a 3.0 and 5.0 earthquake is quite large. That is, the 5.0 earthquake is 10 X 10 (or 100) times more powerful than the 3.0 earthquake.

Similarly, a 6.0 earthquake can cause extensive structural damage and close down buildings for repair; and an earthquake measuring 7.0, if it continues for any length of time, can result in massive destruction of buildings, bridges, and roads. An earthquake measuring 9.0 or 10.0 would level any town.

Most earthquakes occur at a *fault*, which is a crack or fissure in the earth's crust. The majority of earthquakes occur at the boundaries of crustal plates. These plates slip and slide over the earth's inner surface. Sudden motion in these plates is released as an earthquake. Major faults, such as the San Andreas in California, create many thousand "mini-faults" or fractures that spread out like cracks in dried mud. These fractures are also responsible for earthquakes. In fact, two recent large earthquakes in southern California, occurring in 1971 and 1987, were caused by relatively small faults lying outside the San Andreas line.

Though faults are long—some measure thousands of miles—the earthquake occurs at a specific location along it. This location is the *epicenter*. The magnitude of earthquakes is measured at the epicenter (or more accurately, at a standard seismographic station distance of 100 km or 62 miles) where the amount of released energy is the greatest. The shock waves from the earthquake fan out and lose their energy the further they go. Obviously, there can't be a seismograph every 50 or even 100 miles along a fault to measure the exact amplitude of an earthquake. When the epicenter is some distance from a seismograph, its magnitude is inferred, based on its strength at several nearby seismograph stations, past earthquake readings, and the geological makeup of the land in-between.

This accounts for the uncertainty of the exact magnitude of an earthquake immediately after it has occurred and why different seismologists can arrive at different readings. It takes some careful calculations to determine an accurate Richter scale reading for an earthquake, and the precise measurement is sometimes debated for months or even years after the tremor.

HOW ELECTROMAGNETIC SEISMOGRAPHS WORK

The most common seismograph in use today is the electromagnetic variety that uses a sensing element not unlike a dynamic microphone. Basically, the case of the seismograph is a large and heavy magnet. Inside the case is a core, consisting of a spool of fine wire. During an earthquake, the spool bobs up and down, inducing an electromagnetic signal through the wires. A similar effect occurs in a dynamic microphone. Sound vibrates a membrane, which causes a small voice coil (spool of wire) to vibrate. The voice coil is surrounded by a magnet, so the vibrational motion induces a constantly changing alternating current in the wire. As the sound varies, so does the polarity and strength of the alternating current.

To prevent accidental readings of surface vibration, the seismograph is buried several feet into the ground and is sometimes attached to the bedrock. In other cases, it is encased in concrete or secured to a cement piling sunk deep into the ground. Wires lead from the seismograph to a reading station, that might be directly above or several miles away. Telephone lines, radio links, or some other means connect the distant seismographs to a central office location.

The signal from the seismograph is amplified and applied to a galvanometer on a chart recorder (the galvanometer is similar to the movement on a volt-ohmmeter). The galvanometer responds to electrical changes induced by the moving core of the seismograph. The bigger the movement, the larger the response. Attached to the galvanometer is a long needle that applies ink to a piece of paper wound around a slowly rotating drum.

An advance over the chart recorder is the computer interface. The pulses from the seismograph are sent to an analog-to-digital converter (ADC), which connects directly to a computer. The ADC transforms the analog signals generated by the seismograph into digital data for use by the computer. Software running on the computer records each tremor and can perform mathematical analysis.

Laser/Fiberoptic Seismograph Basics

The laser/fiberoptic seismograph (hereinafter referred to as the laser/optic seismograph) doesn't use the electromagnetic principle to detect movement in the earth. Though there are several ways you can implement a laser/optic seismograph, we'll concentrate on just one that offers a great deal of flexibility and sensitivity. The system detects the change in coherency through a length of fiberoptics.

As discussed in Chapter 15, "Lasers and Fiberoptics," when a laser beam is transmitted through a stepped-index optical fiber, some of the waves arrive at the other end before others. This reduces the coherency of the beam in proportion to the design of the fiber, its length, and the amount of curvature or bending of the fiber. Given enough of the right optical fiber, a laser beam could emerge at the opposite end that is totally incoherent.

It is not our intent to completely remove the coherency of a laser beam, but just to alter it slightly through a length of 10 or 20 feet of fiber. Movement or vibration of the fiber causes a displacement of the coherency, and that displacement can be detected with a phototransistor. You can even hear this change in coherency by connecting the phototransistor to an audio amplifier. The "hiss" of the light coming through the fiber changes pitch and makes odd thuds, pings, and thrums as the fiber vibrates. The sound settles as the fiber stops moving or vibrating. In a way, the optical fiber makes a unique form of interferometer that settles quickly after the external vibrations have been removed.

Reducing Local Vibrations

The laser/optic seismograph is susceptible to the effects of local vibrations, movement caused by people walking or playing nearby, passing cars, trucks, and trains, even the vibration triggered by the sound of a jet passing overhead.

To be most effective, the seismograph should be placed in an area where it won't be affected by local vibrations. Those living on a ranch or the outskirts of town will have better luck at finding such a location than city dwellers or those conducting earthquake experiments in a school or other populated area.

Even if you can't move away from people and things that cause vibration, you can reduce its effects by firmly planting the seismograph in solid ground. Avoid placing it indoors, especially on a wooden floor. Most buildings are flexible, and not only do they readily transmit vibrations from one location to the next, they act as a spring and/or cushion to the movement of an earthquake, improperly influencing the readings.

The cement flooring or foundation of the building is only marginally better. Small vibrations easily travel through cement, so if you attach your seismograph to the floor in your room, you are likely to pick up the movement of people walking around in the living room and kitchen.

The best spot for a seismograph is attached to a big rock out in the back yard, away from the house. Lacking a rock, you can fasten the seismograph to a cement piling, and then bury some or all of the piling into the ground. You can also spread out four to eight cement blocks (about 75 cents each at a builder's supply store), and partially bury them in the ground. Fill the center of the blocks with sand and mount the seismograph on top. Other possible spots include (test first):

* ★ The base of a telephone pole.
* ★ A heavy fence post.
* ★ A brick retaining wall or fence.
* ★ The cement slab of a separate garage, work shop, or tool shed.

CONSTRUCTING THE SEISMOGRAPH

Cut a piece of optical fiber (jacketed or unjacketed) to 15 feet. Polish the ends as described in the last chapter. Using a small bit (to match the diameter of the fiber), drill a hole in the top of a phototransistor. Be sure to drill directly over the chip inside the detector, but do not pierce through to the chip. Epoxy the fiber in place. Alternatively, you can terminate the output end of the fiber using a low-cost FLCS-type connector. You can also use a modified FLSC connector for the emitter end of the fiber (as detailed in Chapter 15) or one of the other mounting techniques described.

240

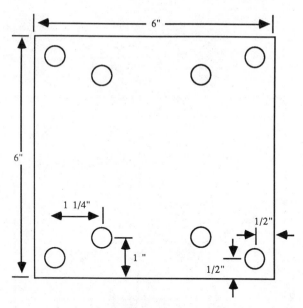

FIG. 16-1. *Drilling and cutting guide for the laser/optic seismograph base.*

Put the fiber aside and construct the base following the diagram in FIG. 16-1. A parts list is included in TABLE 16-1. You can use metal, plastic, or wood for the base, but it should be as dimensionally sturdy as possible. The prototype used ³⁄₁₆-inch-thick acrylic plastic. Cut the base to size and drill the post and mounting holes as shown. Insert four ¼-inch-20-by-3-inch bolts in the inside four holes. Starting at one post, thread the fiber around the bolts in a counterclockwise direction (see FIG. 16-2). Leave 1 to 2 feet on either end to secure the laser and photodiode. If the fiber slips off the bolts, you can secure it using dabs of epoxy.

Mount the photodetector and laser diode (on a heatsink, as shown in FIG. 16-3), in the center of the platform. Alternatively, you can use a He-Ne laser as the coherent light source. Mount the laser tube securely on a separate platform and position the end of the optical fiber so that it catches the beam.

You can use the universal laser light detector presented in Chapter 13 to receive and amplify the laser light intercepted by the phototransistor. You can use either a pulsed or cw drive power supply for the laser diode. Schematics for these drives appear in

Table 16-1. Fiberoptic Seismograph Parts List

1	6-inch-square acrylic plastic base (³⁄₁₆-inch thick)
4	3-inch by ¼-inch 20 carriage bolt, nuts, washers
1	Laser diode on heatsink
1	Sensor board (see Fig. 16-4)
Misc.	15 feet (approx.) jacketed or unjacketed fiberoptics, connector for receiver photodiode, connector or attachmnent for laser to fiber

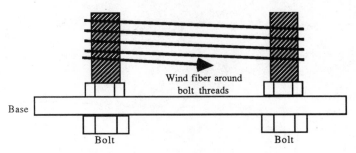

FIG. 16-2. *How to wind the optical fiber around the bolts.*

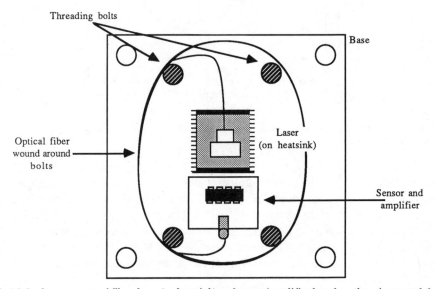

FIG. 16-3. *Arrangement of fiber, laser (on heatsink) and sensor/amplifier board on the seismograph base.*

Chapter 11, "Laser Power Supplies." In either case, be sure that the laser diode does not receive too much current. In pulsed mode, the laser doesn't operate at peak efficiency, but it is not required in this application.

Note that when using a He-Ne tube, you don't need excessive power—a 0.5 or 1 mW He-Ne is more than enough. This simplifies the power supply requirements and allows you to operate the seismograph for a day or two on each charge on a pair of lead-acid or gelled electrolyte batteries.

Test the seismograph by turning on the laser and connecting the output of the amplifier to a speaker or pair of headphones. A small vibration of the base will cause a noticeable thrum or hiss in the audio output. You might need to adjust the control knob on the amplifier to turn the sound up or down.

The mounting holes allow you to attach the seismograph to almost any stable base. Whatever base you use, it should be directly connected to the earth. Use cement or masonry screws to attach the seismograph to concrete or a concrete pylon. When you get tired of listening to earthquake vibrations, connect the output of the amplifier to a volt-ohmmeter. Remove the ac coupling capacitor on the output of the amplifier.

Table 16-2. Commodore 64 Seispograph ADC Parts List

Basic Setup

IC1	TLC548 serial ADC
R1	10 kilohm potentiometer

Light-Dependent Resistor Setup

IC1	TLC548 serial ADC
R1,R2	1 kilohm resistor
R3	22 kilohm resistor

Phototransistor Setup

IC1	TLC548 serial ADC
R1,R2	1 kilohm resistor
R3	100 kilohm potentiometer
Q1	Infrared phototransistor
Misc.	$^{12}/_{24}$ pin connector for attaching to Commodore 64 User Port

Computer Interface of Seismograph Sensor

A small handful of readily available electronic parts are all that is necessary to convert the voltage developed by the solar cell or phototransistor into a form usable by a computer. The circuit shown in FIG. 16-4 uses the TLC548 serial ADC, connected to a Commodore 64. The Commodore 64 provides the timing pulses, so only a minimum number of parts are required. See TABLE 16-2 for a parts list for the circuit.Construct the TLC548 circuit on a perforated board using soldering or wire-wrapping techniques.

Software

The software is relatively simple. You might want to collect a number of samples and either print them out for future reference or graph them in a chart. Such programs are beyond the scope of this book, but if you are interested in pursuing the subject, you can find suitable charting programs using the Commodore 64 in *Practical Interfacing Projects with the Commodore Computers* (Robert Luetzow, TAB BOOKS, catalog #1983), as well as a number of other publications. Check back issues of magazines that cater to owners of the Commodore 64.

LISTING 16-1

```
10  POKE 56579, 255
20  POKE 56577, 0
30  POKE 56589, 127
40  FOR N = 0 TO 7
50  POKE 56577, 0
60  POKE 56577, 1
```

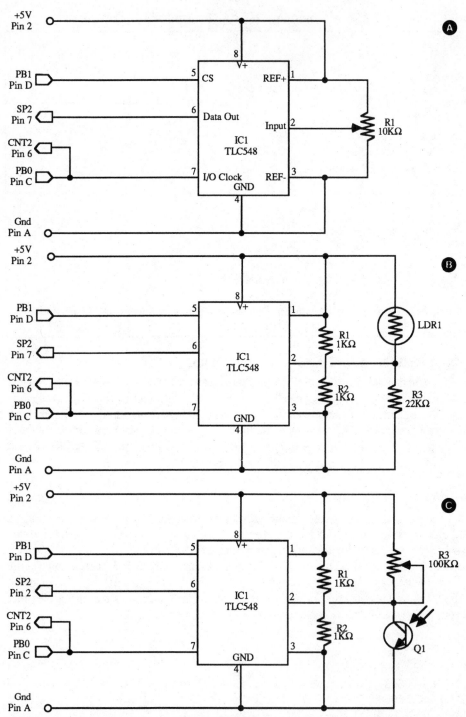

FIG. 16-4. *Hookup diagrams for connecting the TLC548 serial analog-to-digital converter to a Commodore 64 computer. (A) Test circuit (vary R1 and watch change in values); (B) Interfacing the circuit with a photoresistor; (C) Interfacing the circuit with a phototransistor. Adjust R3 to vary the sensitivity.*

```
 70  NEXT N
 80  IF (PEEK (56589) AND 8) = 0 THEN 80
 90  N = PEEK (56588)
100  PRINT N;
110  POKE 65677, 2
120  GOTO 40
```

CORRELATING YOUR RESULTS

I had just put the finishing touches on the coherency change seismograph prototype using a Commodore 64, when Los Angeles, my home town, was racked by a 6.1 Richter-scale earthquake (October 1, 1987). Although the epicenter of the earthquake was more than 50 miles distant with the Hollywood mountains in between, the entire area still shook violently. The earthquake could be felt for about 45 seconds but the remains of the tremor continued for several minutes (with lots of aftershocks).

I immediately checked the computer and found that it had recorded the full duration of the quake. I jotted down the results (I had not yet implemented a recording feature to save the data on disk or tape), and then waited until the seismologists in southern California could settle on an accurate magnitude for the tremor. At the epicenter, the earthquake measured 6.1 on the Richter scale. In my area, however, it was calculated that the earthquake measured only about 5.6 on the Richter scale. I used that information to "calibrate" the results from the computer. That way, with the digital data I recorded more accurately compared to a known value, I could better estimate the magnitude of future quakes.

17

Beginning Holography

Remember the first time you saw a real hologram? The image seemed to float in space, as if a piece of invisible film hung in mid-air in front of you. Moving your head back and forth verified the image was more than two-dimensional. You could actually see around the object to examine its top, bottom, and sides.

If experimenting with fiberoptics, modulators, and power supplies is the technical side of lasers, then holography is the artistic side. Armed with a helium-neon laser, some assorted optics, and a pack of film, you can create your own holograms. Your subjects can be anything that is small enough to illuminate with the laser beam and patient enough to sit still for the exposure.

After development of the film, you can display your holograms for others to see and appreciate. As you gain experience, you can tackle more complex forms of the laser holographic art—rainbow holograms, holographic interferometry, and even motion picture holograms.

This chapter introduces you to the art and science of making holograms, including how holograms work and what you need to set up shop. You'll learn how to build a holographic table and create, and later view, a transmission hologram. The next chapter continues with plans and procedures for a more advanced holographic table as well as how to create elaborate two- and three-beam reflection holograms.

A SHORT HISTORY OF HOLOGRAPHY

The idea of holography is older than the laser. Dr. Dennis Gabor, a researcher at the Imperial College in London, conceived and produced the first hologram in 1948.

Because Gabor was of a scientific bent, he published his ideas in a paper titled "Image Formation by Reconstructed Wavefronts," a phrase that aptly describes the technology behind holography. Gabor was later awarded the Nobel prize for Physics in 1971 for his pioneering work in holography.

Dr. Gabor's first holograms were crude and difficult to decipher. Compared to regular photography, the first Gabor holograms were scratchy throwbacks to pre-Civil War Daguerreotypes. Part of the problem was the lack of a sufficiently coherent source of light. No matter how complex the setup, the images remained fuzzy and indistinct. The introduction of visible-light lasers in the early sixties provided the final ingredient required to make sharp and clear holograms.

Gabor left much of the exploration of holography to other pioneers such as Lloyd Cross, T.H. Jeong, Emmett Leith, Pam Brazier, Juris Upatnieks, Fred Unterseher, and Robert Schinella. Much of the original developments of laser holography—first developed by Leith at the University of Michigan—had scientific and even military applications. Today, holography is used in medicine, fluid aerodynamics, stress testing, forensics, and even art.

In the 1960s and most of the 1970s, the equipment required to make professional-looking holograms was beyond the reach of most amateur experimenters. Now, however, with the wide proliferation of visible-light helium-neon lasers and the availability of surplus optical components, it's possible to build a workable holography setup for less than a few hundred dollars. And, that includes the materials required to construct a vibration-free isolation table.

If you're apprehensive of holography because you are afraid you won't understand how it works, consider that you don't need to know how the silver halide crystals in photographic emulsion react to light to take a snapshot of your family. You just pick up the camera, set some dials, focus, and press the shutter release.

Although you are encouraged to understand how coherent laser light is used to create, and later reconstruct, a multi-dimensional image, you needn't have a college degree in physics to make a hologram. Photographic experience is also not required, though it comes in handy. You will find the process of developing holographic films similar to processing ordinary black-and-white films and papers. If you've never been in a dark-room before, you might want to pick up a basic book on the subject and read up on the various processes and procedures involved. Obviously, you can't learn everything you need to know about holography in the two chapters in this book, and you are urged to expand your knowledge by further reading. A partial list of titles on holography is in Appendix B.

WHAT A HOLOGRAM IS—AND ISN'T

Many people hold misconceptions about the nature of holograms. A hologram is a photographic plate that contains interference patterns that represent the light waves from a reference source as well as from the photographed object itself. The patterns, like those in FIG. 17-1, contain information about the intensity of the light (just as in regular photography), as well as its instantaneous phase and direction. These elements together make possible the three-dimensional reproduction of the hologram.

The interference patterns constitute a series of diffraction gratings, and form what is, in effect, an extremely sophisticated lens. The orientation of the gratings, along with

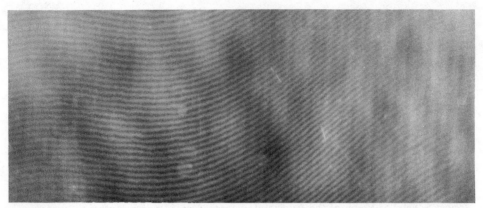

FIG. 17-1. *A thumbprint? No, it's a close-up of the interference fringes of a hologram, captured on high-resolution film.*

their width and size, determine how the image is reconstructed when viewed in light. In most types of holograms, the image becomes clear only when viewed with the same wavelength of coherent light used to expose the film.

Not all holograms are the same. As you'll see later in this chapter and in the next chapter, there are two general forms of holograms: transmission and reflection. The terminology refers to how the hologram is viewed, not exposed. The typical transmission hologram needs a laser for image reconstruction. The reflection hologram, while made with a laser, can be viewed under ordinary light.

The hologram stores an almost unlimited number of views of a three-dimensional object. Put another way, specific areas on the surface of the film contain different three-dimensional views of the object. You see these views by moving your head up and down or right and left. This effect is most readily demonstrated by cutting a hologram into small pieces (or simply covering portions of it with a piece of cardboard). Each piece contains the entire image of the photographed object but at a slightly different angle. As you make your own holograms, you get a clearer view of how the image actually occupies a three-dimensional space.

Some other types of images impart a sense of multi-dimensional imagery, but they are not true holograms. The coin-in-the-bowl scientific novelty, popular in mail-order catalogs, uses concave lenses to project an image of a coin into your eyes. The image appears closer than it is and when you reach out for it, it's not there. Another example: The famous shots of Princess Leia projecting out of the R2-D2 robot in the movie *Star Wars* was created with the aid of photographic trickery.

WHAT YOU NEED

You need relatively few materials to make a hologram:

★ Stable isolation table (often nothing more than a sandbox on cement pillars)
★ Laser—a 1 mW, 632.8 nm helium-neon is perfectly suited for holography, though other visible-light lasers will do as well; it must operate in TEM_{00} mode.
★ Beam expander lens
★ Film holder

★ Film and darkroom chemicals
★ Object to holograph

Other materials are needed for more advanced holographic setups, but this is the basic equipment list. These and other materials are detailed throughout this chapter.

THE ISOLATION TABLE

If you built the Michelson interferometer described in Chapter 9, you know how sensitive it is to vibration. The circular fringes that appear on the frosted glass shake and can even disappear if the interferometer is bumped or nudged. Depending on the table or workbench you use, vibrations from people walking nearby or even passing cars and trucks can cause the fringes to bob.

Because the interference fringes create the holographic image in the first place, you can understand why the table you use to snap the picture must be carefully isolated from external vibrations. If the laser setup moves at all during the exposure—even a few millionths of an inch—the hologram is ruined. Optical tables for scientific research cost in excess of $15,000 and weigh more than your car. But they do a good job at damping vibrations in the lab and preventing them from reaching the laser and optical components.

Obviously, a professional-grade optical table is out of the question, but you can build your own with common yard materials. Your own isolation table uses sand as the heavy damping material. The sand is dumped in a box that is perched on cement blocks. The homemade sandbox table is not as efficient as a commercially made isolation table, but it comes close. Add to your sand table a reasonably vibration-free area (California during an earthquake is out!) and you are on your way to making clear holograms.

Selecting the Right Size

The size of the table dictates the type of setups you can design as well as the maximum dimensions of the holographed object. The ideal isolation table measures 4-by-8 feet or larger, but a more compact 2-by-2-foot version can suffice for beginners. You will be limited to fairly simple optical arrangements and shooting objects smaller than a few inches square, but the table will be reasonably portable and won't take up half your garage.

The design for the sand table is shown in FIG. 17-2. The table consists of four concrete blocks (the kind used for outside retaining walls), a 2-by-2-foot sheet of ¾-inch plywood, some carpeting, four small pneumatic inner tubes, and an 8-inch-deep box filled with sand. You can build this table using larger dimensions (such as 4-by-4 or 4-by-8 feet), but it requires considerably more sand. You can figure about one 75- or 100-pound bag of fine sand for each square foot of table area.

Building the Table

First, build the box using the materials indicated in TABLE 17-1. Use heavy-duty construction—heavy screws, wood glue, and battens. Make the box as sturdy as possible—strong enough for an adult to sit or stand on. Next, lay out the blocks in a 2-foot-square area, inside the house or shop. You'll need a fairly light-tight room to work in, so a garage with all the cracks, and vents, and windows covered up is a good area. In addition, be sure that the room is not drafty. Air movement is enough to upset the fringes. You need a solid, level floor, or the sand table might rock back and forth.

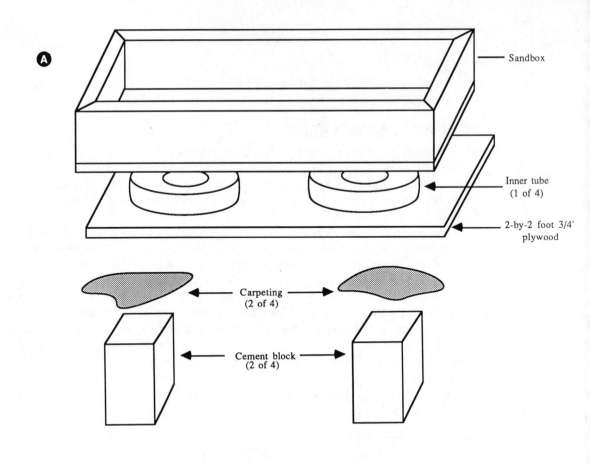

A Sandbox

Inner tube
(1 of 4)

2-by-2 foot 3/4'
plywood

Carpeting
(2 of 4)

Cement block
(2 of 4)

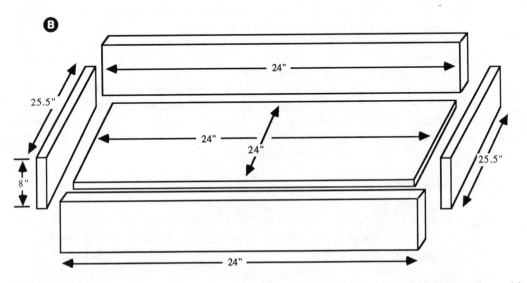

B

24"

25.5"

24"

24"

8"

25.5"

24"

FIG. 17-2. *Design of the 2-by-2-foot sand table. (A) Blocks, carpet, plywood base, inner tubes, and sandbox and how they go together. (B) Construction details of the sandbox. Use ¾-inch plywood for the sandbox.*

250

Table 17-1. Small Sand Table Parts Lists

Base and Pedestal

4	Cement building blocks
4	8-inch-square pieces of carpet
1	2-foot-square piece of carpeting
4	10- to 14-inch inner tubes
1	2-by-2-foot sheet of ¾ inch thick plywood

Sand Box

1	2-by-2-foot sheet of ¾-inch thick plywood
2	8- by 25.5-inch, ¾-inch thick plywood
2	8- by 24-inch, ¾-inch thick plywood
4	75- or 100-pound bags of washed, sterilized, and filtered sand

Place the 2-by-2-foot sheet of plywood over the blocks, then cover the plywood with one or two layers of soft carpeting (plush or shag works well). Partially inflate the four inner tubes (to about 50 to 60 percent) and place them on the carpet. Don't over-inflate the tubes or fill them with liquid. If possible, position the valves so that you can reach them easily to refill the inner tubes. You may need to jack up the sand box to access the valves, but if the tubes are good to begin with, you won't need to perform this duty often.

Carefully position the box over the inner tube. Don't worry about stability at this point: the table will settle down when you add the sand. Be sure to evenly distribute the sand inside the box as you pour out the contents of each bag. There should be little dust if you use high-quality, pre-washed and sterilized play sand. Fill the box with as much sand as you can, but avoid overfilling where the sand spills over the edges.

Testing the Table

You can test the effectiveness of the sand table by building a makeshift Michelson interferometer, as shown in FIG. 17-3. The optical components are mounted on PVC pipe, as described more fully in Chapter 7, "Constructing an Optical Bench." You can place the laser inside the box, on the top of the sand, or locate it off the table. As long as the beam remains fairly steady, any slight vibration of the laser will not affect the formation of the interference fringes necessary for successful holography.

Once you have the optics aligned and the laser on, wait for the table to settle (5 to 15 minutes), then look at the frosted glass. You should see a series of circular fringes. If the fringes don't appear, wait a little longer for the table to settle some more. Should the fringes still not materialize, double-check the arrangement of the laser and optics. The paths of the two beams must meet exactly at the frosted glass, and their propagation must be parallel—vertically as well as horizontally.

Determining Settling Times

Test the isolation capabilities of the table by moving around the room and watching the fringes (you might need someone to stand by the glass and closely watch the fringes).

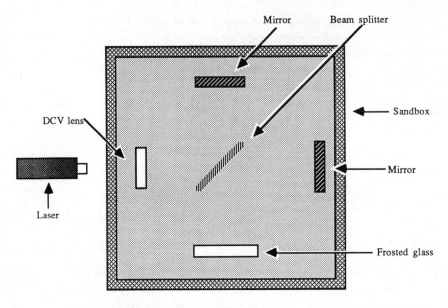

FIG. 17-3. *Optics arranged as a Michelson interferometer for testing the stability of the table.*

The fringes shouldn't move at all, but if they do, it should be slight. More energetic movement, such as bouncing up and down or shaking the walls, will greatly affect the formation of the fringes.

Watch particularly for the effects of vibrations that you can't control, such as other people walking around or cars passing by. Adequate isolation might require a more advanced table (such as the one detailed in the next chapter), more sand, or a change in location. You might also want to wait until after hours, where activity of both people and automobiles is reduced. Keep in mind that you will have better luck when the sand table is located on a ground floor and on heavy wood or cement.

Note the time it takes for the table to settle each time you change the optics or disturb the sand. You will need to wait for at least this amount of time before you can make a holographic exposure. Also note how long it takes for the table to recover from small shocks and disturbances; recovery time for most tables should be a matter of seconds.

THE DARKROOM

Part of holography takes place in the darkroom, where the film is processed after exposure. Although it's most convenient to place the isolation table and processing materials in the same room, they can be separate. As with the isolation table, the darkroom must be reasonably light-tight. It should have a source of running water to rinse processed film and a drain to wash away unwanted chemicals. A bathroom is ideal, as long as you have room to move and can work uninterrupted for periods up to 15 to 20 minutes.

If the bathroom has windows, cover them with opaque black fabric or painted cardboard. If you plan on making many holograms, arrange some sort of easily installed curtain system. Drape some material over the mirror and other shiny objects, like metal towel racks and fixtures. You'll need access to the sink and faucet, so leave these clear.

Only two chemicals are needed for basic holography (advanced holographers use many more chemicals, as detailed in the next chapter). You need to set out two shallow plastic bowls or trays large enough to accommodate the film. If space is a premium inside your bathroom, place a wooden rack over the tub and set the trays on it. The bathtub makes a good location for the trays, because the processing chemicals might stain clothing, walls, floors, and other porous materials. If any drips into the tub, you can promptly wash it away.

OPTICS, FILM, FILM HOLDERS, AND CHEMICALS

Here's a run-down of the basic materials you need to complete your holographic setup.

Optics

Basic holography requires the use of a plano-concave or double-concave lens and one or more front-surface mirrors. The lens can be small—on the order of 4 to 10 mm in diameter. Focal length is not a major consideration but should be fairly short. Mount the lens in 1- or 1¼-inch PVC pipe, painted black, as detailed in Chapter 7.

The purpose of the lens is to expand the pencil-thin beam of the laser into an area large enough to completely illuminate the object being photographed. The area of the beam is expanded proportionately to the distance between the lens and object, and the basic 2-by-2-foot table doesn't allow much room for extreme beam expansion. If the beam is not adequately expanded, use a lens with a shorter focal length, or position two negative lenses together as shown in FIG. 17-4.

The direct one-beam transmission hologram setup described below does not require the use of mirrors or beam splitters, but multiple-beam arrangements do. The size of the mirror depends on the amount of beam spread and the size of the object, but in general, you need one or more mirrors measuring 2-by-3 inches or larger. Some setups also require beam steering or transfer mirrors. These are used before the beam has been expanded, so they can be small. See Chapter 7 for ideas on how to mount mirrors and other optics.

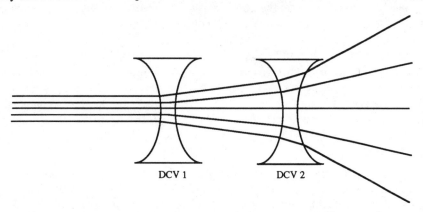

DCV 1   DCV 2

FIG. 17-4. *How to obtain maximum beam spread by using two double-concave lenses.*

Holographic Film

Holography requires a special ultra-high resolution film emulsion that is most sensitive to the wavelength of the laser you are using. Kodak Type 120 and S0-173 or Agfa-Gevaert Holotest 8E75 and 10E75, for example, are made for 632.8 nm helium-neon lasers and are most sensitive to red light (a relatively insensitive wavelength for orthochromatic film). You can handle and develop this film using a dim (7-watt or less) green safelight—a small green bug light works well as long as it is placed 5 feet or more from the sand table and film handling areas. Holographic film comes in various sizes and base thicknesses. Film measuring 2¼ inches square to 4-by-5 inches is ideal for holography.

Holographic films are available with and without antihalation backing. This backing material, which comes off during processing, prevents halos during exposure. The backing is semi-opaque, which prevents light from passing through it. Transmission holograms are made with the laser light striking the emulsion surface, so you can use a film with antihalation backing. Reflective holograms are made with light passing *through* the film and striking the emulsion on the other side. Unless you are eager to make exposures lasting several minutes or more, you'll want a film without the antihalation backing when experimenting with reflection holography.

When buying or ordering film, be sure to note whether the stuff you want comes with antihalation backing. If you want to try both basic forms of holography (transmission and reflection) but don't want to spend the money for both kinds of films, use stock without the backing. Note that this film does not make the best transmission holograms, but the results are more than adequate. Be sure to place a black matte card (available at art supply stores) in back of the film when making the exposure. This prevents light from entering the film through the back and fogging the emulsion.

Another alternative is to use glass plates for your first holograms. These consist of a photographic emulsion sprayed onto a piece of optically clear glass. The benefit of the glass is that it is easier to handle for beginners and is not as susceptible to buckling and movement during the exposure.

Where do you get holographic film? Good question. Start by opening the Yellow Pages and looking under the Photographic heading. Call the various camera stores and ask if they carry it or can special-order it for you. But in the interest of saving you the disappointment, be aware that the personnel at most camera stores are not even aware of such a thing as holographic film. You will probably have better luck calling those outlets that specialize in professional photography or darkroom supplies.

You might also happen to live in an area close to a Kodak or Agfa-Gevaert field office. Call and ask for help. A few companies, such as Metrologic, offer film suitable for holography; check Appendix A for the address. Lastly, write directly to the film manufacturers and ask for a list of local dealers that handle the materials you need. You might even be able to order holographic film through the mail.

Film Holder

You will need some means to hold the film in place during the exposure. One method is to sandwich the film between two pieces of glass held together by two heavy spring clips, as shown in the photograph in FIG. 17-5. The glass must be spotlessly clean. The disadvantage of this method is that the contact of the glass and film can create what's known as Newton's Rings, a form of interference patterns.

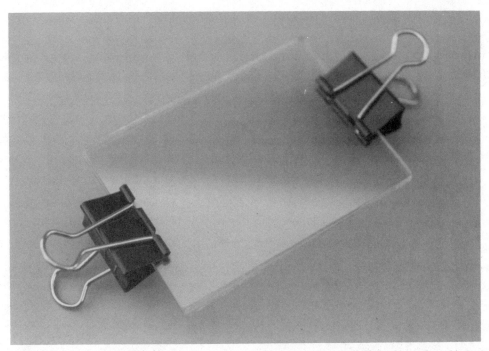

FIG. 17-5. *A homemade film pressure plate consisting of two pieces of optically clear glass and two binder clips. The film is inserted between the glass and pressed tightly together to eliminate air bubbles.*

If you plan on using a glass-plate film holder, remember to press the plates together firmly and keep constant pressure for 10 to 20 seconds (use wooden blocks for even pressure). This removes all air bubbles trapped between the glass and film. Snap the binder clips around the glass and position the holder in front of the object. The handles of the clips can be secured to the table by butting them against two pieces of PVC pipe. Depending on the arrangement of the optics, you might be able to locate the film holder between the PVC pipes that contain the expansion lenses.

Yet another approach is to use a commercially-made film holder, the kind designed for processing plate film. The holders come in a variety of sizes and use novel approaches to loading and holding the film. Most any camera store that carries professional darkroom supplies has a variety of film holders to choose from. You can also make your own following the diagram in FIG. 17-6. Construct the holder to accommodate the size of film you are using. You might need to make several holders if you use different sizes of film.

To use any of the holders, load the film (in complete darkness or under the dim illumination of a small green safelight) into the holder, and stick the holder in the proper location in the sand. If the holder does not allow easy mounting in the sand, attach it to small wood or plastic pieces, as illustrated in FIG. 17-7.

Processing Chemicals

You can use ordinary film developer to process most holographic film emulsions. A good choice is Kodak D-19, available in powder form at some photographic shops. If D-19 is not available, you may use most any other high-resolution film developer. Agfa,

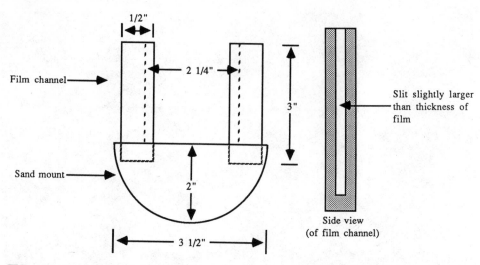

FIG. 17-6. *Construction details for a homemade wooden film holder. The holder shown is made for 2¼-by-2¼ film.*

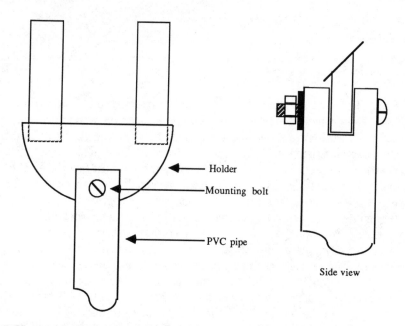

FIG. 17-7. *You can raise the distance between film and sand by mounting the holder on a piece of PVC pipe.*

Nacco, and Kodak make a variety of high-quality powder and liquid developers that you can use.

Both transmission and reflection holograms require bleaching as one of the processing steps. If you can't buy pre-made bleach specifically designed for holography, you can mix your own following the directions that appear below. Note that transmission and reflection holograms need different bleach formulas. Chemicals for making the bleach

256

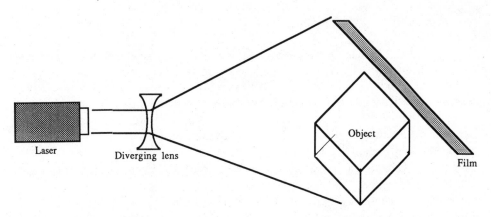

FIG. 17-8. *The basic arrangement for making a single-beam transmission hologram.*

are available at many of the larger industry photographic outlets. If you can't find what you want, dial up a chemical supply house and see if you can buy what you need in small quantities.

The remaining, optional chemicals are available at most camera stores. These include acetic acid stop bath, fixer (with or without hardener), photoflo, and hypo clear.

SINGLE-BEAM TRANSMISSION SETUP

The layout in FIG. 17-8 shows how to arrange the laser, lens, photographic object, and film plate for a direct, one-beam transmission hologram. A parts list is provided in TABLE 17-2. Note that the term "transmission" has nothing to do with the arrangement of the film, optics, or object for making the hologram but rather the method of viewing the image after it is processed. You look through transmission holograms to see an image; you shine light off the emulsion side of a reflection hologram to see the picture.

You can use almost anything as the object, but for your first attempt, choose something small (about the size of a pack of cigarettes or less), with a smooth but not highly reflective surface. A pair of dice, a coffee mug, a chess piece, a small electronic circuit board, or an electronic component make good subjects. Things to avoid include silverware, highly reflective jewelry, the family TV set, and your cat. These and other reflective, large, and/or animate objects can be subjects for holograms, but only after you gain experience.

Making the Exposure

After you arrange the laser and optics as shown above, turn the laser on. The distance between the lens and object/film should be 1.5 to 2 feet. Be sure the beam covers the

Table 17-2. **Single-Beam Transmission Hologram Setup Parts List**	
1	He-Ne laser
1	8 to 10 mm bi-concave lens
1	Holographic film in film holder
1	Object

object and the film, as shown in the figure. Now block the beam with a black "shutter" card.

Turn off all the lights and switch on the safelight. The safelight should emit only a tinge of illumination, hardly enough for you to see your hand out in front of you. The idea is to help you see your way in the dark, not provide daylight illumination. You can test for excessive safelight illumination by placing a piece of film on the sand table for 5 minutes with half of it covered with a piece of black cardboard. Process the film and look for a darkening on the uncovered side. A dark portion means that the film was fogged by the safelight.

Load the film into the holder, emulsion side towards the laser, and place the holder at a 45-degree angle to the laser beam (you can tell the emulsion side by wetting a corner of the film; the emulsion side becomes sticky). The film should be close to the object but not touching it. Wait 10 minutes or so for the table to settle, then quickly but carefully remove the card covering the laser. After the exposure is complete, replace the card. *Do not* control the exposure by turning the laser on and off. The coherency of lasers improves after they have warmed up. For best results, allow the laser to warm up for 20 to 30 minutes before making any holographic exposure.

Exposure times depend on the power output of the laser. As a rule of thumb, allow 3 to 4 seconds for a 1 mW laser and 1 to 2 seconds for a 3 mW laser. You might want to make a test exposure using the technique shown in FIG. 17-9. Place a card in front of the film and make a 1 second exposure. Move the card so that it exposes a little more of the film and make another 1-second exposure. Repeat the process four to six times until you have a series of strips on the film. Each strip along the length of the film denotes an increase in exposure of 1 second.

Develop the test strip film following the steps below, then choose the exposure used for the best-looking strip. In the average transmission hologram, the film should be dark gray with distinct fringe patterning. When viewed with laser light, underexposure causes

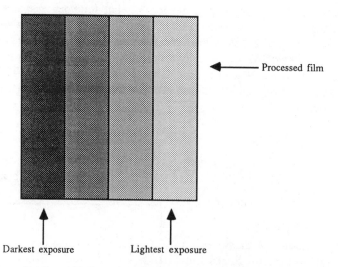

Processed film

Darkest exposure   Lightest exposure

FIG. 17-9. *The results of a sample test strip with increasingly dark stripes of gray across the film. Choose the exposure (and development time) that yields the proper density, as noted in the text.*

the film to be excessively light and lack contrast. Overexposure makes the image hard—if not impossible—to see.

Processing Steps

Once the hologram is exposed, place it in a light-tight container or box. Mix the processing chemicals (if you haven't already) as follows:

✶ D-19 developer—Full strength as detailed in the instructions printed on the package (when using another developer, mix according to instructions to obtain the highest resolution).

✶ Bleach—One tablespoon of potassium ferrocyanide mixed with one tablespoon of potassium bromide in 16 ounces of water. Alternatively, you may dissolve 5 grams of potassium dichromate in 1 liter of water. Then add 20 ml of diluted (10 percent) sulfuric acid.

Use water at room temperature (68 to 76 degrees F), and be sure the temperature is roughly the same between each chemical bath. High temperatures or wide variations in temperature can soften the film and introduce reticulation—two side effects you want to avoid. Place the developer and bleach in large plastic bowls or processing trays (available at photo stores). The chemicals have a finite shelf life of a few months when stored in stoppered plastic bottles. Use small or collapsible bottles to minimize the air content, and store the bottles in a cool place away from sunlight.

In the dark or with a green safelight, remove the film from the light-tight box and place it in the developer. Use plastic tongs to handle the film (avoid using metallic implements and your fingers). Swish the film in the developer and knock the tongs against the side of the tray or bowl to dislodge any air bubbles under or on the film. Maintain a slow and even agitation of the film. After about 2 to 5 minutes, the image on the film will appear and development is complete.

If you have a safelight, you can see the image appear and can better judge when the development is complete. As a rule of thumb, when held up to a safelight, a properly exposed and developed transmission hologram allows about 70 to 80 percent of the light to pass through it (reflective holograms pass about 20 to 30 percent of the light).

After development, rinse the film in running water or a tray filled with clean water. You can also use an acetic stop bath, mixed according to directions. The stop bath more completely removes traces of developer and helps prolong the life of the bleach used in the next step.

After 15 seconds or so of rinsing, place the film in the bleach. Keep it there for 2 to 3 minutes. After a minute or two of bleaching, you can turn on the lights. Follow bleaching with another rinse. Use lots of water and keep the film in the wash for 5 to 10 minutes. A cyclonic rinser, used in many professional darkrooms, does a good job of removing all the processing chemicals.

The hologram is almost ready for viewing. Before drying it, dip the film in Kodak Photoflo solution—a capful in a pint of water is sufficient. The Photoflo reduces the surface tension of the water and helps promote even drying without spotting. It also removes minerals left from the water bath. You may also use a squeegee to remove excess liquid. Darkroom squeegees cost $5 to $10 and are good for the purpose, or a new car windshield wiper blade costs $2 and is just as good.

Use clothespins or real honest-to-goodness film clips to hang up the film to dry. Place the film in a dust-free area. A specially made drying cabinet is the best place, but the shower or bathtub is another good choice.

Wait until the film has completely dried before trying to view it (the fringes might not be clear until drying is complete). Depending on the temperature, humidity, and other conditions, it takes 30 to 120 minutes for the film to dry completely. Glass plate film dries more quickly. Note that when wet, photographic emulsion becomes soft and scratches easily. Be sure to handle the film with care and avoid touching it (with hands or tongs) except by the edges.

Let's recap the processing steps:

- ✰ Develop for 2 to 5 minutes.
- ✰ Rinse for 15 seconds (water or stop bath).
- ✰ Bleach for 2 to 3 minutes.
- ✰ Rinse for 5 to 10 minutes in constantly running water.
- ✰ Dip in Photoflo for 15 seconds (optional step).
- ✰ Squeegee to remove excess liquid (don't scratch the emulsion!).
- ✰ Let dry at least 30 minutes.

Viewing the Hologram

Part of the fun of holography is making the picture; the rest is seeing the result with your own eyes. If you haven't disturbed the setup used to take the picture, simply replace the film in the film holder. The emulsion should face the laser. View the image by looking through the film. The image might not be noticeable unless you move your head from side to side or up and down.

Note that if you have dismantled the sand box setup, you must replace the laser, optics, and film holder in the same arrangement to reconstitute the image. The simple layout used in the single-beam direct transmission exposure makes it relatively easy to reconstitute the image.

Problems

Having problems? The image in the hologram isn't clear or the exposure just isn't right? Most difficulties in image quality are caused by motion. The fringes must remain absolutely still during the exposure or the hologram could be ruined. Consider that a more powerful laser decreases the exposure time, reducing the problems of vibration. For example, a 5 to 8 mW laser might require an exposure of only 0.5 to 1 second. If at all possible, use the most powerful laser you can get your hands on, but don't give up on holography if all you own is a tiny 0.5 mW tube.

Like regular photography, it takes time, patience, and practice to make really good holograms. Don't expect to make a perfect exposure the first time around. Odds are the exposure will be too short and the development too long, or vice versa. If your first attempt doesn't come out the way you want it (or doesn't come out at all!), analyze what went wrong and try again.

Check the layout and be sure that the film is inserted in the holder with the emulsion side out (that is, towards the laser). If you use a film with an AH backing (recommended for transmission holograms), the light won't pass through to the emulsion if the film is inserted backward in the holder. Be sure that the chemicals are mixed right and that you are following the proper procedure.

INTERMEDIATE HOLOGRAPHY

You can learn more about holography by actually making a hologram than reading an entire book on the subject (however, that *is* recommended, see Appendix B for a list of books on holography). Below are some observations that will help you better grasp the technology and artistry of holograms.

Field of Vision

Here's an important point to keep in mind as you experiment further. You might have realized that the film holder represents the frame of a window onto which you can view the subject. Place your head against the film holder, close one eye, and look at the object from up close. Without actually moving your eye, scan your head vertically and horizontally and note the different views you can see of the subject. They are the same views you see in a hologram. Use this technique to view the perspective and field of vision for your holograms. You can then adjust the position of the object, film, or even the optics to obtain the views you want.

Varying Exposure

Unless you have a calibrated power meter and lots of experience, expect mistakes in estimating exposure times when experimenting with different types of holographic setups. The direct, one-beam setup detailed previously conserves laser light energy, thus reducing exposure time. Some multi-beam setups (see below and in the next chapter) require exposure times of 15 to 25 seconds, assuming a 3 to 5 mW laser.

Light Ratios

All holograms are made by directing a reference beam and an object beam onto a film plate. The reference beam comes directly from the laser, perhaps after bouncing off a beam splitter and a mirror or two. The object beam is reflected off the object being photographed. The ratio between the reference and object beams is a major consideration. If the ratio isn't right, the hologram becomes "noisy" and difficult to see. The reflectivity of the object largely determines the ratio, but with most commonly holographed subjects, the reference-to-object light ratio for a transmission hologram is about 4:1 (four parts reference to one part object).

Intermediate and advanced holography requires careful control over light ratios, which means you must take readings using a calibrated power meter (available through Metrologic) or a light meter.

You may adjust the light ratios by repositioning the optics or by using a variable-density beam splitter. A variable-density beam splitter is a wheel (or sometimes a rectangular piece of glass) with an anti-reflective coating, applied in varying density, on one side. An area with little or no anti-reflection coating reflects little light and passes much. The opposite is true at an area with a high amount of AR coating. You can "dial" in the ratio of transmitted versus reflected light by turning the wheel. Note that basic holography doesn't require a variable-density beam splitter, which is good because they're expensive.

Reverse Viewing

If you flip the hologram over top to bottom, you'll see the image appear in front of the film, looming toward you. You see the image inside-out as you view it from the back side. This effect is created by light focusing in the space in front of the hologram. Although the entire image can't be focused onto a plane, you can see the formation of three-dimensions by placing a piece of frosted glass at the apparent spot where the hologram appears. As you move the plate in and out, different portions of the picture will come in and out of focus.

Projected Viewing

The image in a hologram can be projected in a variety of ways. Try this method. Replace the hologram so that its emulsion faces the laser. Remove the beam-expanding lens so that just the pencil-thin beam of the laser strikes the center of the hologram film. Place a white card or screen behind the film to catch the light going through the hologram. The screen or card should be located where your eyes would be if you were viewing the hologram. Watch the image that appears. You'll see the complete object, but the image will be two-dimensional.

Now move the film up and down and right and left. Notice how the picture of the object remains complete but the perspective changes. This is the same effect you get if you cut a hologram into many pieces. Each piece contains a full picture of the subject but at slightly different views.

Transfer Mirrors

Some holographers prefer to place the laser in the sand table. This is perfectly acceptable, as long as you mount the laser on a wood or plastic board. Position it along one side of the table and direct the beam diagonally across the table with a transfer mirror. Aiming the beam diagonally across the surface of the sand table gives you more room and allows you to create more elaborate setups. The general idea is shown in FIG. 17-10.

Alternate Single-Beam Transmission Setups

FIGURE 17-11 shows the setup for an alternate single-beam transmission hologram. This arrangement requires a white card measuring about 4-by-4 inches and a 5- or 6-inch square mirror. Position the film holder as shown and locate the mirror at about an 80-degree angle to the film. Place the card so that no direct light from the laser strikes the film. All of the light used for exposing the film should be reflected off the mirror.

Yet another single-beam transmission hologram can be created by using the simple setup shown in FIG. 17-12. It is similar to the first arrangement, but the film is positioned at almost a 90- degree angle to the path of the laser light. Here, no white card or mirror is used. TABLES 17-3 and 17-4 provide the parts lists for these two setups.

The film can actually be beside, under, or over the object. Place the film angled downward on a hill of sand. Underneath, place a piece of painted airport runway from

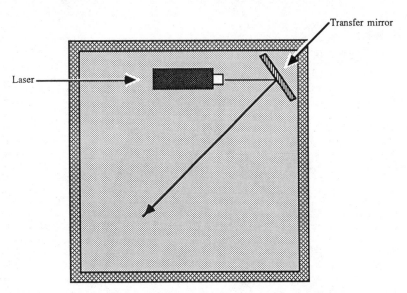

FIG. 17-10. *The idea of transfer mirrors.*

Table 17-3. Alternate #1 Single-Beam Transmission Hologram Parts List

1	He-Ne laser
1	8 to 10 mm bi-concave lens
1	Holographic film in film holder
1	5- or 6-inch-square front-surface mirror
1	Black blocking card
1	Object

Table 17-4. Alternate #2 Single-Beam Transmission Hologram Parts List

1	He-Ne laser
1	8 to 10 mm bi-concave lens
1	Holographic film in film holder
1	Object

a model airplane kit. The finished hologram will have a startling 3-D image of the runway that will appear as if you are actually landing an airplane. Try this technique the next time you exhibit your best model airplane. You're sure to win first place!

SPLIT-BEAM TRANSMISSION HOLOGRAM

The visual effect of a single-beam hologram is limited due to the single source of light. Objects photographed in this manner can look dark or lack detail in shadowed areas. Only one side of the hologram might be illuminated, and as you move your head to see

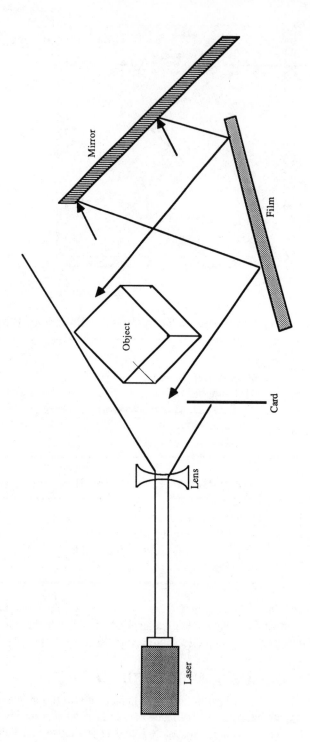

FIG. 17-11. A single-beam transmission hologram using a large mirror. The mirror provides the reference beam as well as a large portion of light for the illumination of the object.

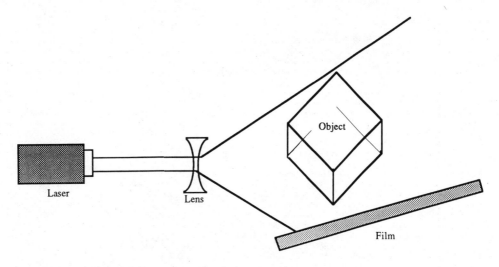

FIG. 17-12. *A simple single-beam transmission hologram arrangement, shown with film set at a 75 to 80 degree angle to the beam path.*

different views, the image grows dim. In portraiture photography, two, three, and sometimes four different light sources are used to illuminate different parts of the head or body. The amount of illumination from each light is carefully controlled to make the picture as pleasing as possible.

A similar approach can be used in holography by splitting the laser beam and providing two or more sources of light to illuminate the object (there is still only one reference beam). The setup in FIG. 17-13 shows how to illuminate an object by splitting the laser light and directing two expanded beams to either side of the object. See TABLE 17-5 for a list of required materials.

The main consideration when splitting the light is that the reference and object beams should travel approximately the same distance from the laser. That is, the reference beam should not travel 2.5 feet and the object beam(s) only 1.2 feet. Use a fabric tape to measure distances. Accuracy within one or two inches will assure you of good results. Longer distances might exceed the coherency length of the laser, and your holograms might not come out right. Note that high-power lasers generally have longer coherency lengths than low-power ones.

The arrangement requires three diverging lenses, three mirrors, and two beam splitters. A slightly different variation, using fewer components, is presented in the holography gallery found in the next chapter. The first beam splitter reflects the light to a reference mirror. This light is then expanded by a bi-concave lens and directed at the film. The transmitted light through the first beam splitter is divided again by a second beam splitter. These two beams are bounced off positioning mirrors, expanded by lenses, and pointed at the object.

It's very important that no light from the two object mirrors strikes the film. It is a good idea to baffle the light by placing black cards on either side of the film holder, as illustrated in the figure.

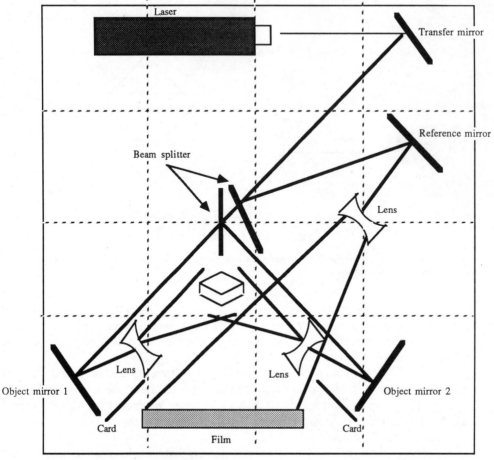

FIG. 17-13. *Optical arrangement (shown in the sandbox) for a multiple-beam transmission hologram. The object is illuminated on two sides, providing even lighting and better contrast.*

Exposure time should be similar to the one-beam method described earlier in this chapter, but you should probably make another test strip to make sure. Although roughly the same total amount of light is striking the film in both single- and multiple-beam setups, there is inherent light loss through the beam splitters as well as additional light scattering through the three lenses.

Table 17-5. Split-Beam Transmission Hologram Parts List

1	He-Ne laser
3	8 to 10 mm bi-concave lens
1	Holographic film in film holder
4	1- to 2-inch-square front-surface mirror
2	Plate beam splitter (50:50)
2	Black blocking card
1	Object

Process the film in the usual manner, and when dried, place it back in the film holder, emulsion side facing the reference beam. Remove the object you photographed as well as the object mirrors and second beam splitter because you need only the reference beam to reconstitute the image. An image should now appear.

You can visually see how the processed hologram must be placed in the exact same position relative to the reference beam, or the image won't appear. Try turning the film in the holder. Notice that the image disappears if you rotate the film more than a few degrees in either direction.

18

Advanced Holography

Most people have enjoyed looking at a hologram, but few people have made one. But if just looking at a hologram is such fun, imagine what its like to make your own. The last chapter introduced the basics of holography and showed you how to make your own transmission-type holograms. This chapter takes you several steps further, revealing how to make colorful reflection holograms, circular holograms, or even holographic movies. There are also several alternative setups for both transmission and reflection holograms you might like to try.

ADVANCED SANDBOX

The 2-by-2-foot sandbox described in the last chapter is fine for tinkering around, but serious holography requires a bigger work area. FIGURE 18-1 shows a design for a 4-by-4-foot sandbox using plywood, concrete blocks, poured concrete, and automobile inner tubes. A parts list is provided in TABLE 18-1. The table uses basic masonry and concrete-pouring techniques, so if you aren't familiar with these, pick up any good book on the subject at the library. Although the procedure of pouring and leveling concrete might seem too involved, it's actually a simple process that takes just one or two hours.

Start the table by laying a 4-by-4-foot carpet on the floor. Place five concrete blocks, each measuring 8 by 8 by 16 inches, over the carpet, as shown in FIG. 18-2. Add small pieces of carpet on top of the blocks, then set a ¾-inch piece of 4-by-4-foot plywood squarely over the blocks. Add yet another piece of carpet over the plywood.

Inflate four small automobile or motorcycle inner tubes to 50 to 60 percent full. Don't overfill—just inflate enough so that the inner tube starts to expand and that a heavy weight

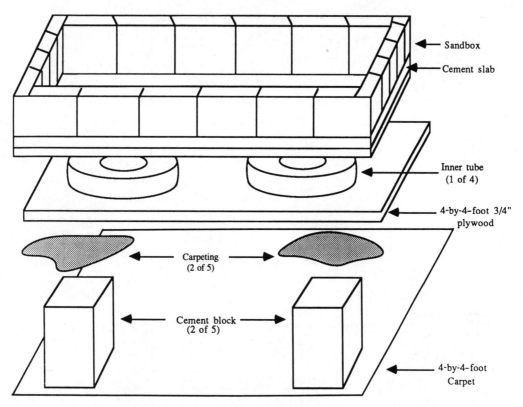

FIG. 18-1. *Materials required to make the 4-by-4-foot concrete sand table.*

Table 18-1. Large Sand Table Parts Lists

Base and Pedestal

5	Cement building blocks (8 by 8 by 16 inches)
5	8-inch-square pieces of carpet
1	4-foot-square piece of carpeting
4	10- to 14-inch inner tubes
1	4-by-4-foot sheet of ¾-inch-thick plywood

Sand Box

1	4-by-4-foot sheet of ¾-inch-thick plywood
16	12-by-12-inch-square stepping stones (or equivalent)
4	48-inch lengths of 2-by-4 framing lumber (for cement slab)
2	Bags Redi-Mix (or similar) cement
1	Bag mortar mix
1	4-by-4-foot piece of chicken wire
4	75- or 100-pound bags of washed, sterilized, and filtered sand

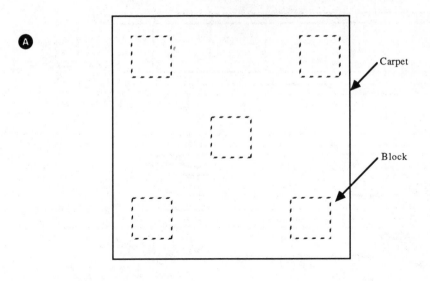

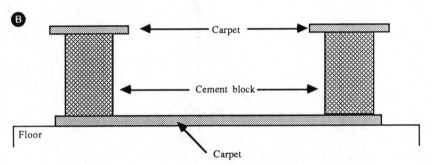

FIG. 18-2. *Placement of the carpet and blocks. (A) Top view; (B) Side view.*

placed on top will not squeeze the rubber together. Inner tubes for small 12- or 13-inch car wheels work well. Alternatively, you can use wide motorcycle-tire inner tubes. Use a fifth tube in the middle of the table if the tubes are very small.

Cut a piece of plywood to 4 by 4 feet. Nail 2-by-4 framing to the outside edge of the plywood. Use only a few nails for each side; the framing is temporary. Mix two bags of Redi-Mix (or similar) cement according to instructions. In most cases, all you add is water. An old washbucket makes a good mixing bin. Lay a piece of plastic tarpaulin in the box to prevent cement from oozing out the sides. Fill the box about ⅓-full with cement, then place a piece of 4-by-4-foot chicken wire (trim the chicken wire so that it fits) on top of the cement. Pour the remaining concrete over the chicken wire.

The concrete should reach the top of the box. If it doesn't, mix a little more, but you don't need to use a whole bag. Use a piece of scrap 2-by-4 as a leveling board to smooth out the concrete. Once the cement is leveled, leave it alone overnight. Read the instructions that came with the cement to see if you should water the slab as it dries to prevent cracking. If watering is recommended, sprinkle a small amount of water on the slab every few hours. Should the slab be cracked after drying, fill it using concrete filling cement, available at most building supply stores. After the slab has dried (the cement is now concrete), remove the 2-by-4 frame pieces and trim away the tarpaulin.

Now arrange a series of stepping stones around the perimeter of the slab as shown in FIG. 18-3. The size of the stepping stones doesn't matter as long as they are about 1 inch thick. A common stepping stone size is 12 by 12 inches; you don't need anything fancy (don't buy colored or fluted stones).

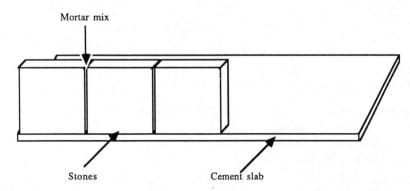

FIG. 18-3. *Adding square stepping stones to the perimeter of the concrete slab. Use mortar to set the stones.*

Prepare a bucket of mortar mix according to instructions. With a trowel, glop the mortar onto the bottom and sides of the stones and apply them to the perimeter of the slab (you might need to wet the slab and stones first with clean water). Work slowly and try to use just the right amount of mortar, but you can scrape off mortar that oozes between the stones with the trowel.

If the stones you use won't fit evenly around the perimeter of the slab—the corners don't meet, for example—you can cut them into smaller pieces by first making a scoring line. Score the stone with a hacksaw at the place where you want to make the cut. After the score is ⅛- to ¼-inch deep, place the stone at a curb or other ledge. Press down with both hands to break the stone at the score. Your first attempt might not work, but with practice, you'll get the hang of it. You can also use smaller stepping stones or even red brick to fill in corners that don't meet.

The mortar takes about a day to dry and might also require a sprinkle now and then with water to prevent cracking. After the mortar has set, add 16 to 18 100-pound bags of sand. Fill to about 2 inches from the top of the table. At this point, you may want to remember that the table will weigh about a ton when filled with sand, so if you put the table on a rickety floor, you'll soon regret it.

The sand table is now complete. Some holographers like to paint the concrete flat black to reduce light scatter and to improve the looks. Painting is not necessary, but do as you please.

MAKING A REFLECTION HOLOGRAM

Transmission holograms require you to shine the expanded beam of a laser through them in order to see an image. Another type of hologram that doesn't require a laser for viewing is the reflection hologram. These work by shining light (white or colored)

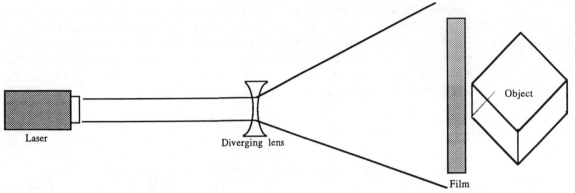

FIG. 18-4. *Basic single-beam reflection hologram arrangement.*

off the diffraction grating surface of the hologram. If viewed under white light, the hologram gives off a rainbow of dazzling colors.

Mixing the Chemicals

Reflection holograms are no more difficult to make than transmission holograms, although the processing chemistry is a bit different. White-light reflection holograms can use Kodak D-19 developer but should be bleached using one of the following formulas:

★ Carefully mix 20 grams (or about one tablespoon) of potassium bromide and 20 grams mercuric chloride *(fatal if swallowed)* in one liter of water. This stuff is dangerous and highly caustic, so never touch it with bare fingers or allow it to splash on skin, clothes, or eyes.

Or,

★ Dissolve two grams of potassium dichromate with 30 grams of potassium bromide in one liter of water. After these have thoroughly mixed, carefully add two cm³ of concentrated sulfuric acid (always add acid to water, not the other way around). This is also nasty stuff and will burn skin if you touch it. Wear gloves and safety goggles when mixing, and use gloves or tongs when processing the film.

The best reflection holograms need a fixing step, just like regular photographic film and paper. Fixer comes pre-made (in powder or liquid), making it easy to use. Kodak Rapid Fixer with hardener is a good choice. Mix according to instructions.

Setting Up

FIGURE 18-4 shows the most rudimentary arrangement for making a reflection hologram. See TABLE 18-2 for a parts list. Position the laser, lens, film, and object in a direct line. For best results, the object used in reflection-type holography should be relatively small in comparison to the film and should be placed within a few inches of the film. Larger images or those placed far away tend to be dark and fuzzy.

As you read in the last chapter, reflection holograms require a film without the antihalation backing; be sure to use this type. When you are ready to make the exposure, turn out the lights, remove the film, and place the film in the holder.

If you are using a glass-plate film holder, remember to press the plates together firmly and keep constant pressure for 10 to 20 seconds to remove air bubbles trapped between the glass and film. Use wood blocks to apply even pressure. Snap the binder clips around the glass and position the holder in front of the object. If necessary, mount the film holder between two PVC or dowel pillars.

As with the transmission hologram, the beam should be expanded so that the outer ⅓ of the diameter falls off the edge of the film. You want the inner ⅔ of the beam, which is the brightest portion. Unlike transmission holograms, reflection holograms call for a ratio between reference and object at a more even 1:1 or 2:1. Use a photographic light meter or power meter to determine proper beam ratios.

Exposure time depends on the power output of the laser as well as the size of the film (or more precisely, the amount of beam spreading), but you might have luck using trial exposures of 3 to 5 seconds with a 1 mW laser and a 1 to 2 seconds for a 3 mW laser.

Processing the Film

In dim or green-filtered light, dip the film in the developer tray and process for 2 to 5 minutes. As a general rule of thumb, a reflection hologram should pass about 20 to 30 percent of the light when held up to a green safelight. After developing, rinse in water or stop bath for 15 seconds.

Dip the film in the first fixer bath for 2 to 3 minutes. After fixing is complete, the room lights can be turned on (fixer renders the film insensitive to further exposure to light). Wash again in water for 15 seconds.

Place the film in the bleach mixture for 1 to 2 minutes or until the film clears. Rinse once more in water for 15 seconds. Finally, place the film in a second fixer bath for 3-5 minutes or until the hologram turns a brown color.

Wash all the chemicals away by rinsing the film under running water for at least five minutes. Then, dip the hologram in Photoflo, squeegee it, and hang it up to dry (details on these last steps are in the previous chapter). Be aware that you can't see a holographic image until the film is completely dried, so don't judge your success (or failure) at this point.

If you are impatient and can't wait for the film to dry on its own, you can hurry up the process by blow-drying the film with a hair dryer. Set the dryer on no heat (air only) and gently waft it 6 to 8 inches in front of the film. The backing will dry quickly but the emulsion takes 5-10 minutes. Always remember to dry film in a dust-free place. Amateur photographers like to use the tub or shower in the bathroom, a place where airborne dust usually doesn't stay for long.

To sum up:

★ Develop in D-19 (or similar) developer for 2 to 5 minutes.

Table 18-2. Single Beam Reflection Hologram Setup Parts List		
	1	He-Ne laser
	1	8 to 10 mm bi-concave lens
	1	Holographic film in film holder
	1	Object

* Rinse in water or stop bath for 15 seconds.
* Fix in first fixer for 2 to 3 minutes (light on after fixing).
* Rinse in water for 15 seconds.
* Bleach for 1 to 2 minutes (or until the film clears).
* Rinse in water for 15 seconds.
* Fix in second fixer for 3 to 5 minutes.
* Wash thoroughly under running water for 5+ minutes.
* Dip in Photoflo for 15 seconds; squeegee.
* Dry for at least 30 minutes.

Alternate Method

A slightly less complex processing method can be used to make reflection holograms. Develop the film in D-19 for 2 to 5 minutes, then wash in running water for 5 minutes. Bleach the film using the sulfuric acid bleach described above for 2 minutes or until the film clears (becomes transparent). Wash another 5 to 10 minutes and dry it.

VIEWING REFLECTION HOLOGRAMS

Reflection holograms don't require a laser for image reconstruction. Just about any source of light will work, including sunlight or the light from an incandescent light. Avoid greatly diffused light such as that from a fluorescent lamp, or the hologram will look fuzzy. The ideal light source is a point-source, such as an unfrosted filament bulb. You will see the image as you tilt the hologram at angles to the light.

Note the many colors in the picture, particularly green. Although made with a red helium-neon light, the film shrinks after processing, so it tends to reflect shorter wavelength light. The amount of shrinkage varies depending on the film, but it often correlates to 50 to 100 nanometers, reducing the red 632.8 nm wavelength of the a helium-neon laser to about 500 to 550 nm.

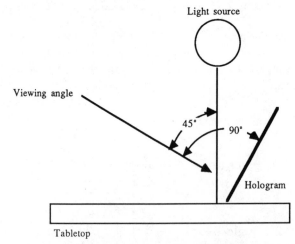

FIG. 18-5. *How to view the processed (and dried) reflection hologram in white light.*

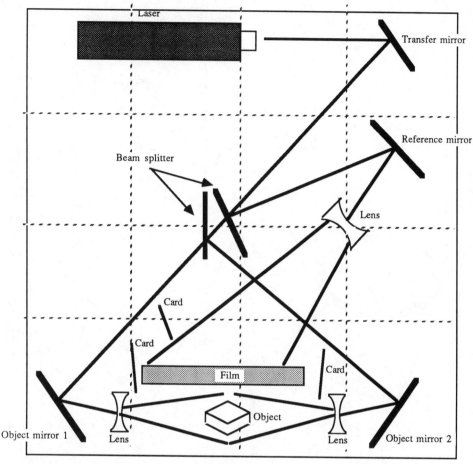

FIG. 18-6. *Optical arrangement (shown in the sandbox) for a multiple-beam reflection hologram. The object is illuminated on two sides, providing even lighting and better contrast.*

The best viewing setup for a reflection hologram is shown in FIG. 18-5. Place the light from a desk lamp straight down at the table. Tilt the hologram toward you until the image becomes clear (about 45 to 50 degrees).

IMPROVED REFLECTION HOLOGRAMS

There are a number of methods to make better and more interesting reflection holograms. One is to use a casting or mold for an object and place the mold directly behind the film. The hologram that results has an eerie 3-D quality to it where you can't tell if the object is concave or convex. Try it.

A split-beam reflection hologram provides more even lighting and helps improve the three-dimensional quality. FIGURE 18-6 shows one arrangement you can use. Except for the angle of some of the objects, it is nearly identical to the split-beam transmission hologram setup described in the last chapter.

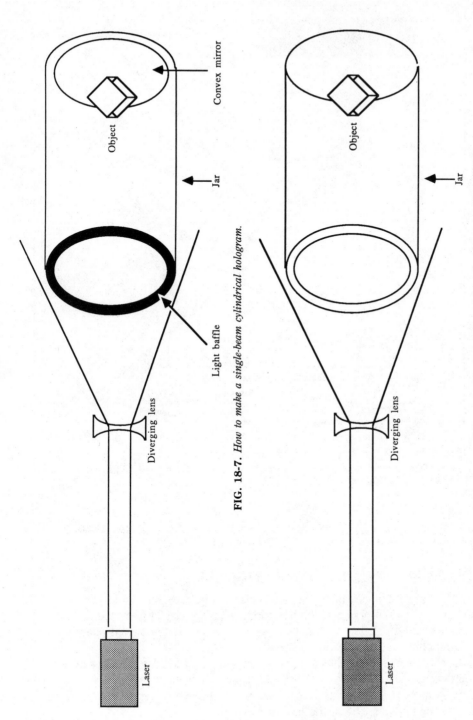

FIG. 18-7. *How to make a single-beam cylindrical hologram.*

Convex mirror

Object

Jar

Light baffle

Diverging lens

Laser

FIG. 18-8. *Another approach to the cylindrical hologram, without the mirror and light baffle.*

Object

Jar

Diverging lens

Laser

Table 18-3. Split-Beam Transmission Hologram Parts List

1	He-Ne laser
3	8 to 10 mm bi-concave lens
1	Holographic film in film holder
4	1- to 2-inch-square front-surface mirror
2	Plate beam splitter (50:50)
3	Black blocking card
1	Object

Table 18-4. Cylindrical Hologram Setup Parts List

1	Laser
1	8 to 10 mm bi-concave lens
1	3-inch diameter (approximately) clear glass or plastic jar (length 2 to 4 inches)
1	3-inch diameter (approximately) convex mirror
1	Object on pedestal
1	Film Strip
1	Light baffle

ADVANCED HOLOGRAPHIC SETUPS

A one-mirror cylindrical transmission hologram can be made using the basic arrangement depicted in FIG. 18-7 (refer to TABLE 18-4 for a parts list). You need wide film for this project; you can either buy it in 70-by-250-mm strips, or cut even larger film to size. The setup is relatively simple. A glass or plastic jar (approximately 3 inches in diameter) holds a convex mirror and the object. Inside the jar, wrap a strip of film so that the emulsion is facing inward. A light baffle made of black artboard, placed at the neck of the jar, prevents laser light from spilling directly onto the film.

In operation, the laser light directly illuminates the object as well as the mirror. The light from the mirror acts as the reference beam. The light from both object and mirror meets at all areas around the circumference of the jar, exposing the entire piece of film. The object shows up better if it is mounted on a pedestal. A small piece of wood painted flat black works well as a pedestal.

Process the film as usual, but be sure the trays are large enough to accommodate the film. To view the hologram, wrap it in a circle and shine an expanded laser beam through the top and against one inside wall. View the film by rotating it. You can support the film by placing it back inside the jar you used for making the hologram.

A reflection hologram is made by changing the convex mirror for a concave one (many such mirrors are coated on both sides so that they are concave/convex). To view the hologram, shine a point source of white light through the top and against one inside wall.

Any object that you can fit in the jar and place on the pedestal can be used for a cylindrical hologram. Good choices for first attempts include chess pieces, old coins (held upright on one edge), and playing dice. If the object doesn't sit the way you want on the pedestal, secure it in place using black modeling clay (not the best method) or Super Glue. Most objects will snap off even with Super Glue, particularly if you apply just a dab on top of the pedestal.

The biggest problem with this method is movement of the film during exposure. Wait at least 5 minutes after loading the film to expose it to allow time for the film to shift and settle. You might want to tap the jar once or twice after placing the film to help set it in position. Leave it alone while you take a coffee break (or better yet, take in a half-hour sitcom).

Direct Beam Cylindrical

Another type of cylindrical hologram doesn't use a mirror. The direct beam cylindrical arrangement shown in FIG. 18-8 makes transmission-type holograms of objects you've mounted on a black pedestal. See TABLE 18-5 for a parts list. The setup requires a well-expanded beam of laser light, so you might need to use two bi-concave lenses, positioned one after the other, to achieve the desired effect.

If you use non-AH film (film without an antihalation backing), you might want to cover the outside of jar with black construction paper to prevent the film from being fogged by stray light that strikes the outside of the jar. Once again, be sure to place the film securely in the jar. Place tape on the top portion of the film to keep it from moving; wait at least 10 to 15 minutes for the film to settle.

Directing the Beam

It is easiest to expose objects by placing the jar sideways on the sand table. That requires you to mount the pedestal and object to the mirror and secure the mirror to the base of the jar. When this isn't practical or desirable, you may place the jar upright and direct the beam into the jar with a large front-surface mirror.

While this approach works, it's not highly recommended, because dust, fingerprints, and other contamination on the mirror can upset the exposure. Reflected laser light is never as pure when it is expanded to cover a wide area. Whenever possible, laser light should should be unexpanded when bounced off mirrors.

Multiple-Channel Holograms

A transmission hologram must always be positioned so that the illumination from the laser is at the same angle as the reference beam when the picture was first taken.

Table 18-5. Cylindrical Hologram Setup Parts List

1	Laser
1	8 to 10 mm bi-concave lens
1	3-inch diameter (approximately) clear glass or plastic jar (length 2 to 4 inches)
1	Object on pedestal
1	Film strip

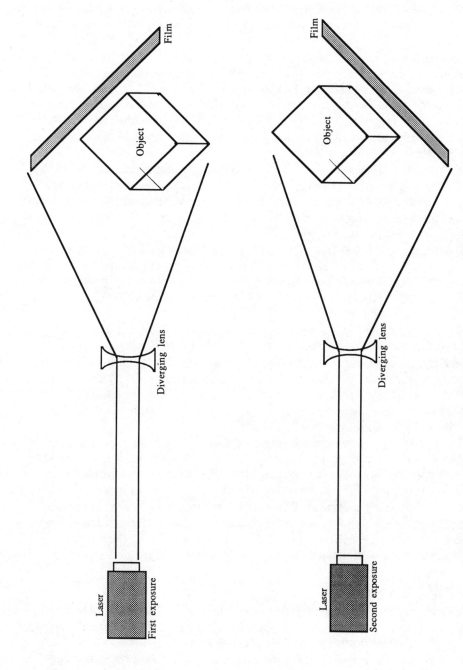

FIG. 18-9. *How to make a multiple-channel hologram. Shown are locations for the film holder for the first and second exposures of the same object.*

Canting the film at an angle makes the image disappear. While this is often considered a nuisance by beginning holographers, it's actually a considerable benefit because it means you can expose the same piece of film with an almost unlimited number of scenes! You see each scene by rotating the film or by shifting its angle relative to the illumination beam. This process is most often called multiple-channel holography, and it takes many forms.

In practice, you want to limit the number of exposures to two or four. The more scenes recorded on the film, the more chance they will interfere with one another. The objects for multiple-channel holography should be relatively small in comparison to the size of film you are using. A good rule of thumb for a two-channel hologram (one that has two separate images on it) is that the subjects should be about half the size of the film. In a four-channel hologram, the subjects should be approximately one-quarter the size of the film.

Other forms of multiple channel holograms, including multiplex still-film holography, can accommodate hundreds and even thousands of separate images when the area of exposure is controlled. *Multiplex still-film holography* (or *holographic stereograms*) use frames of movie film recorded as slits on the hologram. After processing, the film is wrapped in a circle and you see each "frame" as the slits roll by.

A famous example is the "Kiss," created by Lloyd Cross and Pam Brazier, where a woman blows a kiss, then winks. The technique is fascinating but beyond the scope of this book. For more information on this holographic technique (as well as numerous others), consult *Holography Handbook*, by Fred Unterseher, et al (Ross Books, 1982; see Appendix B for other titles).

To make a simple multiple-channel hologram, arrange the laser and optics as shown in FIG. 18-9. Position the film to produce a standard direct-beam transmission hologram. Make the first exposure but at about half the normal time. Shift the position of the film and change subjects (you can also use the same subject, if desired). Make the second exposure, again at about half the normal time. Be sure that the emulsion faces the object and reference beam for both exposures. For best results, the angle of the film should change by 60 to 90 degrees.

After processing, illuminate the hologram in the usual manner. Note how you see object A when the film is tilted one way, then object B when the film is tilted the other way. If you used the same object for both exposures and the angle of the film was roughly 90 degrees, rotating the film will reveal almost 180 degrees of the object.

Instead of physically moving the film, you can rotate it in its holder. For example, you can make a four-channel hologram by rotating the film 90 degrees for each new exposure (exposure time about one-fourth of normal). When viewing the hologram, you see the different views of different objects by spinning the film.

HOLOGRAM GALLERY

Below are several setups you might want to use to create a wide variety of holograms. Included are both transmission and reflection types, using single and multiple beams. In all cases, remember to follow standard holographic practices:

★ Allow time for the table and film to stabilize before taking the exposure.

★ When using glass-plate film holders, be sure the glass is perfectly clean. Press both pieces firmly together for about 30 seconds. Use blocks of wood to exert even

280

pressure. Remove all the trapped air, or the film might move during the exposure.

★ Be sure to place the film so that the emulsion faces the subject and/or reference beam. This isn't always necessary for reflection-type holograms, but it is a good habit.

★ Observe proper lighting ratios between reference and object beams. Generally, transmission holograms have a 3:1 or 4:1 ratio between reference and object beams (but up to 10:1 is sometimes required to eliminate noise); reflection holograms have a 1:1 or 2:1 ratio.

★ Measure distances for reference and object beams to ensure they are approximately equal. Use a cloth or flexible tape.

★ Use the proper chemicals mixed fresh (or stored properly), as per directions. Throw out exhausted chemicals—flush them down the sink and run plenty of water to wash away the chemical residue.

Holographic Interferometry

FIGURE 18-10 shows the basic setup for experimenting with holographic interferometry, a type of metrologic study where you can visually see how an object moves under stress. An interferometric hologram is made with two exposures: one where the object under test is "at rest," and other when it is "under stress." Slight differences in shape and structure can occur between the two states, causing movement that shows up in the two exposures. The amount and type of movement is clearly visible in the form of interference lines.

To make an interferometric hologram, take one exposure of the object (a ruler in the example) at about half-normal time—that is, if the usual exposure is 3 seconds reduce it to 1.5 seconds. Then, without disturbing anything, apply stress (the weight) to the object and after allowing time for settling, take another exposure. It is important that you do not disturb anything in the setup other than to carefully apply the weight.

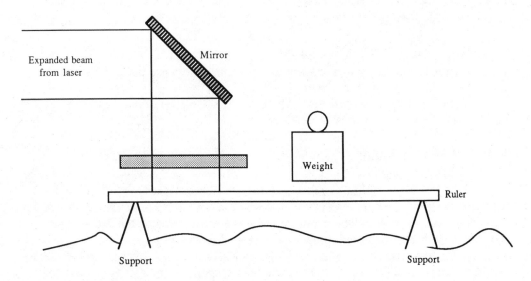

FIG. 18-10. *The basic arrangement for making interference holograms.*

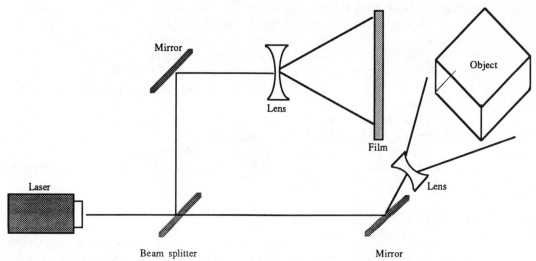

FIG. 18-11. *An alternative scheme for producing a multiple-beam reflection hologram.*

If the ruler, film, or other components shift even a few millionths of an inch, the hologram will be ruined. Don't expect to get this one right the first time. It requires a great deal of patience and careful arrangement.

If the ruler doesn't prove a cooperative subject, try another. One good choice is a small C-clamp pressed tightly around a steel or plastic block. Take one picture with the clamp tightened, but not overly so, on the block. Cinch down on the clamp and take another exposure. You will see stress marks around the C-shaped mouth of the clamp as well as at the point of contact on the block. Other possible subjects include a light bulb in both cold and hot states (turn the bulb on for a few minutes between exposures, but be sure to block the film from exposure), and the strain on a piece of metal from an increased electromagnetic current.

Multiple-Beam Reflection Hologram

The setup in FIG. 18-11 shows another arrangement you can use to make a multiple-beam reflection hologram. It is a more simple approach to the multiple beam plan described earlier in this chapter and doesn't provide the same even lighting of the object, but it's easy to arrange for classroom study.

Soft Lighting Technique

Transmission holograms often suffer from high-contrast lighting where the shadows have little detail. While this can enhance the three-dimensional quality of the image, the lighting effect is unnatural. FIGURE 18-12 shows how you can soften the lighting to achieve a less dramatic appearance. The reference beam is directed towards the film as usual, but the object beam is diffused using opal or frosted glass.

Note that the glass acts as a mirror on the back side so you should add black cards as necessary to avoid light spills. Expand the beam slightly before striking the back of the glass to avoid a hot spot in the center. Try exposure times slightly longer than normal.

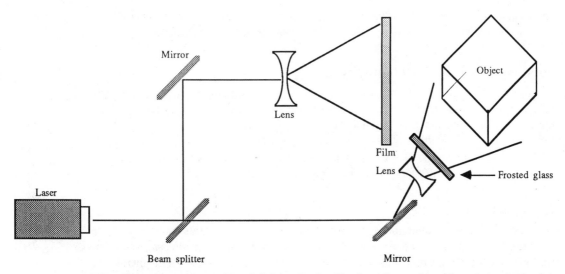

FIG. 18-12. *One way to provide soft lighting for the object is to place a piece of frosted glass in the path of the expanded object beam.*

Downward Facing Hologram

A simple form of single-beam transmission holography entails pointing the film toward the sand. The object must be relatively small and should be lightly colored, because shadows will be heavy (the object is back-lit only).

MISCELLANEOUS IDEAS AND TIPS

Here are a collection of ideas and tips you might find useful.

Holography Groups

Holographers are generally a friendly group and like to share ideas and techniques. If you don't know other holographers in your area, ask at your local camera store or college (the photography, chemistry, or physics departments are good choices). One of the best ways to learn about holography is to watch someone else do it. Offer to help out on the next shoot. You'll gain useful practical knowledge that could otherwise take you years to uncover by yourself.

You will also find it less expensive if you join a holography group. If the group is well organized, it might have community-property equipment such as spatial filters, collimating mirrors, or even argon and krypton lasers. Or, look for someone who is willing to loan you the equipment you need for your next holographic masterpiece. Rental should either be free or modest, but in any case it will be considerably less than if you rented the items from a commercial outfit. You might even have gear that's worth sharing as a way of returning the favor.

Holography groups often put on their own public shows where others can come to wonder in astonishment "How do they do it?" With experience comes more professional results, and you might find yourself submitting your holograms for judging at regional shows and fairs (photo contents often have a special holography category). If prize money is involved, you can use it to help pay for your holographic habit.

Making Money With Holography

Besides contests (which are rare and usually don't provide much money), you might be able to turn your holographic expertise into cash. Consider making holograms of jewelry, gemstones, and other artifacts and selling them at swap meets or flea markets. You don't need to tend the booth yourself; be a distributor and sell them to other retailers. Of course, you can offer your wares to gift and card shops, but you are competing against national companies that specialize in low-cost, "general-purpose" holograms.

Holograms as Gifts

Though it doesn't offer monetary pay, you can make holograms for everyone for Christmas or other gift-giving seasons. Because you pick the subject, you can personalize the holograms for each recipient. Unless your friends and family own their own laser, however, you'll want to make reflection-type holograms so that they can be viewed under white or filtered light. You can complete your holographic offerings by placing them in frames or even housings that include a point light source (an unfrosted 15-watt nightlight works well).

Adding an Aperture Card

Most low-cost helium-neon lasers emit a faint glow around the periphery of the beam. Because this glow is low power, it doesn't often interfere with your holograms. But to ensure that the glow doesn't wreck a hologram that you've spent hours preparing, place a card with a hole in it in front of the laser. Make the hole large enough for the entire beam to pass through (about 1mm in diameter), but nothing else. Avoid making the hole too small or the beam will diffract as it passes through. A better approach to cleaning up the laser beam is to use a spatial filter. See Chapter 3 for details on what a spatial filter is and how it works.

Using View-Camera Film Backs

A view camera is a little more than a lens attached to a bellows. At the focal point of the lens is a removable ground glass. The photographer frames and focuses the subject on the ground glass and then takes it out and replaces it with a film back. A black slide inserted in the film back prevents the film from being fogged in daylight. The slide is removed, the exposure taken, and the slide is returned.

Film backs for view cameras come in various sizes and styles, and many can be adopted for use for transmission holograms. The backs are unique in that they use a pressure plate to hold the film steady. This is a real boon to the holographer who is constantly struggling to keep the film from moving during an exposure. Because of the pressure plate, however, film backs can't be used for reflective holography were the expanded laser beam must pass through the film. You can find film backs in larger photographic stores that cater to professionals. A used film back is just as good as a new one, so don't be shy about saving a few dollars.

Chemical Alternatives

Some holographers like to dilute the D-19 developer 4:1 (four parts water, one part developer) in order to improve the resolution of the film. Because the chemical is diluted, development takes about four times longer. Dilute the developer only as you need it.

If potassium ferrocyanide is not available for making bleach for transmission holograms, you may substitute cupric bromide. An alternative bleach for reflection holograms consists of 30 grams potassium bromide, 15 grams of borax, two grams of potassium dichromate, and two grams of p-benzoquinone (check photo or industrial chemical houses for the last ingredient). The mixture, although still poisonous, is safer than the mercuric chloride bleach mixture described earlier in this chapter.

Displaying Your Holograms

You don't need the sand table to view your finished holograms, so if you want to show your work to others, there is no need to lug around 2,000 pounds of luggage with you. Presentations and shows with transmission holograms can be simplified by building a portable laser hologram table. This table, which can be set up in 10 minutes or less, houses the laser, beam expanding optics, and hologram.

The table can simply be an optical breadboard (see Chapter 7) engineered for permanent use to display holograms. One way of arranging the laser and optics is shown in FIG. 18-13. A parts list for the table appears in TABLE 18-6. The beam-steering mirrors are used to "fold" the light path, enabling you to use a smaller table. The laser is placed on one side and the beam diverted to the middle rear of the table. One or two bi-concave lenses are used to expand the beam, which is directed toward the hologram. The amount of expansion depends on the size of the finished holograms. If your holograms are different

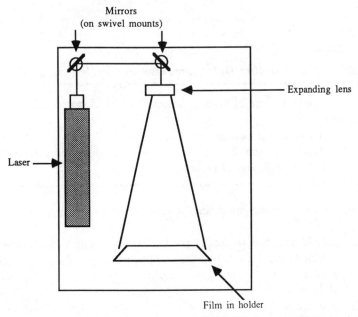

FIG. 18-13. *Top view of the laser, mirrors, lens, and film holder for a portable transmission hologram viewing table.*

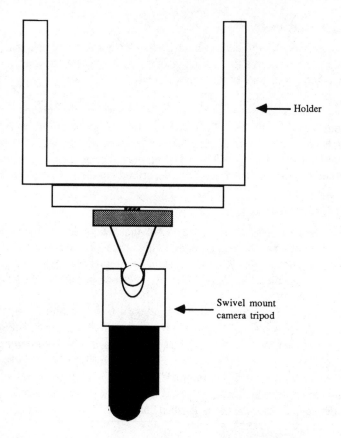

FIG. 18-14. *A swivel mount from a portable camera tripod can be used to position the film holder at angle to a laser or white light source.*

Table 18-6. Portable Hologram Viewing Table Parts List

1	18-by-24-inch piece of ¼-inch Masonite or ⅜-inch plywood
1	Laser
2	Mirrors on swivel mounts (see Chapter 7)
1-2	8 to 10 mm diameter bi-concave lenses, in mounts (see Chapter 7)
1	Film holder, with swivel mount

sizes, you can control the beam spreading by making the second lens removable and adjustable.

The film holder is designed so that it can be tilted in relation to the laser beam. A ball-and-socket joint from a low-cost portable tripod can be used to adjust the angle. Simply loosen the knob and tilt the film holder into the position you want. Lock the knob and the holder stays in place.

Although some people might want to see the inner workings of your holographic setup, you'll probably want to enclose it to block stray light. A frame made of plastic or metal can be constructed and placed over the table. Alternatively, you can position the table behind a partition and expose just the holographic film for viewing. Use pieces of black velvet or other material as drapes. This technique can be used when your holograms are on display at the fair, photo show, or museum.

An even simpler arrangement can be built for reflection holograms. The laser is not necessary, nor are the beam-expanding optics. You need a fairly bright point source of light, positioned above the hologram. The hologram should be mounted in a tiltable holder, as illustrated in FIG. 18-14, to allow you to set the proper angle of view. You can also motorize the holder so that the angle of the hologram shifts slowly, revealing its 3-D nature. An animation motor (designed for point-of-sale promotional signs), hooked up to a cam for the forward and backward motion, should do nicely.

Manual and Automatic Beam Shutter

Using a card as a shutter has its disadvantages: it's clumsy to use and you could actually disturb the table as you lift it out of the sand. Just being close to the sand table during the exposure can cause unwanted air movements that can upset the optics, beam, and film. And if you use a high-powered 8 to 10 mW He-Ne for your holograms, the light might be so strong that you only need exposure times of less than 1 second. With short exposure times come inaccuracy: it is hard to manually control the shutter card with accuracy of better than ½ second.

One way of providing more accurate shutter control is to use the shutter mechanism on a discarded camera. Use an old camera with a leaf-type shutter—the kind resembling an iris. Don't use a focal plane shutter such as those used in most 35 mm SLRs. You can find old cameras at flea marts and garage sales.

Inspect the camera to make sure that the shutter is still working, then take the camera home and dismantle it. Remove the shutter and mount it on a stand where you can easily reach the shutter speed control, release, and cocking mechanism (some shutters are self-cocking).

With the shutter, you can set the desired short-exposure time by dialing it in on the speed control knob, cocking the shutter, and pushing the shutter release. Most of the old shutters open and close with an unmistakable clang that can vibrate through the table and disturb the optics and film. Therefore, for best results, locate the laser and shutter off the table. As long as the laser is on a fairly sturdy mount, any slight movement will not affect the hologram (assuming the laser is well made and doesn't loose coherency when vibrated; a check with the Michelson interferometer can visually show you how well the laser maintains coherency).

Old camera shutters usually don't have settings longer than one second. For time exposures, you place the speed control ring in ''bulb'' position and manually control the duration that the shutter is open. Use a cable release and a darkroom timer to manually control the shutter.

A fully electronic shutter system is shown in FIG. 18-15. Here, a 12-volt solenoid is mounted on a piece of ⅛-inch thick acrylic plastic. The plunger of the solenoid connects to the shutter plate as well as a spring. When the solenoid is not energized, the shutter is at rest and blocks the beam. Energize the solenoid and the shutter moves, unblocking the beam. The finished solenoid shutter is shown in the photo in FIG. 18-16. A parts list for the solenoid shutter is provided in TABLE 18-7.

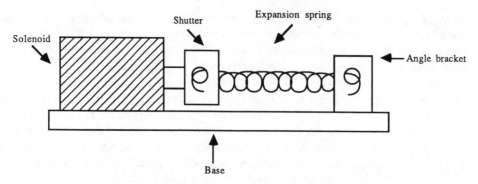

FIG. 18-15. *Layout for the solenoid shutter.*

FIG. 18-16. *The completed solenoid shutter.*

POWER/LIGHT METER

A photographic light meter can be used to test the output of a laser or to make sure that the beam ratios are correct for a given holographic setup. Best results are obtained using a well-made meter with a white diffusing filter placed over the sensing element. The filter prevents the laser light from forcing the sensor into non-linear operation. The filter isn't strictly needed when measuring expanded laser light.

You can make your own power/light meter using an ordinary solar cell and a volt-ohmmeter (VOM). Connect the solar cell to the voltmeter as shown in Chapter 4. Place a small piece of white diffusion over the cell. To use the meter, dial the VOM to the mV range, and position the cell in the path of the beam.

Table 18-7. Solenoid Shutter Parts List

1	12 Vdc solenoid, with mounting bolts or screws
1	½-inch-long expanded spring (length when relaxed)
2	1-by-½-inch corner angle irons
1	½-by-¾-inch plastic (for shutter)
1	⁶⁄₃₂-by-½-inch bolt, nut, washer

These past two chapters have only lightly covered the art and science of holography, leaving many topics untouched. If you would like to learn more about holography, refer to Appendix B for a list of selected books on the subject. One in particular, *Holography Handbook* (Unterseher, Hansen, and Schlesinger) offers an excellent tutorial in making many types of basic and advanced holograms.

19

Basic Laser Light Shows

Over 16 million people have seen the Laserium light show, and just about every one of those 16 million have gone home afterward wishing they could create the same kind of mind-boggling special effects. If you have a laser, you are already on the road to producing your own laser light shows. A small assortment of basic accessories is all you need to make dancing, oscillating shape on the ceiling, wall, or screen.

This chapter details some basic approaches to affordable laser light shows. You'll learn how to produce light shows using dc motors and mirrors that make interesting and controllable "Spirograph" shapes, how to make a laser beam dance to the beat of music, and how to make "sheet" and "cone" effects using mirrors and lenses.

The Laserium "laserists" use advanced components and lasers costing many tens of thousands of dollars. A few of these more sophisticated components are detailed in the next chapter (there, you'll discover the use of servos and galvanometers to control the laser beam, how to make exciting smoke effects, and ways to use argon, krypton, and other laser types to add more colors to the show).

THE "SPIROGRAPH" EFFECT

Imagine your laser drawing unique, "atom-shaped," repeating spiral light forms, with you adjusting their size and shape by turning a couple of knobs. The "Spirograph" light show device (named after the popular Spirograph drawing toy made by Kenner) uses three small dc motors and an easy-to-build motor speed and direction control circuit.

Depending on how you adjust the speed and direction of the motors, you alter the shape and size of the spiral light forms. And because the motors used are not constant

Table 19-1. "Spirograph" Light Show Device Parts List

3	Small 1.5 to 6 Vdc hobby motor
3	Lincoln penny
3	1-inch diameter or square, thin, front-surface mirror
3	¾-inch electrical conduit pipe hanger
3	$^{10}/_{24}$-by-¾-inch bolt, flat washer, tooth lock washer
6	$^{10}/_{24}$ nut
1	8-by-24-inch pegboard (¼-inch thick)
2	24-inch lengths of 2-by-2-inch framing lumber
2	4-inch lengths of 2-by-2-inch framing lumber

speed, slight variations in rotation rate cause the light forms to pulse and change all on their own. A complete parts list for the "Spirograph" light show device is in TABLE 19-1.

Mirror Mounting

Got a penny? That and a little bit of glue is all you need to mount each mirror to a motor. The best motors to use are the 1.5- to 6-volt dc hobby motors that are made by Mabuchi, Johnson, numerous other companies and are sold by Radio Shack and most every other electronics outlet in the country. Measure the diameter of the shaft; it can vary depending on the manufacturer and original application for the motor. Then drill a hole in the exact center of a penny using a bit just slightly smaller than the motor shaft.

Use a drill press to hold the penny in place and to prevent the bit from skipping. You'll find drilling easier if you turn the coin over and position the bit in the middle column of the Lincoln Memorial (for a penny less than about 25 years old). Note that the newest pennies are easiest to drill. Don't worry; the hole can be off a few fractions of an inch, but it should not be larger than the motor shaft. If anything, strive for a press fit. File away the flash left by the bit so the surface of the penny is smooth.

Next, apply a drop of cyanoacrylate adhesive (Super Glue) to a 1-inch-square or diameter mirror to the center of the penny. Best results are obtained when using a fairly thin mirror and gap-filling glue. The Hot Stuff Super "T" glue made by HST-2 (available at hobby stores) is a good choice. Wait an hour for the adhesive to dry and set. Repeat the procedure for the other three mirrors.

Avoid gaps between the mirrors and pennies. Although a small amount of misalignment is desirable, a large gap will cause excessive beam displacement when the motor turns. You'll see exactly why this is important once you build the Spirograph light show device.

Finally, mount the penny and mirror on the end of the motor shaft, as depicted in FIG. 19-1. Apply several drops of adhesive to the shaft and let it seep into the hole in the penny. Wait several hours for the adhesive to set completely before continuing. Alternatively, you can solder the penny to the shaft. This requires a heavy-duty soldering iron or small, controllable torch. Mount the penny on the motor shaft first, then tack on the mirror.

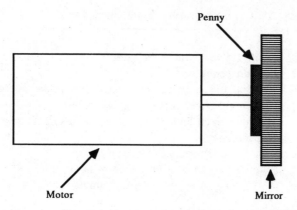

FIG. 19-1. *Mirror and penny mounting detail for the dc motors used in the "Spirograph" light show device.*

Mounting the Motors

Ideal motor mounts can be made with 1-inch plumbing pipe hangers, sold at a plumbing supply outlet or hardware store. The hanger is made of formed, u-shaped metal with a mounting hole on one end and an adjustable open end at the other (many other styles can also be used). Secure the motor in the hanger by loosening the bolt on the end, slipping the motor in, and then finger-tightening the bolt.

Secure the hanger on an 8-by-24-inch piece of ¼-inch hardwood pegboard, as shown in FIG. 19-2. Add wood blocks to the underside of the pegboard to make an optical breadboard, as explained in Chapter 7, "Constructing an Optical Bench." Arrange the hangers as shown in FIG. 19-3, and lightly secure the hangers to the pegboard using ¹⁰⁄₂₄-by-½-inch bolts and matching hardware. Use flat and split washers as indicated in FIG. 19-2 to prevent movement when the motors are turning (and vibrating).

Building the Motor Control Circuit

The motor control circuit allows you to individually control each motor. You have full command over the speed and direction of each motor by flicking a switch and turning a dial.

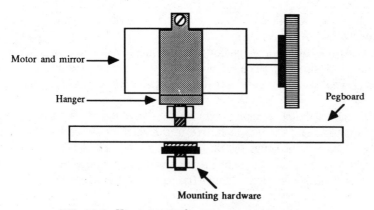

FIG. 19-2. *How to mount the motors to a pegboard base.*

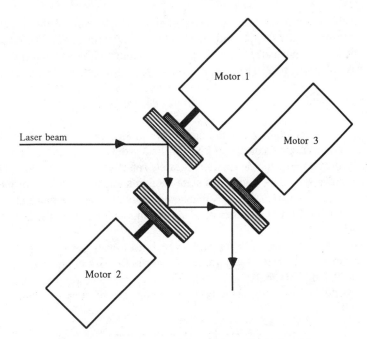

FIG. 19-3. *Arrangement of the motors on the pegboard base.*

The schematic for the motor control circuit is shown in FIG. 19-4. The illustration shows the circuit for only one motor; duplicate it for the remaining two motors. The prototype used a 3½-by-4-inch perforated board and wire-wrapping techniques. Your layout should provide room for the electronics, switches, potentiometers, and transistors on heatsinks (the latter are very important). Lay out the parts before cutting the board to size. Power is provided by a 6 Vdc battery pack consisting of four alkaline "D" cells.

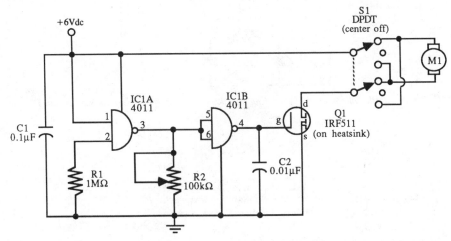

FIG. 19-4. *An easy-to-build pulse width modulation speed control and direction switch for a small dc motor. Q1 must be mounted on a heatsink.*

The double-pole, double-throw switches allow you to control the direction of the motors or turn them off. The potentiometers let you vary the speed of the motors from full to about ½ to ⅔ normal. Different speeds are obtained by varying the "on" time, or duty cycle, of the motors. The more the duty cycle approaches 100 percent, the faster the motor turns. The design of this circuit does not allow the motors to turn at drastically reduced speeds which, in any case, is not really desirable to achieve the spiral light-form effects.

Alternative speed control circuits are shown in FIGS.19-5 through 19-7. FIGURE 19-5 details a similar control circuit using 2N3055 heavy-duty transistors. This circuit also works on the duty-cycle principle (more accurately referred to as pulse width modulation). These transistors should be placed in a suitable aluminum heatsink with proper case-to-heatsink electrical insulation. The 2N3055s and heatsinks require more space than the IRF511 power MOSFET transistors used in the schematic outlined earlier, so make the board larger.

FIGURE 19-6 shows a basic approach using 2- to 5-watt power potentiometers. Be sure to use pots rated for at least 2 watts, or you run the risk of burning them out. Bear in mind that the potentiometer approach consumes more power than the pulse width modulated systems. No matter how fast the motors are turning, a constant amount of

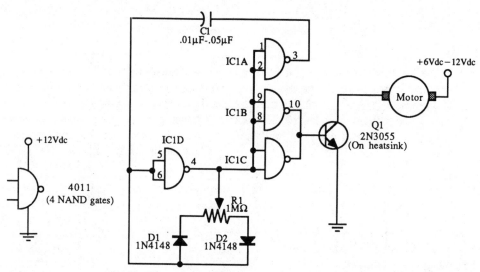

FIG. 19-5. *An alternate method of providing pulse width modulation using a 2N2055 power transistor.*

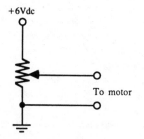

FIG. 19-6. *How to connect a high-wattage rheostat or potentiometer to control the speed of a dc motor.*

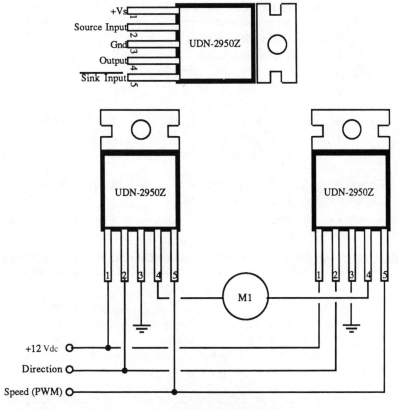

FIG. 19-7. *Hookup diagram for using a Sprague UDN-2950Z half-bridge motor driver IC. Use the output of the circuits in* FIG. 19-4 *or* 19-5 *to supply the Speed (PWM) signal.*

current is always drawn from the batteries. Current not used by the motors is dissipated by the potentiometer as heat.

FIGURE 19-7 shows speed and direction control using a Sprague motor control IC (others are available). You can obtain the IC through most Sprague reps as well as from Circuit Specialists (see Appendix A for sources). Use the pulse width modulation circuit shown in FIG. 19-4 or 19-5 and apply to the speed pin (pin 5) of the UDN-2950Z IC. Parts lists for the three alternative speed control circuits are provided in TABLES 19-2 through 19-4.

Mount the circuit board on one end of the optical breadboard using ¹⁰⁄₂₄ by ½ bolts and ¹⁰⁄₂₄ nuts. Make sure there is sufficient space between the bottom of the board and the wire-wrap posts and top of the optical breadboard.

Mounting the Laser

Cut a 3-inch length of 2-inch plastic PVC pipe lengthwise in half. Using ¹⁰⁄₂₄-by-½-inch bolts and hardware, mount the two halves on the optical breadboard. Insert a 2½- to 3-inch diameter worm-gear pipe clamp under the PVC half before tightening the nuts and bolts. Use silicone adhesive to attach rubber feet above the four bolt heads. The

Table 19-2. Motor Control Circuit Parts List (Each Motor)

IC1	4011 CMOS NAND gate IC
R1	1 megohm resistor
R2	100 kilohm potentiometer
C1	0.1 μF disc capacitor
C2	0.01 μF disc capacitor
Q1	IRF511 MOSFET power transistor (or equivalent)
S1	DPDT switch, center off
Misc.	Heatsink for Q1

All resistors are 5 to 10 percent tolerance, ¼ watt. All capacitors are 10 to 20 percent tolerance, rated 35 volts or more.

Table 19-3. Alternate #1 Motor Control Circuit Parts List (Each Motor)

IC1	4011 CMOS NAND gate IC
R1	1 megohm potentiometer
C1	0.01 or 0.05 μF capacitor
D1,D2	1N4148 diode
Q1	2N3055 power transistor
Misc.	Heatsink for Q1

All resistors are 5 to 10 percent tolerance, ¼ watt. All capacitors are 10 to 20 percent tolerance, rated 35 volts or more.

Table 19-4. Alternate #2 Motor Control Circuit Parts List (Each Motor)

IC1,2	Sprague UDN-2950Z half-bridge motor driver
1	Speed control circuit (minus drive transistor) from Fig. 19-4 or 19-5

rubber feet serve as a cushion between the laser tube and bolts, as well as to increase the height of the beam over the optical bench. More details on this and other laser-mounting methods can be found in Chapter 7.

The PVC pipe laser holder is designed for a cylindrical laser head. If you are using a bare laser tube, install it in a suitable enclosure as detailed in Chapter 6, "Build a He-Ne Laser Experimenter's System."

Aligning the System

Slip the laser into the holder and tighten the two clamps. The distance between the front of the laser and the motors is not critical, but be sure that there is no chance that the mirror on the first motor will touch the end of the laser. Turn the laser on but do not apply power to the motor control circuit. Rotate the hangers on each motor so that the beam is deflected from mirror 1 to 2 to 3.

Fine-tune the alignment by rotating each motor 90 degrees. The misalignment inherent in the mirror mounting should displace the beam on the mirrors. Avoid fall-off where the beam skips off the mirror. Beam fall-off causes a void in the spiral when the motors turn.

If you cannot align the motors so that the beam never falls off the mirrors, check the gap between the mirrors and pennies. Place the motor with the largest gap at the end of the chain as motor number 3. If beam fall-off is still a problem, try mounting a mirror and penny on a new motor.

Place all three switches to their center position and apply power to the motor control circuit. Flick switch #1 up or down and rotate the potentiometer. The motor should turn. If the motor whines but refuses to turn, flick the switch off, turn the pot all the way on, and reapply power. The motor should turn.

Test the speed control circuit by turning the pot. The motor should slow down by an appreciable amount (you'll be able to hear the decrease in speed). If nothing happens, double-check your work. A motor that won't change speed could be caused by improper wiring or a bad transistor. A blown transistor could cause the motor to spin at about a constant 80 percent of full speed. Next reverse the motor by moving the switch to the opposite position. The motor should momentarily come to a halt, turn in its tracks, and go the other way.

Turn the first motor off and repeat the testing procedure for the other two. After all motors check out, turn them back on and point the #3 mirror so that the beam falls on a wall or screen. Watch the spiral light form as all three motors turn. Do you notice any beam fall-off—if so, *stop all the motors* and readjust them. Note that the motors vibrate a great deal at full speed, and that can cause them to go out of alignment. When you get the motors aligned just right, tighten the hangers to prevent them from coming loose.

Test the different types of light forms you can create by turning off the #1 motor and using just #2 and #3. Depending on how the direction is set on the motors, you should see an "orbiting atom" shape on the screen, as depicted in FIG. 19-8. If the form looks more like constantly changing ellipses, reverse the direction of one of the motors. Adjust the speed control on both motors and watch the different effects you can achieve. Now try the same thing with motor #1 and #3 on. Try all the combinations and note the results.

What happens if the light form doesn't show up or appears very small, even when the screen is some distance from the light show device? This can occur if the mirror is precisely aligned with the rotation of the motor. Although this is rare using the construction technique outlined above, it can happen. You can see how much each motor contributes to the creation of the light form by turning on each one in turn. You should see a fairly well-formed circle on the screen. The mirror is too precisely aligned if a dot appears instead of the circle. Replace the mirror and motor with another one and try again.

Note that the size of the circle does not depend on where the beam strikes the mirror. The circle is the same size whether the beam hits the exact center of the mirror or its edge.

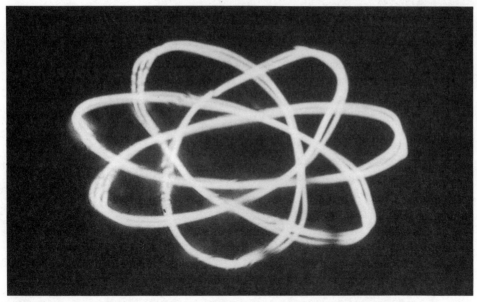

FIG. 19-8. *The "atom" laser lightform made with the "Spirograph" device.*

Notes On Using and Improving the "Spirograph" Device

Here are some notes on how to get the most from the spiral light-form device:

★ Keep the mirrors clean and free of dust or the light forms will appear streaked and blurred.

★ *Never* adjust the position of the motors when they are turning. The mirrors are positioned close together, and moving the motors could cause the glass to touch. The mirrors will then shatter and fragments of glass will fly in all directions. It is a good idea to use protective goggles when adjusting and using the spiral light show device.

★ Cheap dc motors like those used in this project make a lot of noise. You might want to use higher quality motors if you plan on using the "Spirograph" maker in a light show. Get ones with bearings on the shaft. You can also place the device in a soundproof box. Provide a clear window for the beam to come out.

★ The "Spirograph" device is designed for manual control. With the right interface circuit, you can easily connect it to a computer for automated operation. The motor direction and speed control circuit used in this project is similar to the robotic control schemes outlined in my book *Robot Builder's Bonanza* (TAB BOOKS, catalog number 2800). Refer to it for ideas on how to control motors via computer.

★ You can obtain even more light forms by adding a fourth motor. Try it and see what happens.

★ Don't be shy about turning some of the motors off. Some of the most interesting effects are achieved with just two motors.

SOUND-MODULATED MIRRORS

In the early seventies, during the psychedelic light show craze, Edmund Scientific Company offered an unusual device that transformed music into a dancing beam of light.

The system, called MusicVision, was simple: Thin front-surface mirrors were attached to a sheet of surgical rubber. The rubber sheet was then pulled taut across the front of an 8- or 10-inch woofer. A projector was positioned off to one side so that it cast one or more beams on the mirrors mounted on the rubber.

When the hi-fi was turned on, the speaker would move, vibrating the rubber sheet and causing the mirrors to bob up and down. The beam of light from the projector would follow the mirror, projecting an undulating and constantly changing pattern on a wall or the back side of a rear-projection screen. A filter wheel added color to the light shapes, which then colorfully bounced and jumped in time to the music.

Imagine what would happen if you replaced the projector and color wheel with a laser. Point the thin beam of a laser at the mirror and you get a projected image of it on the wall doing a dance. There are numerous ways to build sound-modulated mirror systems. Below are just a few of them; you are free to experiment and come up with some of your own.

The Old Rubber Sheet Over the Speaker Trick

A simple yet effective light show instrument can be made using a small mirror, a sheet of rubber, and a discarded peanut can. This design, with parts indicated in TABLE 19-5, comes from laser light show designer and consultant Jeff Korman, who calls it "PeanutVision." Using an all-purpose adhesive, mount a thin front-surface mirror—measuring approximately ½ inch in diameter—onto a 6-inch square sheet of surgical rubber.

Surgeon's gloves (available at many surplus stores) are a good source of surgical rubber, but a better choice is to use flat squares of the stuff. Check an industrial supply outlet and don't be afraid to improvise. If the rubber dates some time back, however, check to be sure it's still in its protective wrapper. The rubber dries out in time when exposed to air. In a pinch, you can get by using the rubber from balloons, but if possible, use thin-walled balloons.

While waiting for the adhesive to dry, drill 5 to 10 small holes in the metal end of a Planter's Peanuts can. Mount a 3-inch round speaker over the holes. Use small hardware, epoxy, or glue to hold the speaker in place.

Stretch the rubber over the open end of the can. Pull the rubber tight while making sure that the mirror is placed in the approximate center of the opening. Wrap one or more rubber bands around the sheet to secure it, as indicated in FIG. 19-9. Solder a pair of wires to the speaker terminals and connect the leads to a low-wattage stereo or hi-fi. Unless you use a high-capacity rated speaker, don't connect it to a stereo system that delivers more than a few watts—otherwise you'll burn out the voice coil in the speaker.

Table 19-5. PeanutVision Parts List

1	Peanut can measuring approx. 4 inches in diameter by 3¾ inches tall
1	2- to 3-inch diameter speaker
1	Latex rubber; approximately 6 inches square
1	Thin, ½-inch diameter, front-surface mirror
1	Rubber band

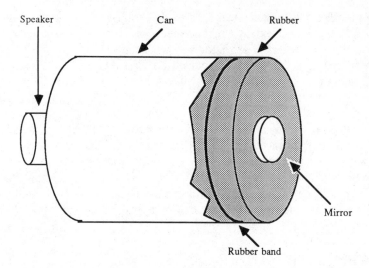

FIG. 19-9. *How to build the PeanutVision, using a peanut can, rubber sheet, mirror, and speaker.*

Connecting the light show speaker in parallel with the main speakers of the hi-fi reduces the chance of burnout, but it also changes the output impedance and might affect the sound. Instead of the usual 8 ohms of impedance of regular hi-fi speakers, adding the light show instrument in parallel reduces the output impedance to 4 ohms (assuming you use an 8-ohm speaker in the light show instrument). This generally causes no damage, but the sound quality of the stereo could be affected.

Use the sound-modulated mirror system by attaching it to a swivel mount and frame. Place the frame in front of the laser and adjust the swivel until the beam strikes the mirror and is reflected to a back wall or screen. The further the wall is from the instrument, the larger the beam pattern will be. You can also control pattern size by adjusting the volume. Again, be careful you don't turn up the volume too much, or the speaker could be ruined.

Enhancing the Light Show

Your light shows look even better if you are the proud owner of an argon, krypton, green He-Ne, helium-cadmium, or other non-red, visible-light gas plasma laser tube. The argon and krypton lasers produce light with many distinct wavelengths. These are mainlines and can be separated by using an equilateral prism or dichroic filters. Send each beam to a different PeanutVision system. An argon laser with its 488.0 and 514.5 mainline beams separated can be used with two sound-modulated mirrors. Alter the appearance of the light forms by feeding the right channel to one mirror and the left channel to the other.

Alternatively, you can add a filter to the sound output of your hi-fi and separate the highs from the lows. Route the high-frequency sounds to one mirror and the lows to the second mirror. Position the mirrors so that the beams converge on the screen. The colors will appear as if they are dancing with one another. Each has its own dance steps, but both are moving together to the beat of the music.

Direct Mirror Mounting

Instead of mounting the mirror on a sheet of surgical rubber, mount it directly onto the speaker cone. The best mounting location is in the center, above the voice coil. If the mirror is thin enough (0.04 inch or so), its mass won't overload the speaker and it should vibrate in unison with the sound.

Frequency response using this mounting technique is excellent—almost the frequency response of the speaker itself. Note that higher frequency sounds don't move the speaker cone and mirror as much as low-frequency sounds so the visual beam pattern effect is more marked at low frequencies.

If you want the visuals only and don't want the speaker to emit sound, you can reduce its audio output by carefully cutting away the cone material. Use a razor blade to cut the cone at the outer edges. Next, cut out the inside where the cone attaches to the voice coil, but keep the *spider*—the portion attaching the voice coil to the frame of the speaker—intact. The electrical connection for the voice coil might be physically attached to the outside of the cone, so be sure to leave this part attached.

The speaker will still produce sound, even with all or most of the cone removed. However, the sound level will be low and not generally audible when the room is filled with music from the main sound system. If sound from the speaker is a problem, mount it in a small wooden box. Fill the box with fiberglass padding (the kind made for speaker stuffing), and provide a clear window for the laser beam. Or, you can make the speaker and laser self-contained by making the box large enough for the tube. Keep the fiberglass away from the tube and add one or two small holes for ventilation.

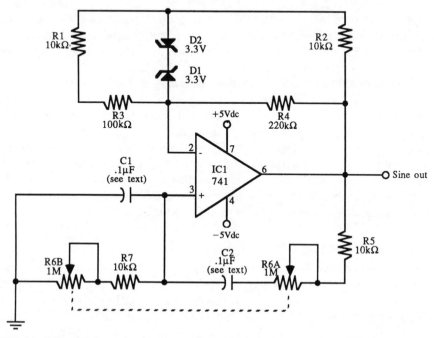

FIG. 19-10. *Adjustable-frequency sine-wave oscillator. For lower frequency tones, increase value of C1 and C2 (make them the same). R6 is a dual-ganged 1-megohm precision potentiometer.*

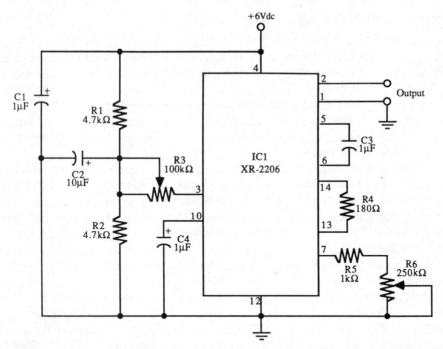

FIG. 19-11. *How to build a sine-wave generator using an XR-2206 function generator IC.*

The soundproof box comes in handy if you are controlling the beam with an audio oscillator. The oscillator, operating under your control, produces a buzzing or whining noise that is distracting when accompanied with a music soundtrack. With the speaker systems stuffed in the box, the oscillator noise will be largely inaudible (unless you are standing right next to the box).

Schematics for useful audio oscillators appear in FIGS. 19-10 and 19-11. Both oscillators produce sinusoidal ac signals that cause the speaker cone to move both in and out from its normal centered position (a waveform that is positive only moves the cone all the way out but not in). The op-amp circuit shown in FIG. 19-10 is the cheapest to build but requires a healthy assortment of parts. The circuit in FIG. 19-11 is based around the versatile Exar XR-2206 monolithic function generator. Many mail-order electronics firms, such as Circuit Specialists, offer this chip. Its higher cost (about $4 to $5) is offset by the minimum number of external components required to make the circuit function. TABLES 19-6 and 19-7 include parts lists for the generators.

Bell-Crank Mounting

Both speaker systems outlined above cause the mirror to bob up and down, thereby moving the beam across a wall or screen. Moving the apex of the mirror back and forth—in an arc motion—produces better pattern effects. You can build a sound-modulated mirror system using a speaker, mirror, and a model airplane plastic bell-crank. The bell-crank and other mounting hardware are available at hobby shops that carry radio control (R/C) model parts.

You can use the speaker as-is or remove the cone, as described above. Devise the crank and mount as shown in FIG. 19-12. Use a gap-filling cyanoacrylate glue to bond the parts to the metal speaker frame. Note: be sure the bell crank and other model parts are not made of nylon; these don't adhere well to any glue.

After the glue used to attach the mirror and mount has set (allow at least 30 minutes), secure the frame of the speaker to a tiltable stand. Position the speaker so that the laser beam glances off the mirror at a 45-degree angle. The mirror should rock the beam right and left (or up and down, depending on how the laser and speaker are arranged). Some "sloppiness" is inherent in this and other sound-modulated mirror systems. The beam will not trace a perfect line as it moves back and forth.

The beam pattern using just one mirror is more-or-less one-dimensional. You can create two-dimensional patterns using two mirror/speaker systems. Position the two speakers at 90 degrees off-axis to one another, and aim the laser so that the beam strikes one mirror and then the next.

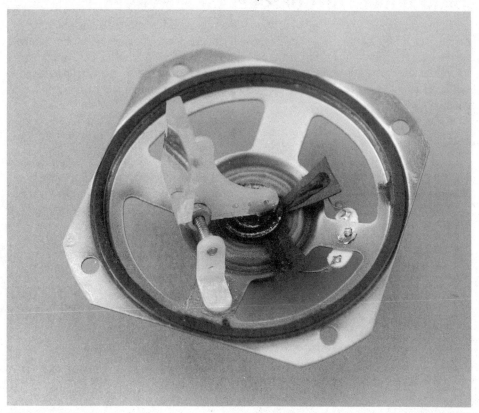

FIG. 19-12. *One way to attach a mirror to a speaker. The bell-crank arrangement, when used with the proper lever/fulcrum geometry, can greatly increase the deflection of the mirror.*

Table 19-6. Op-Amp Sine Wave Generator Parts List

IC1	LM741 op-amp IC
R1,R2	10 kilohm resistor
R5,R7	
R3	100 kilohm resistor
R4	220 kilohm resistor
R6	1 megohm dual precision potentiometer
C1,C2	0.1 μF silvered mica capacitor (capacitors must be closely matched for best operation)
D1,D2	3.3-volt zener diode

All resistors are 5 to 10 percent tolerance, ¼ watt. All capacitors are 10 to 20 percent tolerance, rated 35 volts or more, unless otherwise indicated.

SOUND MODULATION OF AIRPLANE SERVOS

Radio-controlled (R/C) model airplanes use motorized servo mechanisms for controlling such things as the rudder, ailerons, and landing gear. The servos, which connect to a central receiver on board the craft, consist of a miniature dc motor, a control board, a potentiometer, and a gear reduction system. All work together to provide closed-loop feedback, a system where the position of the servo arm is known and maintained at all times.

How Servos Work

The servo operates from a 4.5 to 8 volts dc source. That provides power to the motor and circuitry. To actuate the servo, the receiver (acting under command from the radio control transmitter), sends a series of pulses. The width of the pulses varies from 1.0 to 2.0 milliseconds and determines the direction and distance of travel.

Table 19-7. Function Sine Wave Generator Parts List

IC1	Exar XR-2206 function generator IC
R1,R2	4.7 kilohm resistor
R3	100 kilohm potentiometer
R4	180 kilohm resistor
R5	1 kilohm resistor
R6	250 kilohm potentiometer
C1,C2,C4	1 μF electrolytic capacitor
C2	10 μF electrolytic capacitor

All resistors are 5 to 10 percent tolerance, ¼ watt. All capacitors are 10 to 20 percent tolerance, rated 35 volts or more.

When a pulse is received, the servo circuit actuates the motor, which turns the gearing system as well as an output potentiometer. The position of the potentiometer wiper indicates the position of the servo arm (connected to some linkage on the aircraft). The servo circuit monitors the position of the potentiometer and turns off the motor when the pot reaches a given point. Very fine movement—much less than 1 degree of revolution—is possible with most servo systems.

Although R/C servos are meant to be used with the proper type of receiver, you can rig up your own actuating circuit using only a handful of components. By applying an audio input to the circuit, you can make the servos dance back and forth to the music. Laser light is deflected by a mirror mounted on the end of the servo arm. Note that servos are not as nimble as other light-show devices (such as the galvanometer described in the next chapter), but they can be used to create interesting "sweeping scan" effects. Using two motors placed at right angles to one another lets you create two-dimensional light forms.

Building the Sound Servo Systems

FIGURE 19-13 shows the circuit for the sound-modulated servo (parts list in TABLE 19-8). The project is is designed around the 556 IC, a dual version of the venerable 555 timer chip (two timers in one integrated circuit). One half of the 556 provides a series of pulses, and the other half varies the width of the pulses based on the voltage presented at the modulation input. Potentiometer R2 provides a threshold adjustment that lets you find a suitable "mid-point" where the servo arm swings both clockwise and counter-clockwise when music is applied to pin 3, the modulation input.

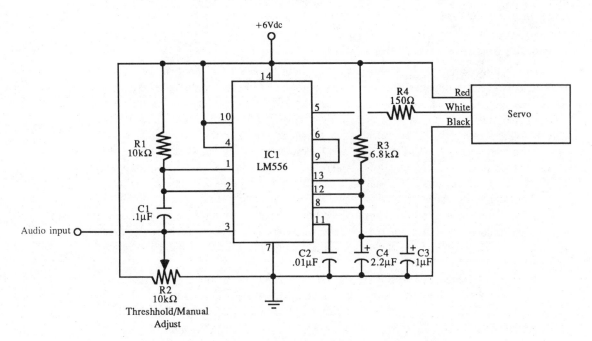

FIG. 19-13. *Schematic diagram for operating a model aircraft servo with amplified audio signals.*

Table 19-8. Sound Modulated Servo Parts List

IC1	LM556 dual timer IC
R1	10 kilohm resistor
R2	10 kilohm potentiometer
R3	6.8 kilohm resistor
R4	150 ohm resistor
C1	0.1 μF disc capacitor
C2	0.01 μF disc capacitor
C3	2.2 μF electrolytic capacitor
C4	1 μF electrolytic capacitor
M1	Servo motor

All resistors are 5 to 10 percent tolerance, $\frac{1}{4}$ watt. All capacitors are 10 to 20 percent tolerance, rated 35 volts or more.

Most R/C servos work the same, but a few odd-balls can present problems. The prototype circuit used Aristo-Craft Hi-Tek HS-402X servos, which are low-cost Korean copies of the popular Futaba servo motor. Capacitors C4 and C3, with resistor R3, determine the pulse width. If you don't get the results you want with the servo you use, try varying the values of these components.

The servo has three color-coded wires: red, white, and black. The red and black wires are the positive and ground leads, respectively. The white wire is the pulse lead, and connects to the output of the circuit. Build two identical circuits if you are controlling a pair of servo motors.

Using the Sound Servo System

Connect the output of an amplified music source to pin 3, the input of the circuit (a 500 mW to 1 watt amplifier provides more than enough power). Connect the circuit to the servo motor as indicated in the schematic. Turn up the volume on the amplifier and watch for a racking motion of the servo. If nothing happens or the servo immediately travels to the far end of its rotation and stays there, adjust R2 to modify the input voltage. If the servo moves to one extreme and makes a chattering noise, disconnect the power immediately. The chattering is caused by the gears in the gear train skipping. If allowed to continue, the gears will strip and the servo will be useless.

When adjusted properly, the servo should move back and forth in syncopation with the music. The amount of movement depends on the relative sound level of the music. The servo tends to react more to low-frequency sounds, which generally have a higher power content than higher frequency ones. The higher the volume, the more the servo will wiggle back and forth.

Note that the frequency response of the servo depends on the amplitude of rotation. The more the servo rotates, the lower the frequency response. If the servo is allowed to swing too far in both directions, the motor won't respond to changes in the music of more than 8 to 10 Hz. When the motor is set so that it slightly vibrates, frequency response is increased to a more respectable 30 to 50 Hz.

Mirror Mounting

Your local hobby store should stock a variety of plastic and hardware items that can be used to mount a suitable mirror on the servo. The output shaft of the servo is designed to accommodate a number of different plastic wheels, armatures, and brackets. You can glue the pieces together or use miniature 4/40 or 3/56 hardware (or whatever happens to be handy). Attach the mirror to the bracket using epoxy.

Positioning the Servos

Mount the servos on an optical breadboard (as discussed in Chapter 7) using the hardware provided with the servo or purchased separately. By mounting two servos at a 90-degree angle (one vertical and one horizontal) and positioning the mirrors so that the beam is deflected off one mirror and then the other, you gain complete control of the X and Y coordinates of the laser beam. If you provide each servo with a slightly different signal (left and right stereo channels, for example), you can create unusual lithesome patterns. Using active or passive filtration you can divert high-frequency sounds to one servo and low-frequency sounds to the other.

Remember that the servo is not really sensitive to frequencies, just the relative amplitude of the music generated by these frequencies. The servos respond best to such sounds as drums and bass and other low-frequency, short-duration instruments. Filter these out with a circuit that rolls off at about 300 to 500 Hz, and the servo will no longer respond to them but act on the amplitude of the remaining frequencies.

One of the best advantages of the sound-modulated servo system is it doesn't reproduce the music—unlike the speaker/mirror light-show instrument detailed earlier. This is especially important if you're putting on a light show. It can be disconcerting to an audience to hear the squeaky, raspy sounds of the speaker/mirror system along with the high fidelity of the auditorium audio system.

SIMPLE SCANNING SYSTEMS

Not all light-show effects are designed to bob with the music. Some effects are made by scanning the beam using prisms, mirrors, mirror balls, and other rotating reflecting optics. Depending on how you arrange the optical components and laser, you can create unique "sheet" and "cone" effects.

A sheet is a one-dimensional scan where the pinpoint laser beam is spread out in a wide arc. When projected on a screen, the beam draws out a long, streaking line. A cone is a three-dimensional scan where the beam is moved both up and down as well as right and left. When projected on a screen, the beam draws a circle or oval.

You need an extremely powerful laser (100 mW or more) to see the scanning effect in mid air, and then the beam is most visible when it shoots towards you, rather than away from you. As an example, light-show experts rig up mirrors or fiberoptics so the rays of laser light are directed toward the audience. Of course, the beams are aimed so that they don't actually strike anybody but are deflected to "beam stops"—flat-black fabric or metal baffles that prevent the beam from bouncing around the room.

Unless you fill the room with smoke or fake fog, the scanned beam from a 10 mW He-Ne is invisible, and even with the smoke it is extremely weak. Details on adding smoke can be found in Chapter 20, "Advanced Laser Light Shows."

Sheet Effects

There are three basic ways to create sheet-effect scanned images (more sophisticated approaches are shown in the next chapter).

☆ Reflect the beam off a mirror attached to the shaft of a motor, as shown in FIG. 19-14. The arc of the scan is approximately 170 degrees with a one-sided mirror (silvered on one side only).

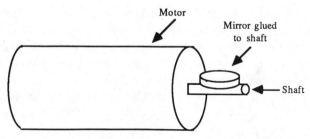

FIG. 19-14. *Mount a mirror on the shaft of a small dc motor as shown to produce a sweeping scan effect.*

☆ Bounce the beam off a holographic scanner, which is a specialized mirrored wheel used in laser-based supermarket checkout systems. The scanner is a wheel with mirrored or flat, polished edges. The number of facets on the outside of the wheel determine the arc of the scan.

☆ Pass the beam through a cylindrical lens. The lens expands the beam in one direction only. The angle of the arc is determined by the focal length of the lens. Most cylindrical lenses expand the beam to cover a 90- to 120-degree arc.

In all approaches, the intensity of the beam is reduced by a factor determined by the arc of the scan as well as any time the beam is stopped or blocked. Beam intensity is reduced the most with the one-sided mirror techniques. If you were to slow down the motor spinning the one-sided mirror, you'd see that the beam is not reflected for half the period of rotation (that is, when the beam strikes the back of the mirror). When the reflective side of the mirror faces the laser, the beam is directed outward in an arc. The beam intensity along the arc is only a fraction of what it is when the beam is stationary.

Holographic scanners must be precisely mounted on the motor shaft. Wobble of the scanning wheel causes multiple scan lines when the beam is projected. The multiple lines might be desirable when using certain smoke effects because the width (not arc) of the scan is increased. The scan width increases from the diameter of the actual beam to the distance between the far right and left lines.

The cylindrical lens does not suffer from excessive reduction in beam intensity or multiple scan lines. The beam is refractively widened into an arc so that no motors or mirrors need be used to provide the scanning action. The only requirement of the cylindrical lens is that its focal length must be carefully chosen if you desire a specific scanning arc.

Cone Effects

A cone effect is made by mounting a mirror off-axis on the shaft of a motor. A similar mounting technique was described for the "Spirograph" laser light show device, detailed at the beginning of this chapter. These mirrors are mounted slightly off-axis to produce a small circle shape on a screen. In the cone-scanning system, the mirror is mounted at a greater off-axis angle to produce a larger circle.

Altering the Speed of the Scan

With the exception of the cylindrical lens system, the scanning systems described here use motors that can be accelerated or decelerated as desired for a particular effect. Beyond a certain speed, the scanning rate is not detectable to the human eye and further speed increase is not necessary. This can help prolong the life of your motors as well as make the light show system quieter.

Both the one-sided mirror and cone systems use optical components that could present an uneven load on the motor shaft. That can lead to excessive noise and wear on motors that are not equipped with shaft bearings. You can reduce wear and noise by decreasing the speed of the motor without adversely affecting the visual effects of the scan. Use the motor speed circuits provided earlier in this chapter.

Slow "sweeping" effects can be achieved by reducing the motor speed to a crawl. Most speed control circuits cannot slow down a motor beyond a certain point without stalling the motor or causing the shaft to jerk instead of turn smoothly. If you can't get the motor to turn slowly enough, consider adding a gear reduction system to decrease the rotation of the mirror.

Sweeping scans can also be created using R/C servos. Even at top speed, the scan of one servo is slow enough to see, so the beam appears as a comet with a streaking tail. For a repetitive sweep, the servo circuit described earlier requires a low-frequency sinusoidal waveform. The oscillators depicted back in FIGS. 19-10 and 19-11 serve as excellent tone sources for the servos.

20

Advanced Laser
Light Shows

Want to put on a neighborhood light show that will dazzle your friends and family? Considering going into the professional laser light show business? Or just interested in experimenting with new art forms? The projects described in this chapter can turn you into a light show wizard. You'll learn how to make complex geometric patterns using devices known as galvanometers. You'll also find details on making your own galvanometers and the basics of how to effectively use argon and krypton lasers in light shows. Finally, this chapter provides important information on restrictions and rules governing public light show performances.

WHERE TO HAVE A LIGHT SHOW

Before discussing the hows of advanced light shows, let's take a moment to examine where to have them. Your choice of location goes a long way toward the overall enjoyment of the show and the ease with which you can produce it. You must consider the size of the room or auditorium, the location of screens or backdrops, the degree of light-proofing for doors and windows, and several other factors.

Even high-powered lasers look dim when they are used to create flashing beams and undulating light forms on the screen. A truly professional light show uses high-powered 2- to 5-watt argon and krypton lasers that cost $15,000 to $35,000. Unless you find one of these that has fallen off some truck (in which case you probably don't want it), you'll be using a trusty red helium-neon laser for your light shows. The higher the output of the laser, the brighter the beam. However, note that a 10 mW laser might not necessarily appear twice as bright as a 5 mW laser. Beyond a certain level the eye can no longer discern brightness. But the higher output tube will deliver more light as the beam is swept across the screen.

Speaking of screens, you must provide some type of light colored background or the laser beam may not be clear or easy to see. A beaded glass screen designed for movie projection is not a good idea because the little beads of glass act as prisms and mirrors. Not only will the beam be reflected back into the audience, it will appear fuzzy due to dispersion inside the glass.

Any light-colored wall, preferably one painted with flat white paint, will do. If such a wall is not handy, bring one in the form of a well-pressed sheet, a piece of photographic background paper (this stuff comes in convenient rolls), or a scenery flat. A flat, used in live theater, is a piece of painted muslin stretched taut in a wood frame. The flat is lightweight, but its size makes it hard to transport.

Obviously, the room or auditorium must be large enough to accommodate the number of people attending the show, but it must also be spacious enough to allow the light show pattern to spread to a respectable size. The distance between the laser and the projection surface is called the *throw*. The longer the throw, the larger the light show image. An image that is 1 foot high at a distance of 6 feet will measure 2 feet high at a throw distance of 12 feet. An easy rule of thumb is that every time you double the throw, the image size increases by 100 percent.

Because the laser beam is so compact, the effects of the inverse square law are minimized (recall from Chapter 3 that the inverse square law requires the intensity of light to fall off 50 percent for every doubling of light-to-subject distance). It doesn't really matter if the laser is 10 feet from the wall or 20 feet, but remember the effects of a long throw. Too much distance can cause the light show patterns to fan out excessively. Beam divergence also becomes a problem at long distances. A laser beam with a divergence of two milliradians will diverge to a spot approximately two inches in diameter at 80 feet.

How will you seat the audience? The conventional chairs- facing-the-screen seating arrangement is only marginally useful in laser light shows. Projecting the laser beam like a movie image requires that the laser projector be placed high above the audience, and your room might not easily allow for this.

EXPERIMENTING WITH GALVANOMETERS

Professional light shows don't use R/C servos or stepper motors as laser beam scanners. Rather, they use a unique electromechanical device called the *galvanometer*. A galvanometer—or galvo for short—provides fast and controllable back-and-forth oscillation. Mount a mirror on the side of the shaft and the reflected light forms a streak on the wall. Position two galvanometers at a 90-degree angle, apply the right kind of signal, and you can project circles, ovals, spirals, stars, and other multi-dimensional geometric shapes.

What Is a Galvanometer?

Most electronics buffs are familiar with the basic galvanometer movement of an analog meter. The design of the movement is shown in FIG. 20-1. A coil of wire is placed in the circular gap of a magnet. Applying current to the coil causes it to turn within the

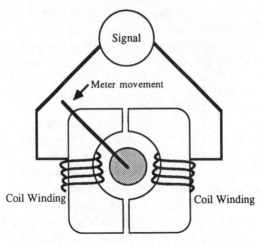

FIG. 20-1. *The basic operation of the galvanometer.*

electromagnetic field of the magnet. The amount of turning is directly proportional to the amount of current that is applied. The needle of the meter is attached to the coil of wire.

Some meter movements are designed so that the needle rests in the center of the scale. Applying a positive or negative voltage swings the needle one direction or the other. Say that at full deflection the meter reads + and −5 volts. When charged with +5 volts, the needle swings all the way to the right. When charged with −5 volts, the needle sways all the way to the left. Current under 5 volts (positive or negative) causes the needle to travel only part way right or left.

Meter movements are designed for precision and are not capable of moving much mass. But by using a stronger magnet and a larger coil or wire, a galvanometer can be made to motivate a larger mass. Heavy-duty scanning galvanometers are often made to actuate the needles in chart recorders and they have more than enough "oomph" to rack a small first-surface mirror back and forth. Some galvo manufacturers, most notably General Scanning, make units specifically designed for high-speed laser light deflection. These are best suited to laser light show applications but their cost is enormous. Surplus galvanometers, available from several sources including Meredith Instruments and General Science & Engineering, cost from $50 to $150 a piece; new high-speed models cost upwards of $350.

You may use either commercially made galvanometers for the projects that follow or make your own using small hobby dc motors. Be aware that commercially-made galvos work much better than homemade types, but if you are on a budget and simply want to experiment with making interesting light show effects, the dc motor version will prove more than adequate.

Using Commercially-Made Galvos

FIGURE 20-2 shows a typical commercially made galvanometer. The General Scanning model GVM-735 galvanometers illustrated in the picture are actually Cadillacs among

FIG. 20-2. *A commercially made precision galvanometer.*

scanners, so the units are not representative of typical quality. There are other makers of fine galvanometers, including C.E.C., Minneapolis-Honeywell, and Midwestern.

Galvanometers carry a number of specifications you can use to judge quality, versatility, and practicality. Among the most useful specifications are:

✷ **Rotation.** The amount of deflection of the shaft, in degrees. Usually stated in both + and − about a center point. A deflection of 10 to 20 degrees is fine for light show applications.

✷ **Natural frequency.** Often stated as the resonance frequency, in hertz, and provides an indication of top speed. Most galvo applications call for a top frequency of 80 percent of the resonance frequency. For example, with a resonance frequency of 225 hertz, top operating frequency is about 180 Hz.

✷ **Rotor inertia.** The measurement of the ability to move mass. A higher inertia means the galvanometer can move a greater amount of mass. The rotor inertia is relatively small—1 to 2 g/cm², but it's sufficient to move a small, mounted mirror.

✷ **Coil resistance.** The resistance, in ohms, of the drive coil. Helpful in designing and applying drive circuits.

✷ **Operating voltage.** Nominal and/or maximum operating voltage, typically 12 volts. Also useful in designing and using drive circuits.

★ **Power consumption**. The power consumption, in milliamps or amps, of the galvanometer, typically under worst-case conditions (full rotor deflection, full voltage, etc.). The drive circuit you use must be able to deliver the required current.

Driving Galvanometers

Galvanometers can be driven in a variety of ways, including power op amps, audio amplifiers, and transistors. A basic, no-frills drive circuit appears in FIG. 20-3 (refer to TABLE 20-1 for a parts list). The input can be an audio signal from the LINE OUT jack of a hi-fi or an unamplified input from a frequency generator (more on these later). You *can* apply an amplified signal to the input of the drive circuit, but the op amp will clip the output if the input is excessively high.

The two drive transistors, a complementary pair consisting of TIP31 and TIP32 power types mounted on heatsinks, interface the output of the op amp to the coil of the galvanometer. The circuit works with a variety of voltages from ±5 volts to ±18 volts. Most scanners operate well with supply voltages of between ±5 and ±12 volts. Check the specifications of your galvanometers and make sure you don't exceed the rated voltage.

If anything, operate the galvos at a reduced voltage. They will still operate satisfactorily but the rotor might not deflect the full amount. This is not a problem for most applications, including laser light shows, where full deflection is not always desired.

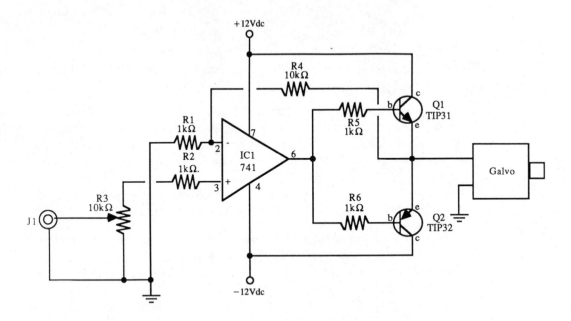

Notes:
Q1 and Q2 must be on heasinks!

Use supply voltage to complement galvanometer; up to ± 18 VDC.

FIG. 20-3. *Driver circuit for operating a galvanometer from line-level audio source. Build two circuits for controlling two galvanometers.*

Table 20-1. Galvanometer Drive Parts List

IC1	LM741 op amp IC
R1,R2 R5,R6	1 kilohm resistor
R3	10 kilohm potentiometer
R4	10 kilohm resistor
Q1	TIP 31 npn transistor
Q2	TIP 32 pnp transistor
J1	⅛-inch jack
G1	Galvanometer
Misc.	Heatsinks for Q1 and Q2.

All resistors are 5 to 10 percent tolerance, ¼ watt.

One by-product of full deflection is a "ringing" that occurs when the rotor hits the stop at the ends of both directions of travel. The ringing appears in the laser light form as glitches or double-streaks.

To make two-dimensional shapes, you need two galvanometers positioned 90 degrees apart, as illustrated in FIG. 20-4. Mount mirrors on the shafts using aluminum or brass tubing. Add a set screw (see FIG. 20-5) so that you can tighten the mirror mounts on the rotor shaft of the galvanometer.

You can use any number of mounting techniques to secure the galvos to an optical breadboard or table, but the mounts you use must be sturdy and stable. Vibrations from the galvos can be transferred to the mounts, which can shake and disturb the light forms.

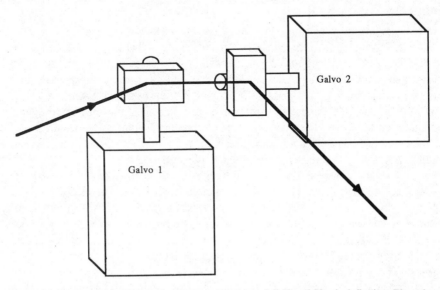

FIG. 20-4. *How to arrange two galvanometers to achieve full X and Y axis deflection. Place the mirrors of the galvanometers close together to counter the effects of beam deflection.*

315

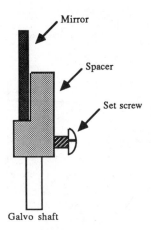

Mirror

Spacer

Set screw

FIG. 20-5. *You can mount a thin, front-surface mirror to the galvanometer shaft using an aluminum spacer that has been filed down.*

Galvo shaft

Build a separate drive circuit for both galvanometers and enclose it in a project box (or you can include the driver circuit in a larger do-everything light-show console). I built the prototype drive circuit on a universal breadboard PCB and had plenty of room to spare. The enclosure measured 4⅜ by 7¾ by 2⅜ inches. Subminiature ⅛-inch phone jacks were provided for the audio inputs and scanner outputs and potentiometers were mounted for easy control of the input level.

To test the operation of the galvanometers and drive circuits, plug in the right and left channels of the hi-fi and turn the gain controls (R1 for both drives) all the way up. The galvos should shake back and forth in response to the music. Shine a laser beam at the mirrors so the light bounces off one, is deflected by the other, and projects on the wall. The light forms you see should undulate in time to the music.

Providing An Audio Source

The test performed above only lets you test the operation of your galvanometer setup. With the arrangement detailed above, the light form will always be squeezed into a fairly tight line that crawls up and down the wall at a 45-degree angle. Full flexibility of a pair of galvanometers, physically set apart 90 degrees, requires an audio source that has two components — both of which are set 90 degrees apart in phase.

Audio signals are sine waves, and sine waves are measured not only by frequency and amplitude but by phase. The phase is measured in degrees and spans from 0 to 360 degrees. FIGURE 20-6 shows two sine waves set apart 90 degrees. Notice that the second wave is a quarter step (90 degrees) behind the first one.

If you could somehow delay the sound coming from one channel of your stereo, you can broaden the 45-degree line into a full two-dimensional shape. The closer the delay is to 90 degrees out of phase, the more symmetrical the light form will be. Imagine a pure source of sine waves—a sine wave oscillator. The oscillator is sending out waves at a frequency of 100 Hz. It has two output channels called sine and cosine. Both channels are linked so they run at precisely the same frequency, but the cosine channel is delayed 90 degrees. The lightform projected on the screen is now a perfect circle.

The circuit in FIG. 20-7 provides such a two-channel oscillator. Two controls allow you to change the frequency and "symmetry" of the sine waves. The symmetry (or

316

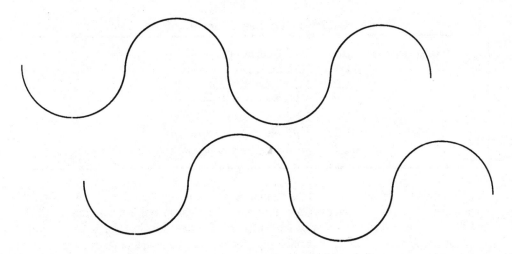

FIG. 20-6. *Two sine waves; the bottom wave is delayed 90 degrees from the top wave.*

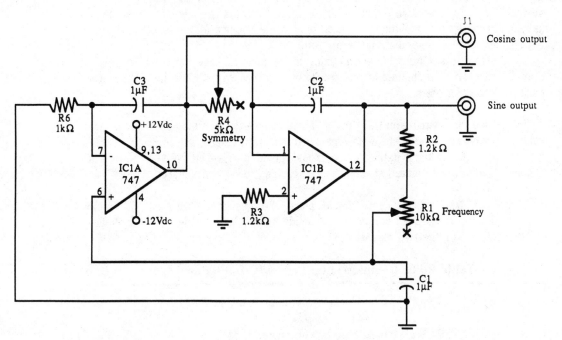

FIG. 20-7. *One way to implement a sine/cosine audio generator for operating two galvanometers.*

phase) control counters the destabilizing effects caused by rotating the frequency knob. This circuit, with parts list provided in TABLE 20-2, is designed so that the resistors and capacitors are the same value. Changing the value of one resistor throws off the balance of the circuit, and the symmetry control helps rebalance it. Note that the frequency and symmetry controls provide a great deal of flexibility in the shape of the projected beam. Fiddling with these two potentiometers allows you to create all sorts of different and unusual light forms.

Table 20-2. Sine/Cosine Generator Parts List

IC1	LM747 dual op amp IC
R1	10 kilohm potentiometer
R2,R3	1.2 kilohm resistor
R4	5 kilohm potentiometer
R5	1 kilohm resistor
C1-C3	1 μF electrolytic capacitor
J1,J2	⅛-inch miniature phone jack

The circuit is designed around the commonly available 747 dual op amp (two LM741 op amps in one package). You can use almost any other dual op amp, such as the LM1458 or LF353, but test the circuit first on a breadboard. For best results, use a dual op amp.

You can obtain even more outlandish light forms when combining the sine and cosine signals from two separate oscillator circuits. On a perf board, combine two oscillators using two LM741 op amps. An overall circuit design is shown in FIG. 20-8. One set of switches allows you to turn either oscillator on and off and the input controls let you individually control the amplitude from each sine/cosine channel (for a total of four inputs). A parts list for the general circuit is in TABLE 20-3.

The switch lets you flip between mixing the sine inputs together or criss-crossing them so that the sine channel of one oscillator is mixed with the cosine of the other oscillator, and vice versa.

★ When the switch is in the "pure" position(A)—sine with sine and cosine with cosine—you obtain rounded-shape designs, such as spirals, circles, and concentric circles.

★ When the switch is in the "cross-cross" position (B)—sine with cosine for both channels—you obtain pointed shapes, like diagonals, stars, and squares.

Like the drive circuit, you should place the oscillator, with all its various potentiometers, in a project box or tuck it inside a console. Provide two ⅛-inch jacks for the outputs for the two galvos.

Table 20-3. Complete Light Show Circuit Parts List

2	Sine/Cosine generators (see Fig. 20-7)
IC1,IC2	LM741 op amp IC
R1,R2 R6,R7	10 kilohm potentiometer
R3,R4 R8,R9	1 kilohm resistor
R5,R10	10 kilohm resistor
J1,J2	⅛-inch miniature phone jack
S1-S3	DPDT switch

All resistors are 5 to 10 percent tolerance, ¼ watt.

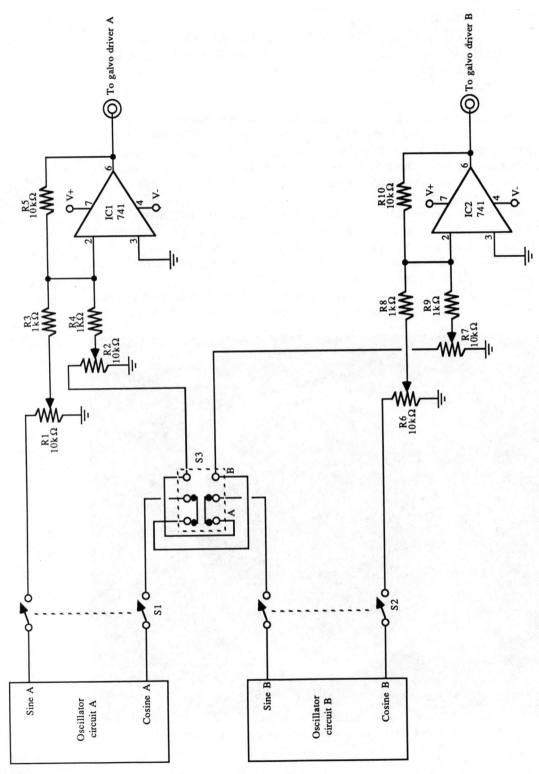

FIG. 20-8. *A schematic for designing a two-channel sine/cosine audio generator, with dual op amp mixers (note: use separate op amps for the mixers).*

319

Using the Oscillator

Connect the outputs of the oscillator to the inputs of the drive circuit. Apply power to both circuits and rotate the mixer input controls (R1, R2, R6, and R7) to their fully on positions. Flick on switch #1 so that only the signals from one oscillator are routed to the mixer amps and turn switch #3 to "A" position. Slowly turn the control knobs until the galvanometers respond.

If the galvanometers don't seem to respond, temporarily disconnect the jumpers leading between the oscillator and drive circuits and plug an amplifier into one of the oscillator output channels. You should hear a buzzing or whining noise as you rotate the frequency and symmetry controls. If you don't hear a noise, double-check your wiring and be sure the mixer controls are turned up. When turned down, no signals can pass through the mixing amps.

Aim a laser at the mirrors and watch the shapes on a nearby wall or screen. Get the feeling of the controls by turning each one and noting the results. With the symmetry control turned down and the frequency control almost all the way down, you should see a fairly round circle on the screen. If the circle looks like an egg, adjust the mixer controls to decrease the X or Y dimension, as shown in FIG. 20-9.

If the egg is canted on a diagonal, the phase of the cosine channel is not precisely 90 degrees. Try adjusting the symmetry control and fine tuning it with the frequency

FIG. 20-9. *The basic circle produced by turning on one sine/cosine generator and adjusting the frequency and symmetry controls to produce a pure sine wave.*

320

FIG. 20-10. *One of many spiral light forms created by turning on both sine/cosine generators.*

control. There should be one or more points where you achieve proper phase between the sine and cosine channels.

Now turn off the first oscillator and repeat the testing procedures for the second oscillator. It might behave slightly different than the first due to the tolerance of the components. If you need more precise control over the two oscillators, use 1-percent tolerance resistors and High-Q capacitors.

For really interesting light forms, turn on both oscillators and adjust the controls to produce various symmetrical and asymmetrical shapes. At many settings, the light forms undulate and constantly change. At other settings, the shape remains stationary and can appear almost three-dimensional. FIGURES 20-10 and 20-11 show sample lightforms created with the circuit and galvanometers described above.

Alternate between "A" "B" settings by flicking switch #3. Note the different effects you create when the switch is in either position.

Powering the Oscillator and Drive Circuits

So far we've discussed using galvo oscillator and drive circuits but have paid no attention to the power supply requirements. Although you can build your own dual-polarity power supply to run the galvo system, I strongly recommend that you use a well-made commercial supply, one that has very good filtering. *Sixty cycle hum*, caused by insufficient filtering and poor regulation, can creep into the op amps and make the galvos shudder continuously.

Output voltage and current depend on the galvanometers you use. I successfully used two General Scanning GVM 734 galvos with a power supply that delivered ±5

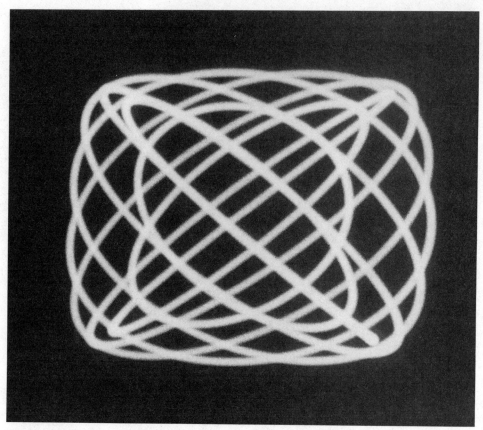

FIG. 20-11. *A clearly definable "Lissajous" figure, made with the galvanometer light show device.*

volts at about 250 mA for each polarity. You might need a more powerful power supply if you use different galvanometers (some require ± 12 at 1 amp or more). The power supply I used for the prototype system was surplus from a Coleco computer system that cost $12.95. Look around and you can find an equally good deal.

Useful Modifications and Suggestions

Not shown in the driver circuit above is an additional switch that allows you to change the polarity to one of the galvanometers. This provides added flexibility over the shape of the light forms. Wire the switch as shown in FIG. 20-12.

The drive circuit detailed above might not provide adequate power for all galvanometers. If the galvos you use just don't seem to be operating up to par, use the drive circuit shown in FIG.20-13, developed by light show producer Jeff Korman (see parts list in TABLE 20-4). This new driver is similar to the old one but provides better performance at low frequencies. Korman also uses a special-purpose sine/cosine oscillator chip made by Burr-Brown. The chip, called the 4423, is available directly from Burr-Brown or from Fobtron Components (see Appendix A).

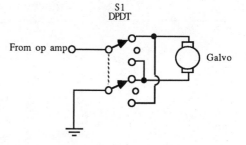

FIG. 20-12. *Wire a DPDT switch to one of the galvanometers to reverse its direction in relation to the other galvanometer.*

Cost is high ($25 to $35) but it provides extremely precise control over frequency without the worry of knocking the signals out of 90-degree phase.

Other useful tips:

★ Whenever possible, try to reduce the size of the light forms; it makes them appear brighter.

★ Connect an audio signal into one channel of the drive circuit and connect the sine/cosine oscillator into the other channel. The light form will be modulated in 2-D to the beat of the music.

★ You can record a galvanometer light show performance by piping the output of the oscillators into two tracks of a four- track tape deck. Use the remaining two tracks to record the music. When played back, the galvanometers will exactly repeat your original recording session. This is how most professional light show producers do it.

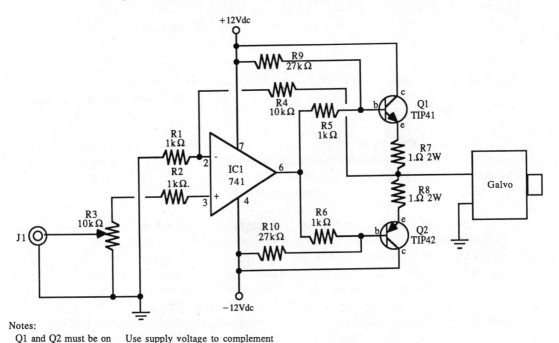

Notes:
Q1 and Q2 must be on heatsinks!   Use supply voltage to complement galvanometer; up to ± 18 Vdc.

FIG. 20-13. *An enhanced high current galvanometer circuit designed by light show consultant Jeff Korman.*

Table 20-4. Enhanced Galvanometer Driver Parts List

IC1	LM741 op amp IC
R1,R2,	1 kilohm resistor
R5,R6	
R3	10 kilohm potentiometer
R4	10 kilohm resistor
R3	10 kilohm potentiometer
R7,R8	0.1 ohm, 2-watt resistor
R9,R10	27 kilohm resistor
Q1	TIP 41 npn transistor
Q2	TIP 42 pnp transistor
J1	⅛-inch jack
G1	Galvanometer
Misc.	Heatsinks for Q1 and Q2.

All resistors are 5 to 10 percent tolerance, ¼ watt, unless otherwise indicated.

✭ Try adding one or two additional oscillators. Combine them into the mixing network by adding a 10k pot (for volume control) and a 1k input resistor. Wire in parallel as shown in the FIG. 20-8 schematic.

✭ Triangle or sawtooth (ramp) waves create unusual pointed-star shapes, boxes, and spirals. You can build triangle and sawtooth generators using op amps, but an easy approach is to use the Intersil 8038 or Exar XR-2206 function generator ICs.

Making Your Own Galvos

A set of commercially-made galvanometers can set you back $100 to $200, even when they are purchased on the surplus market. If you are interested in experimenting with laser graphics and geometric designs but are not interested in spending a lot of money, you can make your own using small dc motors.

Follow the diagram shown in FIG. 20-14 to make your own galvanometers. You can use most any 1.5- to 6-volt dc motor, but it should be fairly good quality. Test the motor by turning the shaft with your fingers. The rotation should be smooth, not jumpy. Measure the diameter of the outside casing. It should be about 1 inch. Refer to TABLE 20-5 for a list of required parts.

Use a high-wattage soldering iron or small brazing torch to solder a penny onto the side of the motor shaft. For best results, clean the penny and coat it and the motor shaft with solder flux. Make sure the penny is thoroughly heated before applying solder or the solder may not stick. You'll need to devise some sort of clamp to hold the penny and motor while soldering—you need both hands free to hold the solder and gun.

Let the work cool completely, then mount a small ½-by-¾-inch front-surface mirror to the front of the penny. You can use most any glue: Duco adhesive or gap-filling cyanoacrylate glue are good choices. Note that the mirror should not be exceptionally thick. You get the best results when the mirror is thin—the motor has less mass to move, so it can vibrate faster.

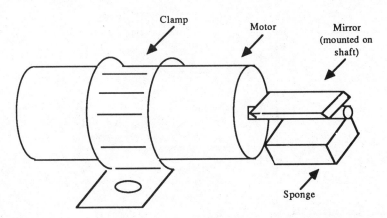

Clamp   Motor   Mirror
(mounted on
shaft)

Sponge

FIG. 20-14. *Basic arrangement for making galvanometers using small dc motors. The sponge prevents the mirror and shaft from turning more than 20-25 degrees in either direction and helps damp the vibrations.*

Table 20-5. Dc Motor Galvanometer Parts List

1	Small 1.5- to 6-Vdc hobby motor
1	¾- to 1-inch pipe clamp
1	2-by-3-inch, ⅛-inch-thick acrylic plastic
1	⁸⁄₃₂-by-½-inch bolt, nut, washer
1	½-by-¾-inch thin, front-surface mirror
1	Lincoln penny
1 ea.	Small piece of sponge, anti-static foam

Snap the motor into a ¾-inch electrical conduit clamp and fit it into position (the clamp will have a 1-inch opening and will hold most small hobby motors). If the clamp is too small, widen the opening by gently prying it apart with a pair of pliers. Mount the motor and clamp to a 2-by-3-inch acrylic plastic base (⅛-inch thickness is fine). Drill holes as shown in FIG. 20-15. Use a ⁶⁄₃₂ by ½-inch bolt and a ⁶⁄₃₂ nut to secure the pipe clamp to the base.

Use a treated, synthetic sponge and cut into two 1-by-½-inch pieces. The sponge should be soft but will dry out when left overnight. After the sponge has dried out, compress it and secure it under the penny using all-purpose adhesive. Now slide a 1-by-½-inch piece of anti-static foam into the gap between the sponge and penny. The fit should be close but not overly tight. If you need more clearance, compress the sponge by squeezing it some more. The finished dc motor galvanometer is shown in FIG. 20-16.

As an alternative, cut a piece of ½-inch foam and stick it under the penny. Try different foams to test their "suppleness." The foam should be soft enough to let the penny and mirror vibrate but not so soft that it acts as a tight spring and bounces the penny back after only a small movement.

Using the Motor/Galvanometers

Attach leads to the motor terminals and connect the homemade galvanometers to the drive circuit and oscillator detailed earlier in this chapter. Repeat the testing

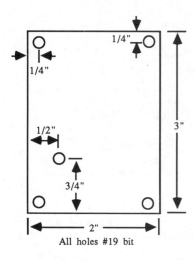

All holes #19 bit

FIG. 20-15. *Cutting and drilling template for the base for the homemade motor galvanometers.*

FIG. 20-16. *A completed motor galvanometer, secured to the base with a ¼-inch electrical conduit clamp.*

procedures outlined for commercially made galvos. Position the motors so that they are 90 degrees off-axis and shine a laser onto both mirrors. You should see shapes and patterns as you adjust the controls on the oscillator.

The motor/galvanometers can be mounted in a variety of ways. One approach is to use metal strips bent 90 degrees at the bottom. Drill matching holes in the strip and attach the base of the motor/galvos to the strips using 6/32 hardware. You can also secure the base of each motor/galvanometer using ½-inch galvanized hardware brackets (available at the hardware store).

The light forms might not be perfectly symmetrical. Depending on the motors you used and the type and thickness of foam backing you installed, one motor might vibrate at a wider arc than the other. Try adjusting the foam and sponge on both motors to make them vibrate the same amount.

SMOKE EFFECTS

Smoke effects are obtained by spreading the laser beam into an arc (the light is projected as a straight line) and filling the room with smoke or vapor. Because you see only a thin slice of the smoke through the arc of laser light, you can clearly view the air currents as they swirl and shift.

Smoke effects in professional light shows require multi-watt lasers—a 2- to 5-watt argon laser makes wonderful smoke effects. The "smoke" is often vapor left by heating dry ice. Dry ice vapor is heavier than air so it must be blown through the stream of laser light. Small blowers keep the air circulating.

You can experiment with laser smoke effects by using a helium-neon laser (1 mW or more). Although the smoke is not visible at any distance, it can be used for small, amateur light shows. You can also use laser smoke effects to view fluid aerodynamics (see Chapter 22) or just to see what happens to smoke particles in a ventilated room. You can use dry ice vapor (buy the dry ice from a local ice-packing company), smoke from a cigarette, incense stick, or match.

In all cases, exercise reasonable care. Dry ice can cause frostbite, so handle it only with gloves and place it in Pyrex or metal containers (plastic and regular glass could shatter). Cigarettes, incense, and matches present a fire hazard. Get help if you are not sure of what you are doing.

Note that smoke—and all particles in the air—are seen when laser light shines through them. You see them best when you are on one side of the smoke particles and the laser is on the other. You see the outline of the smoke particles as they swirl around.

Casting the Arc

Spread the laser beam into an arc using the sheet effects described in the last chapter. As a recap, you can spread light out in an arc by:

* ★ Spinning a mirror on the side of a motor shaft.
* ★ Rotating an off-axis mirror mounted on the end of a motor shaft.
* ★ Spinning a holograph scanner mirror wheel.
* ★ Spreading the light with a cylindrical lens.

These approaches do not allow you to change the spread of the arc. You can obtain more control over the angle—make it narrow or wide—by using a single galvanometer (commercially made or homemade) and by adjusting the amplitude of the drive signal. You can use the oscillator and driver circuits detailed previously in this chapter.

Making the Smoke Effect

Turn off all the lights except for a dim pilot light that you can extinguish remotely. Light the cigarette, incense, or match, and waft the smoke under the laser light arc. Alternatively, dunk the dry ice in a pail of tepid water. Move to a position that allows you to clearly see the smoke and turn off the pilot light. Depending on the spread of the arc, the power of the laser, and the amount of smoke or vapor, you should clearly see particles swirling in the air.

Watch the effect as you add more smoke or vapor or make the spread of the arc smaller. If the air is moving too fast, the smoke/vapor might disperse too quickly. Turn off blowers, air conditioners, and other appliances that may be agitating the air. On the other hand, if the room gets too filled with smoke, you'll see a general haze in the light and the swirling won't be as easy to see. Clear the room by airing it out. Use blowers or fans to speed up the process.

USING ARGON AND KRYPTON LASERS

Most professional laser light shows use argon or krypton lasers. Both of these are manufactured in high-output versions and provide two or more colors. As you learned in earlier chapters, argon lasers emit light at two principle wavelengths or mainlines—488.0 nm and 514.5 nm. Krypton lasers have the unique ability to produce light at just about any wavelength, providing a spectrum of colors. Because they can produce all three primary colors—red, green, and blue—krypton lasers are often used in color holography. The three primary colors can also be created by using an argon and helium-neon laser.

Using a prism or set of dichroic filters allows you to separate the mainlines. The prism disperses the light into its component colors so that each color can be individually manipulated. Dichroic filters let you block all but the color you want. For example, a "red" filter placed in front of an argon laser will block the green 514.5 nm mainline.

Serious light-show applications require a laser with a minimum power output of 100 mW. Better, more dramatic results are obtained with even higher wattages; it's not uncommon to see 3- to 5-watt argon lasers used in professional light shows. These Class IV devices are downright dangerous unless you know precisely what you are doing. They also require water cooling and extensive plumbing, making them a hassle to use.

High-output krypton and argon lasers are unreasonably expensive but if you are serious about laser light shows, you might locate a used specimen at an affordable price ($3,500 or so). If you keep an eye out and make plenty of contacts, you might luck onto a high-output industrial-grade argon laser—used for such applications as medicine, optical disc manufacturing, and forensics—that is no longer functioning but is repairable. Perhaps the tube is gassed out or maybe the power supply is fried, but the cost of fixing the laser will be less than buying a new one. Before you sign the check, make sure you know what the problem is and that you are confident the laser can be fixed.

High-output lasers can't be plugged into the ac socket and aimed at the wall. Most require 220-volt, three-phase ac (wall outlets provide 110-volt single-phase ac). The laser generates too much heat for air cooling and must be shrouded by water to keep it cool. The water supply must be continually circulating from the faucet to a drain. Adequate filtration and pressure regulation is needed to prevent deposits in the cooling jacket or rupturing the tube.

Finally, but most importantly, the power supplies used to operate high-output lasers produce high current at high voltage. Touching a high-voltage component or wire on the power supply or tube will kill you—*no exceptions*. Thoroughly familiarize yourself with the safe operation of the laser before using it.

YOU AND UNCLE SAM

They say the government has its fingers in everything, and lasers are no exception. Uncle Sam's interest in lasers is purely one of safety—the federal government wants you to comply with minimum safety requirements before you put on a light show. As you learned in Chapter 2, ''Working With Lasers,'' the branch of the government that regulates the laser industry is the Center for Devices and Radiological Health, or CDRH (formerly BRH, or Bureau of Radiological Health). The CDRH monitors the manufacture and use of lasers so that harmful laser radiation does not befall unsuspecting people.

Most of the CDRH regulations concern the manufacture of lasers, but some sections deal with the use of lasers in public arenas, including light shows. Briefly stated, anyone wishing to conduct a laser light show for public viewing or otherwise demonstrate the operation of lasers to the public, must fill out forms and submit them to the CDRH. These forms provide necessary information on the type and class of laser and how you intend to use it.

You must also provide details, as precisely as possible, of how the light show equipment will be arranged, where beam stops will be placed, and the number and type of fail-safe mechanisms used. You must also demonstrate an understanding of the regulations and that you intend to comply with them. In the case of a traveling light show, you must indicate how your laser system can adapt to different rooms and auditoriums.

The complete CDRH requirements for laser light shows is too involved to repeat here. You can obtain compliance regulations and application forms directly from the CDRH; their address is given in Appendix A.

GOING PROFESSIONAL

Laser light shows can be both fun and rewarding, both on a personal and financial level. Although permanent laser shows, most notably Laserium, are the most visible, they represent only a small number of light shows conducted in the U.S. Many rock bands like to play to the accompaniment of a light show, especially one that includes lasers. Contact local bands and clubs and ask if they would add a laser show to their gig.

A local non-laser light show producer who has not had the time, inclination, or background to include lasers in his repertoire, might be delighted to have you as a consultant. If you can't find music groups or nearby light show producers, ask at the radio stations in town (call or drop by). You will probably need to start small and work your way to the big time, all the while adding to your laser system.

On a smaller scale are light shows for schools and organizations. What boys' or girls' club wouldn't like to be treated to a light show? These gigs are mostly non-paying, but they are an excellent way to hone your light show talents.

Even if you don't take your light show on the road and perform before a live audience, you can doodle with the artforms created by the assortment of mirrors, motors, servos, galvanometers, and other sundry equipment on film or videotape. A telecine adapter, used primarily for converting Super-8 movies and 35 mm transparencies to videotape, can be used to capture the light-show images on film. Simply replace the film projector with the laser projector. You can use a video camera or still camera to capture the images on the rear-projection plate.

Another alternative is to aim the laser at the wall or screen and photograph or videotape the images directly. This method doesn't yield the best results, because you pick up the pattern on the wall or screen and the images aren't generally as brilliant.

To photograph the light show artforms, place the camera (video or film) on a tripod and focus the lens on the front of the screen. You might need to use the zoom or macro feature of the lens, or else attach supplementary positive diopter lenses in front of the camera in order to take sharp pictures.

Persistence of vision is the capacity of the eye to blend a series of still pictures into smooth motion. A movie is made up of thousands of individual still pictures. These pictures are flashed on the screen faster than our eyes can detect, so the image appears to be in motion. The same technique is used in laser light shows. The scanning of motors, R/C servos, or galvanometers produces a two-dimensional shape on the screen. What appears as a spiral or circle is actually one beam of light, moving so fast that our eyes (and brain) synthesize it into a complete, moving picture.

While your eye smooths the scanning of the laser beam to create the illusion of motion, the eye of the camera may be faster, so the results you see on film might not match those you see in person. When taking still pictures of a laser image, choose a shutter speed $\frac{1}{15}$th of a second or longer. Shorter (higher in number) shutter speeds might result in only partial images.

Television pictures are also created by flashing a series of still pictures on the screen. The video frame rate for a complete picture is $\frac{1}{30}$ of a second. That's faster than your eye can detect, so you don't witness any flicker. Videotaping a laser light show image could result in objectionable flicker. You can see the flicker on your TV set while recording. However, you can often minimize the flicker by adjusting the speed or frequency knob on the light-show motor/galvanometer controller.

21

Experimenting with Laser Weapons Systems

Even before the laser was invented, science fiction writers told of incredible weapons and machines that emitted a bright saber of light, a death ray that disintegrated everything in its path. In the 1951 classic *The Day the Earth Stood Still*, a 7-foot tall "police robot" was equipped with a powerful disinto-ray gun. The gun was mounted behind a visor in the robot's helmet and shot its high-intensity, pencil-thin flame with great precision.

Even today, science fiction movies and books place high emphasis on weapons that use light instead of bullets. But real science hasn't kept up with science fiction; most lasers do little or no harm to human flesh, and many can't cut through a piece of paper, let alone tanks, automobiles, and spaceships.

As unlikely as practical laser-based weapons seem given the current limitations of the state-of-the-art, it's possible that weapons using an intense form of light could someday be developed. In 1983, President Ronald Reagan outlined a plan for outfitting land- and space-born satellites with laser weapons as a defensive measure against ballistic missiles. What are the possibilities of developing powerful laser weapons that thwart nuclear destruction? Can weapons be placed on the ground, installed in tanks or towed on trailers? And what about non-war use of laser guns: could they be used—as they are in the "Star Trek" TV show and movies—to either kill or stun an opponent?

Let's take a brief look at laser weapons and the technology that's currently available. Then read the plans for constructing a useful but relatively harmless laser "gun" using a helium-neon tube. The emphasis of the laser gun is to show you what goes into handheld lightwave weapons, not to make an effective munition. On a more practical standpoint, laser pistols and rifles provide a means for target practice without wasting bullets, pellets, or B-B's and with much less risk of bodily injury.

AN OVERVIEW OF LAND/SPACE LASER WEAPONS

Lasers are already used in the battlefield, but not as offensive weapons. The U.S. Army and North Atlantic Treaty Organization (NATO) use laser rangefinders for determining the distance between the firing line and the target. The laser system can be mounted on a rifle stock and carried by one person. A number of high-caliber cannon, tanks, and helicopters use their own laser rangefinders, controlled by an on-board firing computer.

Laser rangefinders operate in a variety of ways, but many work by means of transmitting short modulated pulses, then waiting for an echo. The timing of the return signal indicates distance. Accuracy is within 10 to 20 feet for most systems and range is up to 7 miles.

Reagan's Strategic Defense Initiative (SDI), which he announced in a widely publicized speech on March 23, 1983, calls for the deployment of land- and space-based weapons using one or all of three possible technologies:

★ Particle beam weapons, shooting atoms of neutral or charged particles from ground- or space-based platforms.

★ Kinetic energy weapons, from ground-based cannon, that fling high-speed projectiles at the target.

★ Laser weapons, which use short-wavelength electromagnetic energy to heat up the target, damaging its electronics or flight control mechanism.

Of the three types, kinetic energy weapons (KEW) have received the greatest attention. Battlefield KEWs are believed to be technologically possible, and may provide the greatest amount of firepower. The most common variety of KEW is the electromagnetic rail-gun, which shoots specially designed heavy metal "bullets" at about six miles per second (the bullet from an M-16 rifle travels at about ⅔ of a mile per second). Rail-guns require a great deal of energy and many designs are one-shot affairs: the gun is severely damaged after just one firing.

To be effective, the wavelength of a laser weapon must be short, at least in the visible band but preferably in the ultraviolet or x-ray band. The greatest difficulty in designing short-wavelength lasers is power—the shorter the wavelength, the more energy that is required. Optical (visible or ultraviolet) lasers work by heating the skin of the target. The beam must remain at the same spot for several seconds until the skin is hot enough to do internal damage to the target. This is tough because the typical ballistic missile travels in excess of 6 miles per second. Imagine focusing on the same 2- or 3-foot spot over a distance of 50,000 feet and you have an idea how accurate such a laser weapon must be.

In addition to the problems of accuracy, laser weapons of any power tend to be monstrous. That limits them to ground-based ray guns, using mirrors to direct the beam to the target. High-powered lasers using turbine-powered chemical jets have been developed and even placed aboard aircraft, but the wavelength of the light is long—6 to 10 micrometers—far in the infrared region. This makes the laser relatively inefficient at destroying their targets.

X-ray lasers, still in the "so secret they don't exist" category, emit an extremely high powered beam that can literally destroy a missile in mid-flight. X-rays can't be deflected by mirrors, however, which means that the weapon must be easily aimed and in a direct line of sight to the target. Fortunately, x-ray lasers can be built small, experts say, making them suitable for space-based operation. The biggest disadvantage to x-ray lasers is that they use an internal atomic explosion to work, so they are essentially one-shot devices.

A relative newcomer to the SDI (or "Star Wars") scene is the free-electron laser, which is being developed at several national laboratories and universities. The free-electron laser uses a stream of electrons that is made to emit photons of light after being oscillated by giant electromagnets. Free-electron lasers (FELs) have been built and they do work. But if put into production, an actual anti-ballistic missile FEL would take up a football field or more. Obviously, such a device would be useful only as a stationary ground-based weapon. It's possible, though unlikely, that it would be built over a span of several years on a low-orbiting space platform.

SMALL-SCALE WEAPONS USING LAB-TYPE LASERS

So far, we've discussed high-power laser weapons designed to counter a nuclear attack. Laser guns in the movies are often hand-held devices, or at most, small enough to prop on a vehicle. Lasers powerful enough to inflict damage but small enough to be carried have been developed, but they are not used in any current military application. It's relatively easy, for example, to build a hand-held ruby laser that puts out bursts of large amounts of light energy. When focused to a point, the light from a ruby laser can cut through paper, cloth, skin, or even thin metal.

Ruby crystals are poor conductors of heat, so ruby lasers emit only short pulses of light to allow the crystal to cool between firings. Nd:YAG lasers operate in a similar fashion as ruby lasers but can produce a continuous beam. Making a hand-held Nd:YAG laser is no easy feat, however. The Nd:YAG crystal must be optically pumped by another high-powered laser or by an extremely bright flash lamp or light source. Though the power output of an Nd:YAG laser is extremely high, considering the current state-of-the-art, a hand-held model is impractical. However, such a weapon could be built as a "laser cannon," transported on an armored vehicle or on a towed trailer.

CO_2 lasers are often used in industry as cutting tools. This type of laser is known for its efficiency—30 percent or more compared to the 1 to 2 percent of most gas and crystal lasers. A pistol-sized CO_2 laser would probably be difficult to design and manufacture because the CO_2 gas mixture (which includes helium and nitrogen) must be constantly circulated through the tube. What's more, the laser requires a hefty electrical power supply. Still, such a weapon could be built in an enclosure about the same size as a personal rocket launcher. These are designed to be slung over a shoulder and fired when standing in an upright position.

BUILD YOUR OWN HELIUM-NEON LASER PISTOL

The small lab-type laser weapons described above would cost several thousand dollars to build and require expert machining and tooling. You can readily build your own low-power, hand-held laser using a commonly available and affordable helium-neon tube and 12-volt dc power supply. The pistol is made from 2- and 1¼-inch schedule PVC plumbing pipe that you can cut with an ordinary hack saw.

Defining Barrel Length

The exact length of the pistol barrel depends on the laser tube and power supply you use. The laser used in the prototype pistol is a common variety 2 mW tube measuring 1½ inches in diameter by 7½ inches. You can obtain even smaller tubes through some laser dealers and make your pistol more compact.

The power supply is one of the smallest commercially made, measuring a scant ⅞-inch in diameter by slightly under 4 inches in length. This particular power supply was purchased through Meredith Instruments (see address in Appendix A) but is also available from Melles Griot and several laser manufacturers.

Cutting the Barrel and Grip

TABLE 21-1 provides a parts list for the laser pistol. Assuming you use the same or similar tube and power supply, cut a piece of 2-inch schedule 40 PVC to 12½ inches. Sand or file the cut ends to make them smooth. Drill a ⅜-inch diameter hole four inches from one end of the tube, as shown in FIG. 21-1. This hole serves as the leadway for the power wires.

Next, cut a length of 1¼-inch schedule 40 PVC to 6¼ inches. Using a wide, round file, shape one end of this piece so that its contour matches the 2-inch PVC. The angle of the smaller length of pipe, which serves as the grip, should be approximately 10 to 15 degrees. The match between the barrel and grip does not need to be exact, but avoid big gaps. Cut the grip a little long so to allow yourself extra room for shaping the contour. The grip is about the right length if it measures 6 inches top to bottom.

Cut a ¼-inch hole 1¼ inches from the bottom of the grip and another ¼-inch hole 90 degrees to the right but at a distance of 4¾ inches from the bottom (see FIG. 21-2). The lower hole is for the ¼-inch phone jack for the power, and the upper hole is for the push button switch. Note that the size of the upper hole depends on the particular switch you use. The switch detailed in the parts list is commonly available at Radio Shack and other electronics outlets. If you use another switch, you should measure the diameter of the shaft and drill a hole accordingly.

Cut two ⅙-inch-wide slits approximately ½ inch from the top of the grip. The slits should be opposite one another and at right angles to the top hole (used for the switch). Unthread the loose end of a 12-inch-long (3½-inch diameter) hose clamp through the slits. Tighten the clamp one or two turns, but not so much that you can't insert the barrel into it. The pistol so far should look like FIG. 21-3.

Table 21-1. Laser Pistol Parts List

1	12½-inch length, 2-inch schedule 40 PVC pipe
1	6¼-inch length, 1¼-inch schedule 40 PVC pipe
1	3-inch adjustable hose clamp
2	2-inch test plugs
1	1¼-inch PVC end cap
J1	¼-inch phone jack
S1	SPST momentary switch (normally open)
Misc.	Miniature 12 Vdc He-Ne power supply (see text), laser tube (see text)

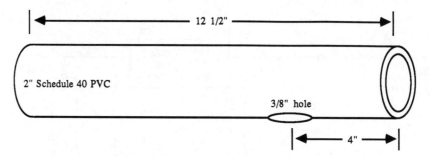

FIG. 21-1. *Cutting and drilling guide for the He-Ne laser pistol barrel.*

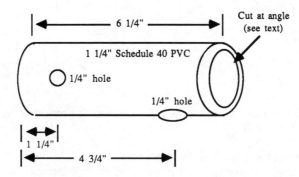

FIG. 21-2. *Cutting and drilling guide for the He-Ne laser pistol grip.*

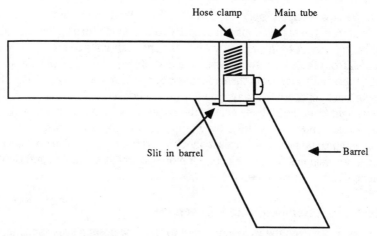

FIG. 21-3. *Barrel and grip held in place with an adjustable car radiator hose clamp.*

Wiring the Jack and Switch

Use 20- or 22-gauge stranded wire to connect the components as shown in FIG. 21-4. The wire lengths are approximate and provide some room for easily fitting the jack, switch, power supply, and tube into the PVC enclosure. After you have soldered the jack and

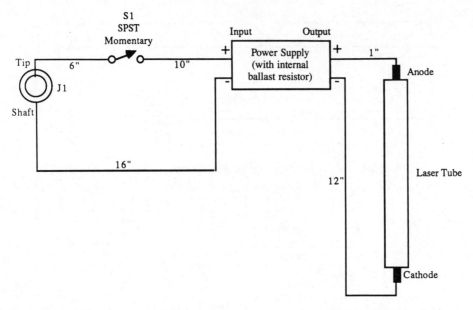

FIG. 21-4. *Wire the He-Ne laser pistol as shown in this diagram. Be aware of the high voltages present at the output of the power supply.*

switch, mount them in the grip, feed the wires through the hole in the barrel, and mount the grip to the barrel. Center the hole in the grip and tighten the clamp.

Mounting the Laser Tube and Power Supply

Attach the power supply leads to the tube. With the power supply and tube used, the anode lead from the supply connects directly to the anode terminal of the tube. This wire must be kept short to minimize current loss and high-voltage arcing. The cathode lead stretches from the supply to the opposite end of the tube. This design works well because the tube emits light from the cathode side, opposite the power supply. Because not all He-Ne lasers operate this way, you'll want to choose the tube carefully. The tube must emit its light from the cathode side or the power supply will get in the way. Loosely attach the power supply to the tube by wrapping them together with electrical tape.

The tube should be protected against shock by wrapping it in a thin styrofoam sheet, the kind used for shipping fragile objects. One or two wraps should be sufficient. Close up the sheet with electrical tape.

Final Electrical Connection and Assembly

Solder the wires from the jack and switch to the power supply. Be sure to observe proper polarity. Slide the tube and power supply into the barrel so that the output end of the laser faces forward. Drill a ⅜-inch hole in the approximate center of a 2-inch knockout "test" plug (available in the ABS plumbing pipe department of most hardware stores). Because the tube might not rest in the exact center of the barrel, determine the proper spot for drilling by first inserting the plug and observing the location of the output mirror. Mark the spot with a pencil, remove the plug, and drill out the hole.

After drilling, replace the plug and push the tube into the barrel so that the output window is just flush with the inside of the plug. Don't allow the window to protrude outside the plug, or you run a greater risk of chipping or breaking the output mirror. Insert another 2-inch plug in the rear of the barrel.

Test plugs are not routinely available for 1¼-inch pipe, so you must use an ordinary end cap for the bottom of the grip. Slip the end cap over the pipe; the fit should be tight. If not, try another end cap.

Building the Battery Pack

The battery pack for the laser is separate, not only because it allows you to build a smaller gun but allows you to power the device from a variety of sources.

The main battery pack consists of two 6-volt 4 AH lead-acid batteries contained in a 4³/₈-by-7³/₄-by-2³/₈-inch phenolic or plastic experimenter's box (available at Radio Shack). The batteries are held in place inside the box with heavy double-sided foam tape. The battery pack contains a fuse and power jack, mounted as shown in FIG. 21-5. A parts list for the pack is provided in TABLE 21-2.

The fuse is absolutely necessary in case of a direct short. Lead-acid batteries of this capacity produce heavy amounts of current that can easily burn through wires and cause a fire. A 5-amp fast-acting bus fuse provides adequate short-circuit protection without burning out during the short-term shorts that can occur when plugging in the battery pack cable.

Wire the battery pack as indicated in FIG. 21-6. When wired in series, the two 6-volt batteries produce 12 volts. The high capacity of the batteries means you can operate

FIG. 21-5. *The internal arrangement of the twin six volt batteries, fuse, and power jack for the battery pack.*

Table 21-2. Battery Pack Parts List

B1,B2	6 Vdc high-output (2 AH or more) gelled electrolyte or lead-acid rechargeable battery
F1	5-amp fuse
J1	¼-inch phone jack
1	Project box, measuring 4⅜ by 7¾ by 2⅜-inches
2	⅝-inch eyelets
1	Camera or guitar strap

the laser for at least an hour before needing a recharge. I have operated the prototype pistol for up to 6 hours before needing a recharge.

To make the battery pack conveniently portable, insert two ⅝-inch eyelet screws near the top of the box. Snap on a wide camera strap and adjust the strap for your shoulder. You can sling the battery pack over your shoulder while holding the pistol in your hands.

Current is delivered from the battery pack to the pistol by means of a 6- to 12-foot coiled guitar extension cord. These are available at Radio Shack and most music stores. Buy the two-wire mono variety; you don't need the three-wire stereo type. Remember to double-check the hookup of the power jacks in the battery pack and pistol. Make sure that the positive terminal connects to the tip of the jack.

Battery Recharger

The battery recharger is a surplus battery eliminator/charger pack designed for 12-volt systems. The pack outputs 13 to 18 volts at about 350 mA. The batteries can be recharged in the box by unplugging the power cord that stretches between the pack and pistol and plugging in the recharger. At 350 mA, recharging takes from 10 to 14 hours. Because the battery eliminator/recharger is not "intelligent," be sure to remove it after the recharge period. Otherwise the batteries could be damaged by overcharging.

ADDING ON TO THE LASER PISTOL

There are several modifications you can make to the laser pistol to increase its functionality and versatility. These include a power indicator and modulation bypass jack.

Power Light Add-on

A power light provides a visual indication that the laser gun is on. The light is especially helpful if you are using the pistol in daylight when the beam is hard to see and you want to be sure that the gun is working.

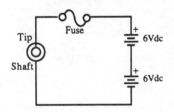

FIG. 21-6. *The wiring diagram for the battery pack. Do not omit the fuse.*

The power light consists of a light-emitting diode and current-limiting resistor. Drill a hole for the LED in the rear test plug and mount it using all-purpose glue. The LED is connected in parallel with the laser power supply: when you pull the switch, current is delivered to the power supply and LED.

Modulation Bypass Add-on

The modulation bypass permits you to easily modulate the beam with an analog or digital signal. The bypass consists of a ⅛-inch miniature earphone jack connected between the cathode lead of the power supply and the cathode terminal of the tube. Both LED and modulation bypass add-ons are shown in FIG. 21-7.

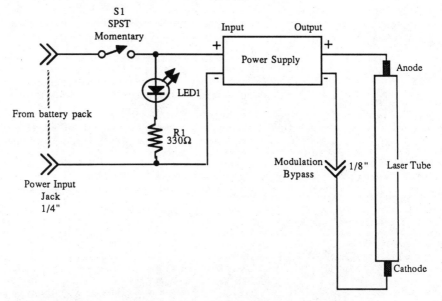

FIG. 21-7. *Schematic diagram for adding the modulation bypass and LED indicator enhancements to the He-Ne laser pistol.*

To add the bypass to the pistol, drill a hole for the jack in the rear test plug. Apply high-voltage putty around the terminals to the jack; be sure that none of the putty interferes with the contacts of the jack. The putty helps prevent arcing due to high voltages. When not using the bypass jack, insert a shorting plug into the jack (the shorting plug has its internal contacts shorted together).

To use the modulation bypass, remove the shorting plug and insert the leads from a modulation transformer, as detailed in Chapter 13, "Free-Air Laser Light Communications." The transformer can be driven by an audio amplifier so signals can be transmitted via the laser beam. Depending on what you connect to the modulation transformer, you can transmit audio signals or digital data over distances exceeding 1 mile.

Aiming Sights or Scope

If you plan on using the pistol for target practice, you'll want to add front and rear sights along the top of the barrel. Because the beam comes out of the barrel almost

1¼ inches from the top, you are bound to experience parallax problems with any type of sight system you use. For best results, mount the sights as close the barrel as possible—the further away they are from the barrel, the more pronounced are the effects of parallax error.

Effective homemade sights can be made inserting two small, flat or pan-head machine screws into the top of the barrel (use flat-blade screws, not hex or Philips). The screws should be short and should not overly extend inside the barrel. If they do, the tips of the screws could interfere with the laser and power supply.

Adjust the leveling of the sights by turning the screws clockwise or counter-clockwise. Position the "slits" of the tops of the screws to that they are parallel to the length of the barrel. By adjusting the height of the screws you compensate for the differences in parallel between the sight and the laser tube. Note that there is no need to adjust for windage because the light is not affected by the wind, nor do you need to compensate for bullet trajectory, because the light beam will continue in a straight path.

Painting

Although painting the laser pistol won't make it work better, it certainly improves its look. Prior to painting, you should remove the end caps, tube, switch, and jack, or else use masking tape to prevent paint from spraying on them. Be particularly wary of paint coming into contact with the output mirror of the tube and the internal contacts of the power plug.

Black ABS plastic pipe does not need to be painted, saving you from this extra step. However, straight pieces of 1¼- and 2-inch black ABS plastic are hard to find. Most plumb-

FIG. 21-8. *The completed He-Ne laser pistol, with modulation bypass shorting plug installed.*

ing and hardware stores carry only pre-formed fittings for drains and other waste-water systems.

Spray on a light coast of flat black paint; Testor's hobby paints are a good choice, and when used properly, won't sag during drying. The paint dries to a touch in 10 to 15 minutes, but it isn't cured until overnight. You can, of course, paint the pistol any color. A complete, painted pistol is shown in FIG. 21-8.

APPLICATIONS FOR THE LASER PISTOL

Let's face it, the He-Ne laser pistol isn't going to make you Luke Skywalker, so there is little chance you'll be going around the galaxy disintegrating bad guys. The output of the tube used in the pistol (like all other helium-neon lasers) isn't enough to be felt on skin, it won't burn holes in anything nor is it capable of any kind of destruction. That leaves rather peaceful applications of the pistol.

Before detailing some of the fun you can have with the laser pistol, I need to stress the safety requirements once more. The hand-held nature and design of the pistol might prompt you to use it in a Laser Tag-like game. Don't. The eyes of you and your opponent (man or beast) could be exposed to the laser beam—a definite health hazard. Point the pistol only at inanimate "blind" objects.

Hand-Held Pointer

Although the He-Ne pistol is rather large for the task, it can be effectively used as a hand-held pointer. Even in a large auditorium with a brightly lit movie, the laser pointer can be easily seen on the screen.

If you don't care for the pinpoint of light, try shaping the beam with optics and a shape mask. Use a double-concave lens to expand the beam from its nominal 0.75 mm to about 10 millimeters. A bi-convex lens collimates the beam—makes the rays parallel again. In front of the collimating lens you place a mask of an arrow, cut from a piece of thick black plastic, aluminum foil or photographic film (your local offset printer can provide a high-contrast mask of any artwork). When using aluminum foil, paint both front and back sides to cut down light reflections.

Because the positioning of the optical components depends on the focal lengths of the lenses, you should experiment with some lenses and try out the system before building it onto the pistol. When you have determined the proper placement of the lens and mask, mount them in PVC pipe or paper tube and attach the pipe to the end of the pistol. A 2-inch coupler can be used to easily attach extensions to the front of the laser.

The Ultimate Cat Toy

Believe it or not, many owners of helium-neon lasers spend countless hours using the bright red beam as a high-tech cat toy. If you have a cat and it's fairly playful, try this experiment: When the animal is least expecting it, shine the beam on the floor (not into its eyes). Most cats will react to the beam by pouncing on it. Of course, they can never get it because it's simply a spot of light on the carpet. Scan the beam at a fairly slow rate and have the cat chase after it.

Dogs don't seem to be much interested in laser beam spots. Mine just licks the end of the laser and wags his tail.

22

Laser Projects Potpourri

In the past several hundred pages, you've discovered numerous applications for your helium-neon and diode laser. Yet there are many, many more, enough to fill several volumes twice the size of this book. This chapter briefly reviews additional applications for lasers and details the components and procedures involved. Feel free to expand upon the ideas presented in the pages that follow; consider the projects as springboards for your own advanced experiments.

OPTICAL SWITCH USING LCD PANEL

An ordinary liquid crystal display (LCD) can be used as a type of laser shutter. Instead of using the display to indicate letters or numbers, you shine a laser through it. When voltage is applied to the terminals of the display, the crystals alternately rotate between their aligned and non-aligned states, breaking and passing the beam.

The basic construction of an LCD is shown in FIG. 22-1. To use the panel as a laser beam switch, remove the back reflective paper. If you accidentally remove the polarizing sheet, replace it with another. Be sure the replacement sheet is oriented in the proper direction.

Connect a 5-volt battery source to the terminals on the display. One or more segments should blink on. You can use any of the segments as long as it is wide enough to block the entire beam. If the segments dim out after you apply current, drive the display with 5- to 10-Hz pulses provided by a 555 timer IC (any book on the 555 can provide details on how to wire it as a pulse generator). You can also drive the panel with an audio source such as a tape recorder, radio, or amplified microphone.

342

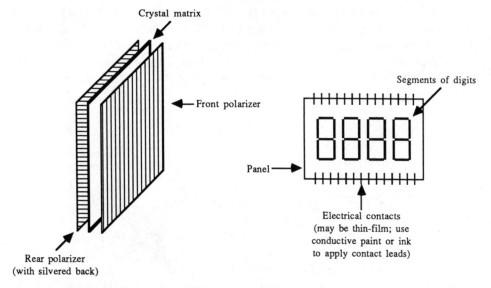

FIG. 22-1. *Construction of a reflective liquid crystal display (LED) panel. To use the panel as a shutter, remove the silver backing on the reverse side of the rear polarizer. Activate the panel by applying low-voltage dc to the electrical contacts. Shine the laser through one or more segments.*

To use the panel, position it in front of the laser and turn the segment on and off. Note that even when the segment is on, some of the laser beam will pass through, so the LCD panel can't be used as an absolute shutter. If you are using the panel as a modulation system, place a phototransistor in the path of the light and amplify the signal using the universal laser light receiver detailed in Chapter 13.

LASER SURVEILLANCE

Not long ago, Radio-Electronics magazine ran a cover story on using a helium-neon laser as a way to intercept the private conversations of others. The article provided step-by-step instructions on building the system, including a sensitive laser light detector. Although the "laser bug" is patently illegal, it's fun to play around with and to use as a way to test positional or geometrical modulation.

The system, as indicated in the article, is comprised of a laser mounted on a tripod and a receiver mounted on another tripod (you can combine the two on the same tripod but alignment is very difficult). The laser is aimed at a window, and the reflected beam is directed to the receiver. Voices or sound inside a building cause the windows to vibrate slightly, which positionally modulates the window pane. The movement is picked up by the receiver and amplified. An identical system is covered in Chapter 13 using a speaker and Mylar foil stretched over an embroidery hoop.

An interesting story about the laser bug is included in Forrest Mims' *Silliconnections* book (see Appendix B for publishing information). Apparently, Mims was asked by the *National Enquirer* to construct and use a laser bug to intercept the private conversations of the late Howard Hughes. Mims finally declined, worrying about the legality of the system, but later went on to demonstrate the bug on television.

You can conduct your own laser bug experiments using a He-Ne laser and the universal laser light receiver found in Chapter 13. You can make the photosensor more directional by mounting it on the back of a riflescope, as shown in FIG. 22-2. Experiment with the placement of the sensor until the received audio signal is the strongest. When using the bug during the daytime, place two polarizing filters in front of the sensor or scope to prevent swamping from sunlight.

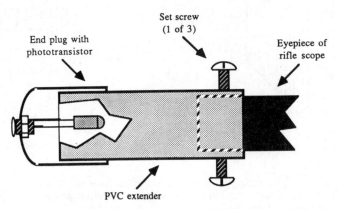

FIG. 22-2. *Use a piece of PVC 1-inch pipe to attach the end plug and phototransistor to the eyepiece of the rifle scope. Three set screws (set in a triangular configuration) clamp the pipe to the eyepiece.*

Depending on the type of window, drapes, and acoustic conditions inside, the sounds you pick up could be faint or overpowering. Strong background noise, like a dishwasher, stereo, or air conditioner, can totally mask any intelligible conversations inside. Of course, if the sound source you want to listen to isn't near a window, you won't hear anything.

Be aware that owning and using a laser bug might be illegal. It's decidedly a bad idea to construct and use a bug to intercept the conversations of others unless you have their permission.

LASER TACHOMETER

Pulses of light can be used to count the rotational speed of just about any object, even when you can't be near that object. The heart of the laser tachometer is the counter circuit shown in FIG. 22-3 (parts A through D). A parts list is provided in TABLE 22-1. In use, shine a helium-neon laser at the rotating object you want to time. If you can, paint a white or reflective strip on the object so the beam is adequately reflected. Place a photosensor and amplifier at the point where the reflected beam lands, and connect the amp to the input of the counter circuit.

At each revolution, the light beam is reflected off the reflective strip on the object and directed into the photosensor, amplifier, and counter. The counter resets itself every second, thereby giving you a readout of the rate of the revolution per second. For example, if a car wheel is turning at 100 revolutions per second, the number 100 appears on the display. To compute revolutions per minute (rpm), multiply the result on the readout by 60.

The maximum count using the three digits of the counter is 999. That equates to 59,940 revolutions per minute, faster than even a jet turbine. You shouldn't run into

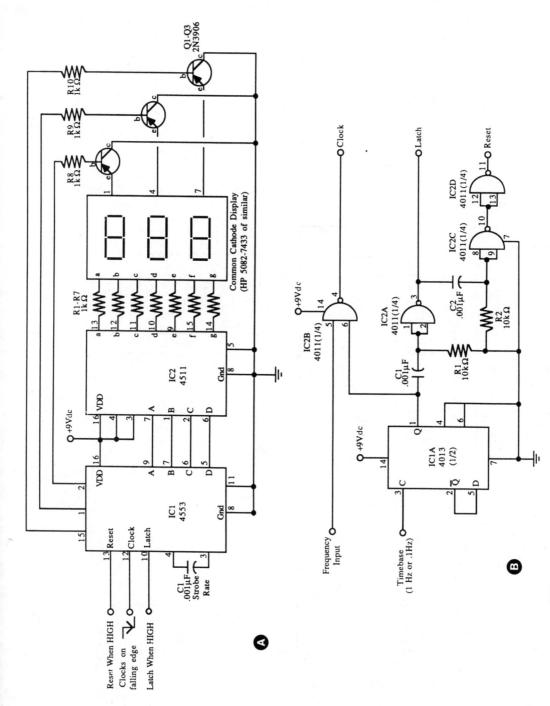

FIG. 22-3. *A schematic diagram for a light-activated tachometer, shown in subsections. (A) Main counter circuit; (B) Trigger logic circuit; (C) Crystal-controlled time base; (D) Phototransistor inputs for trigger reset (optional).*

345

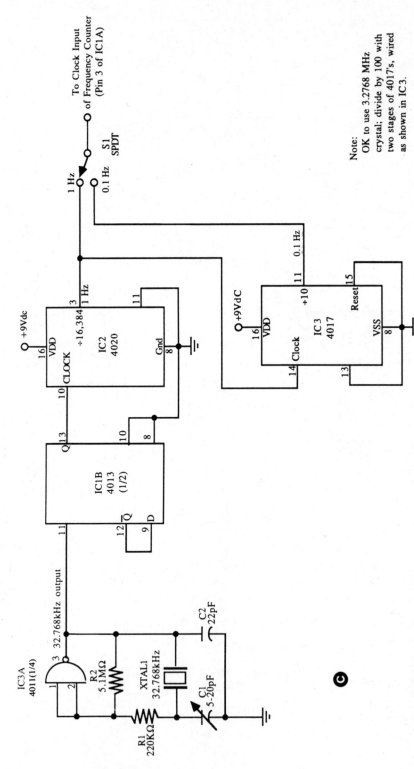

Fig. 22-3. *Continued.*

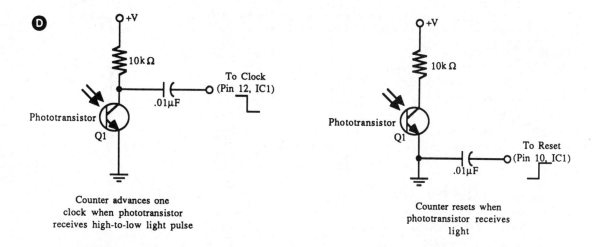

D

+V

10kΩ

To Clock
(Pin 12, IC1)

.01μF

Phototransistor

Q1

Counter advances one
clock when phototransistor
receives high-to-low light pulse

+V

10kΩ

Phototransistor

Q1

To Reset
(Pin 10, IC1)

.01μF

Counter resets when
phototransistor receives
light

Notes:
Use appropriate filters (red, IR)
depending on the laser used.

Adjust sensitivity by replacing 10kΩ resistor
with 250kΩ potentiometer.

Fig. 22-3. *Continued.*

too many things that outpace the tachometer. If the object you are timing is distant, use the riflescope arrangement presented in the previous section that allows you to aim the sensor directly at the reflective strip.

PERIMETER BURGLAR ALARM

Many stores use infrared break-beam systems to detect when a customer walks through the front door. Each time the beam is broken, a bell in the back rings, telling store personnel that someone has entered. A similar approach can be used in a perimeter burglar alarm system. The narrow divergence of the lasers means you can bounce a single beam around the yard, thus protecting a relatively large area. FIGURE 22-4 shows a few ways to position the laser, mirrors, and receiver around the house.

To actuate an alarm, connect the output of the receiver to a relay and bell. When the beam is broken, the voltage present at the output of the receiver drops, the relay kicks in, and the bell sounds off.

BASICS OF THE LASER GYROSCOPE

The latest commercial jets use a laser-based inertial guidance system. Three ring-shaped lasers detect rate and rotation of the aircraft in all three axes (pitch, yaw, and roll) and provide the data to an on-board computer. The advantage of the laser gyroscope is accuracy, even after many thousands of hours of use.

The laser gyroscope, shown schematically in FIG. 22-5, is modeled after a Michelson interferometer. Two beams are passed around the circumference of the gyro and

Table 22-1. Laser Tachometer Parts List

Counter Block (Fig. 22-3A)

IC1	4553 CMOS counter IC
IC2	4511 CMOS LED driver IC
R1-R10	1 kilohm resistor
C1	0.001 μF disc capacitor
Q1-Q3	2N3906 transistor
LED1-3	Common-cathode seven-segment LED display

Trigger Logic Circuit (Fig. 22-3B)

IC1	4013 CMOS flip-flop IC
IC2	4011 CMOS NAND gate IC
R1,R2	10 kilohm resistor
C1,C2	0.001 μF disc capacitor

Crystal-Controlled Time Base (Fig. 22-3C)

IC1	4013 CMOS flip-flop IC (from trigger logic circuit)
IC2	4020 CMOS counter IC
IC3	4017 CMOS counter IC
IC4	4011 CMOS NAND gate IC
R1	220 kilohm resistor
R2	5.1 megohm resistor
C1	5 to 20 pF miniataure variable capacitor
C2	22 pF disc capacitor
S1	DPDT switch

Phototransistor Inputs (Fig. 22-3D)

Q1	Infrared phototransistor
R1	10 kilohm resistor
C1	0.01 μF disc capacitor

All resistors are 5 to 10 percent tolerance, ¼ watt. All capacitors are 10 to 20 percent tolerance, rated 35 volts or more, unless otherwise indicated.

integrated at one or more photosensors and receivers. When at rest, the two beams travel the same distance and there is no change at the photosensors. But when the gyroscope is set into motion, one beam travels a longer distance than the other, and that results in a Doppler shift that causes beat frequencies or optical heterodyning at the detectors. These beat frequencies, which are like low-frequency audio tones, are correlated by a flight computer.

The Michelson interferometer detailed in Chapter 10 can be modified to work as a laser gyroscope. Just mount the base, with laser, on a ball-bearing turntable (lazy Susan).

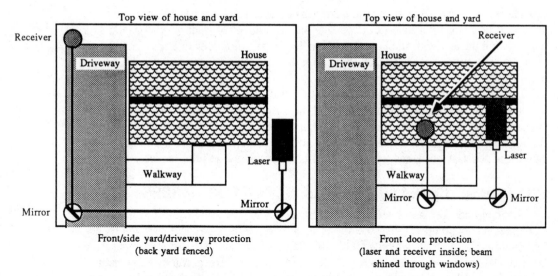

Receiver

Top view of house and yard

Driveway

House

Walkway

Laser

Mirror

Front/side yard/driveway protection
(back yard fenced)

Top view of house and yard

Receiver

Driveway

House

Walkway

Laser

Mirror

Mirror

Front door protection
(laser and receiver inside; beam
shined through windows)

FIG. 22-4. *Two approaches to building a break-beam perimeter laser alarm system. For very long distances, collimate the beam as detailed in Chapter 8.*

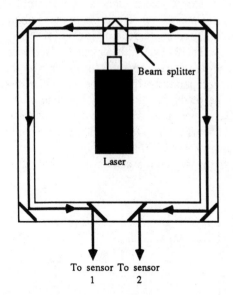

Beam splitter

Laser

To sensor
1

To sensor
2

FIG. 22-5. *Basic configuration of the laser gyroscope. The two beams may also be directed to one photosensor.*

Position a photosensor at the spot where the two beams meet and route the signal to an amplifier. You can hear the effect of movement when the interferometer is rotated.

STUDYING FLUID AERODYNAMICS

You can readily see the effects of wind on airfoils by rigging up your own laser smoke chamber. Build an accurate plastic model and make sure that the wings, fuselage, and other parts are adequately sealed. A coat of paint or lacquer will seal the seams and provide better results. Suspend the airplane in a cardboard box that has its top and bottom cut out. Cut a small portion out of the side and fit a sheet of clear acrylic plastic as a window. Fill the box with smoke and use a motor or galvanometer to scan a laser beam over the wing surface of the aircraft (details on these techniques are covered in Chapter 20, "Advanced Laser Light Shows").

Direct a fan into the box and watch the effect of the smoke as it streams past the airplane. Point the laser at different parts of the airframe to study the aerodynamics at specific points. With enough smoke, you should clearly see the effects of the wings, the turbulence behind the airplane, drag on structural elements, and more. Load your camera with a fast black-and-white film and photograph the results for later analysis.

CRYOGENIC COOLING OF SEMICONDUCTOR LASERS

All semiconductor lasers become super efficient at low temperatures. The first diode lasers could only be operated at the sub-freezing temperature of liquid nitrogen (minus 197 degrees Celsius). Only by the mid 1970's had they perfected the room-temperature laser diode. By dipping a low-power (1 to 5 mW) double-heterostructure cw laser in a glass filled with liquid nitrogen, you can increase its operating efficiency by several hundred percent.

Try this experiment. Connect a laser diode as shown in FIG. 22-6, and dip it into a glass Pyrex measuring cup. Connect a meter to the monitor photodiode to register light output. Note the reading from the photodiode on the meter. Now slowly fill the cup with liquid nitrogen. The liquid will bubble violently as it boils. Watch the reading on the meter jump as the diode is cooled. In my experiments, the reading on the meter was some 300 times higher at cryogenic temperatures than it was at regular room temperature.

This extra light can be readily seen not only because of the higher output of the laser, but because the laser operates at a lower wavelength when cooled. Instead of operating at the threshold of visible light (about 780 nm), the liquid nitrogen brought the operating wavelength down to about 730 nm, in the far red region of visible light. Use a lab-grade spectroscope to measure the exact wavelength of the emitted light.

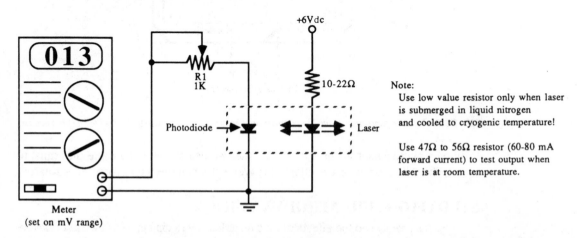

FIG. 22-6. *Use this setup to test the power output of a diode laser in and out of a bath of liquid nitrogen. The laser can be placed directly in the liquid without fear of short circuits because liquid nitrogen is not electrically conductive.*

350

Liquid nitrogen tips and techniques:

★ You can buy liquid nitrogen at welding, hospital, and medical supply outlets. Price is typically between $2 and $3 per liter.

★ Use a stainless steel or glass Thermos bottle to hold the liquid nitrogen (some outlets won't fill these canisters due to breakage and waste; check first). Drill a hole in the top of the Thermos to allow the nitrogen vapor to escape. Without the hole, the Thermos will explode.

★ The liquid nitrogen will last about a day in a Thermos. If you want to keep it longer, use a Dewar's flask or other approved container designed for handling refrigerated liquid gas.

★ Wear safety goggles and waterproof welder's gloves when handling liquid nitrogen. Never allow the liquid to touch your skin or you could get serious frostbite burns.

★ If the liquid touches clothing, immediately grasp the material at a dry spot and pull it away from your body.

★ While experimenting, fill liquid nitrogen only in Pyrex glass or metal containers. Avoid plastic and regular glass containers as they can shatter on contact with the extremely cold temperatures.

ADDITIONAL PROJECTS

There are numerous additional applications of lasers. These include using lasers in surveying as a means to calculate distances, as well as provide straight lines for grading, leveling, and pipe laying. For example, you can use an unexpanded beam to help you dig a straight trench in your backyard for a water pipe. A laser beam can also be used to test for flatness of a foundation or flooring. Place the laser on the ground and either move it manually or scan the beam using one of the techniques detailed in Chapters 19 and 20.

Very small movements in an object can be detected by attaching a mirror to the object and shining a laser on the mirror. Position a card or screen some distance from the object, as shown in FIG. 22-7, and watch the beam scan back and forth as the object is moved.

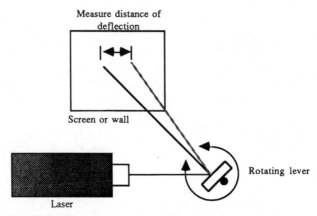

FIG. 22-7. *The basic arrangement for using a laser as a remote lever.*

This chapter has covered only a handful of useful applications for lasers; there are many, many more. Give it some thought, and you can probably come up with dozens of intriguing ways to put lasers to use. A good source for ideas is Metrologic's *101 Ways to Use a Laser*, a short mini-book that is bound into their catalog. In 15 pages, you are provided quick glimpses of ways to put lasers to work and how you can apply laser light to solve even the most demanding problems.

23

Tools and Supplies
for Laser Experimentation

Take a long look at the tools in your garage or workshop. You probably have all the implements necessary to build your own laser systems. Unless your designs require a great deal of precision (and most don't), a common assortment of hand tools are all that's really required to construct all sorts of laser projects.

Most of the hardware, parts, and supplies are things you probably already have that are left over from old projects from around the house. The pieces you don't have can be readily purchased at a hardware store and a few specialty stores around town or through the mail.

This chapter discusses the basic tools and supplies needed for constructing hobby laser systems and how you might use them. You should consider this chapter a guide only; suggestions for tools and supplies are just that—suggestions. By no means should you feel that you must own each tool mentioned in this chapter or have on hand all the parts and supplies.

Some supplies and parts might not be readily available to you, and it's up to you to consider alternatives and how to work these alternatives into the design. Ultimately, it will be your task to take a trip to the hardware store, collect various miscellaneous items, and go home to hammer out a unique creation that's all your own.

CONSTRUCTION TOOLS

Construction tools are the things you use to fashion the frame and other mechanical parts of the laser system. These include hammer, screwdriver, saw, and so forth. Tools for the assembly of the electronic subsystems are discussed later on.

Basic Tools

No workshop is complete without the following:

* **Claw hammer**, used for just about anything you can think of.
* **Rubber mallet**, for gently bashing pieces together that resist going together; also for forming sheet metal.
* **Screwdriver assortment**, including various sizes of flat-head and Philips-head screwdrivers. A few long-blade screwdrivers are handy to have, as well as a ratchet driver. Get a screwdriver magnetizer/demagnetizer to magnetize the blade so it attracts and holds screws for easier assembly.
* **Hacksaw**, to cut anything. The hacksaw is the staple of the laser hobbyist. Get an assortment of blades. Coarse-tooth blades are good for wood and PVC pipe plastic; fine-tooth blades are good for copper, aluminum, and light-gauge steel.
* **Miter box**, to cut straight lines. Buy a good miter box and attach it to your work table (avoid wood miter boxes; they don't last). You'll also use the box to cut stock at near-perfect 45-degree angles, helpful when building optical benches.
* **Wrenches**, all types. Adjustable wrenches are helpful additions to the shop but careless use can strip nuts. The same goes for long-nosed pliers, useful for getting at hard-to-reach places. A pair or two vise-grips (indispensable in my workshop) help you hold pieces for cutting and sanding. A set of nut-drivers make it easy to attach nuts to bolts.
* **Measuring tape**. A six- or eight-foot steel measuring tape is a good choice. Also get a cloth tape at a fabric store for measuring flexible things.
* **Square**, for making sure that pieces you cut and assemble from wood, plastic, and metal are square.
* **File assortment**, to smooth the rough edges of cut wood, metal, and plastic (particularly important when working with metal for safety).
* **Drill motor**. Get one that has a variable speed control (reversing is nice but not absolutely necessary). If the drill you have isn't variable speed, buy a variable-speed control for it. You need to slow the drill when working with metal and plastic. A fast drill motor is good for wood only. The size of the chuck is not important, because most of the drill bits you'll be using will fit a standard ¼-inch chuck.
* **Drill bit assortment**. Good, sharp ones *only*. If yours are dull, have them sharpened (or do it yourself with a drill-bit sharpening device), or buy a new set. *Never* use dull drill bits or your laser systems won't turn out.
* **Vise**, for holding parts while you work. A large vise isn't required, but you should get one that's big enough to handle the size of pieces you'll be working with.
* **Clear safety goggles**. Wear them when hammering, cutting, drilling, and any other time flying debris could get in your eyes.

If you plan on building your laser systems from wood, you might want to consider adding rasps, wood files, coping saws, and other woodworking tools to your toolbox. Working with plastic requires a few extras as well, including a burnishing wheel to smooth the edges of the cut plastic (a flame from a cigarette lighter also works but is harder to control), a strip-heater for bending, and special plastic drill bits. These bits have a modified tip that isn't as likely to rip through the plastic material. Bits for glass can be used as well. Small plastic parts can be cut and scored using a sharp razor knife or razor saw, available at hobby stores.

Optional Tools

There are a number of tools you can use to make your time in the laser shop more productive and less time-consuming. A *drill press* helps you drill better holes, because you have more control over the angle and depth of each hole. Use a drill press vise to hold the pieces; never use your hands.

A *table saw* or *circular* saw makes cutting through large pieces of wood and plastic easier. Use a guidefence, or fashion one out of wood and clamps, to ensure a straight cut. Be sure to use a fine-tooth saw blade if cutting through plastic. Using a saw designed for general wood-cutting will cause the plastic to shatter.

A motorized *hobby tool* is much like a hand-held router. The bit spins very fast (25,000 rpm and up), and you can attach a variety of wood-, plastic-, and metal-working bits to it. The better hobby tools, such as those made by Dremel and Weller, have adjustable speed controls. Use the right bit for the job. For example, don't use a wood rasp bit with metal or plastic, because the flutes of the rasp will too easily fill with metal and plastic debris.

A *nibbling tool* is a fairly inexpensive accessory (under $20) that lets you "nibble" small chunks from metal and plastic pieces. The maximum thickness depends on the bite of the tool, but it's generally about 1/16 or 1/8 inch. Use the tool to cut channels, enlarge holes, and so forth. A *tap and die set* lets you thread holes and shafts to accept standard size nuts and bolts. Buy a good set. A cheap assortment of taps and dies is more trouble than its worth.

A *thread size gauge* made of stainless steel might be expensive ($10 to $12), but it helps you determine the size of any standard SAE or metric bolt. It's a great accessory for tapping and dieing. Most gauges can be used when chopping threads off bolts with a hacksaw, providing a cleaner cut.

A *brazing tool* or *small welder* lets you spot-weld two metal pieces together. These tools are designed for small pieces only; they don't provide enough heat to adequately weld pieces larger than a few inches in size. Be sure that extra fuel and/or oxygen cylinders or pellets are readily available for the brazer or welder you buy. There's nothing worse than spending $30 to $40 for a home welding set, only to discover that supplies are not available for it. Be sure to read the instructions that accompany the welder and observe all precautions.

A *caliper* and *micrometer* let you measure small things such as lenses and other optical components. The projects in this book don't call for an extremely accurate caliper or micrometer, but if you're serious about laser work you'll want the best laboratory-grade instruments you can afford. At the very least, the caliper should be accurate to 1/64 of an inch and the micrometer to 0.001 of an inch.

ELECTRONIC TOOLS

Constructing electronic circuit boards or wiring the power system of your laser requires only a few standard electronic tools. A *soldering iron* leads the list. For maximum flexibility, invest in a modular soldering pencil with a 25- to 30-watt heating element. Anything higher could damage electronic components. (If necessary, a 40- or 50-watt element can be used for wiring switches, relays, and power transistors.) *Stay away from "instant-on" soldering irons*. Supplement your soldering iron with these accessories:

⭐ **Soldering stand**, for keeping the soldering pencil in a safe, upright position.

⭐ **Soldering tip assortment**. Get one or two small tips for intricate PCB work and a few larger sizes for routine soldering chores.

⭐ **Solder**. Resin- or flux-core only. Acid core and silver solder should *never* be used on electronic components.

⭐ **Sponge**, for cleaning the soldering tip during use. Keep the sponge damp and wipe the tip clean after every few joints.

⭐ **Heatsink**, for attaching to sensitive electronic components during soldering. The heatsink draws the excess heat away from the component to help prevent damage to it.

⭐ **Desoldering vacuum tool**, to soak up molten solder. Used to get rid of excess solder, to remove components, or redo a wiring job.

⭐ **Soldering or "dental" picks**, for scraping, cutting, forming, and gouging into the work.

⭐ **Resin cleaner**. Apply the cleaner after soldering is complete to remove excess resin.

⭐ **Solder vise** or "third hand." The vise holds together pieces to be soldered, leaving you free to work the iron and feed the solder.

VOLT-OHMMETER

A *volt-ohmmeter* is used to test voltage levels and the impedance of circuits. This moderately priced electronic tool is the basic requirement for working with electronic circuits of any kind. If you don't already own a volt-ohmmeter, seriously consider buying one. The cost is minimal considering the usefulness of the device.

There are many volt-ohmmeters (or VOMs) on the market today. For work on lasers, you don't want a cheap model, but you don't need an expensive one. A meter of intermediate quality is sufficient and does the job admirably. The price for such a meter is between $30 and $75. Meters are available at Radio Shack and most electronics outlets. Shop around and compare features and prices.

Digital or Analog

There are two general types of VOMs available today: digital and analog. The difference is not that one meter is used on digital circuits and the other on analog circuits. Rather, digital meters employ a numeric display not unlike a digital clock or watch, and analog VOMs use the older fashioned—but still useful—mechanical movement with a needle that points to a set of graduated scales.

Digital VOMs used to cost a great deal more than the analog variety, but the price difference has evened out recently. Digital VOMs, such as the one shown in FIG. 23-1, are fast becoming the standard; in fact, it's becoming difficult to find a decent analog meter anymore.

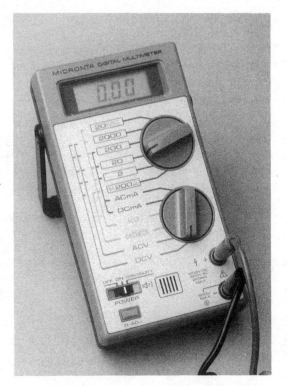

FIG. 23-1. *A typical digital volt-ohmmeter.*

Analog VOMs are traditionally harder to use, because you must select the type and range of voltage you are testing, find the proper scale on the meter face, then estimate the voltage as the needle swings into action. Digital VOMs, on the other hand, display the voltage in clear numerals and with a greater precision than most analog meters.

Automatic Ranging

As with analog meters, some digital meters require you to select the range before it can make an accurate measurement. For example, if you are measuring the voltage of a 9-volt transistor battery, you set the range to the setting closest to, but above, 9 volts (with most meters it is the 20- or 50-volt range). Auto-ranging meters don't require you to do this, so they are inherently easier to use. When you want to measure voltage, you set the meter to volts (either ac or dc) and take the measurement. The meter displays the results in the readout panel.

Accuracy

Little of the work you'll do with laser circuits requires a meter that's super accurate. A VOM with average accuracy is more than enough. The accuracy of a meter is the minimum amount of error that can occur when taking a specific measurement. For example, the meter may be accurate to 2,000 volts, ±0.8 percent. A 0.8-percent error at the kinds of voltages used in most laser experiments—typically 5 to 12 volts dc—is only 0.096 volts!

Digital meters have another kind of accuracy. The number of digits in the display determines the maximum resolution of the measurements. Most digital meters have 3 and one-half digits, so it can display a value as small as .001 (the half digit is a "1" on the left side of the display). Anything less than that is not accurately represented.

Functions

Digital VOMs vary greatly in the number and type of functions they provide. At the very least, all standard VOMs let you measure ac volts, dc volts, milliamps, and ohms. Some also test capacitance and opens or shorts in discrete components like diodes and transistors. These additional functions are not absolutely necessary for building general-purpose laser circuits, but they are handy to have when troubleshooting a circuit that refuses to work.

The maximum ratings of the meter when measuring volts, milliamps, and resistance also varies. For most applications, the following maximum ratings are more than adequate:

dc Volts	1,000 volts
ac Volts	500 volts
dc Current	200 milliamps
Resistance	2 megohms

One exception to this is when testing current draw for motors and other high-demand circuits. All but the smallest dc motors draw an excess of 200 milliamps, and an entire laser system for a light show is likely to draw 2 or more amps. Obviously, this is far out of range of most digital meters. You might need to get a good assessment of current draw, especially if your laser projects are powered by batteries, but to do so, you'll need either a meter with a higher dc current rating (digital or analog) or a special-purpose ac/dc current meter. You can also use a resistor in series with the motor and apply Ohm's Law to calculate the current draw.

Meter Supplies

Meters come with a pair of test leads—one black and one red—each equipped with a needle-like metal probe. The quality of the test leads is usually minimal, so you might want to purchase a better set. The coiled kind are handy. They stretch out to several feet yet recoil to a manageable length when not in use.

Standard leads are fine for most routine testing, but some measurements require the use of a clip lead. These attach to the end of the regular test leads and have a spring-loaded clip on the end. You can clip the lead in place so your hands are free to do other things. The clips are insulated to prevent short circuits.

Meter Safety and Use

Most applications of the meter involve testing low voltages and resistance, both of which are relatively harmless to humans. Sometimes, however, you might need to test high voltages—like the input to a power supply—and careless use of the meter can cause serious bodily harm. Even when you're not actively testing a high-voltage circuit, dangerous currents might still be exposed.

Proper procedure for meter use involves setting the meter beside the unit under test, making sure it is close enough so that the leads reach the circuit. Plug in the leads and test the meter operation by first selecting the resistance function setting (use the smallest scale if the meter is not auto-ranging). Touch the leads together: the meter should read 0 ohms.

If the meter does not respond, check the leads and internal battery and try again. If the display does not read 0 ohms, double-check the range and function settings and adjust the meter to read 0 ohms (not all digital meters have a 0 adjust, but most analog meters do).

Once the meter has checked out, select the desired function and range and apply the leads to the circuit under test. Usually, the black lead is connected to ground, and the red lead is connected to the various test points in the circuit.

When testing high-voltage circuits, make it a habit to place one hand in a pants pocket. With one hand out of the way, you are less likely to accidentally touch a live circuit.

LOGIC PROBE

Meters are typically used for measuring analog signals. *Logic probes* test for the presence or absence of low-voltage dc signals that represent digital data. The 0s and 1s are usually electrically defined as 0 and 5 volts, respectively, with TTL ICs. In practice, the actual voltages of the 0 and 1 bits depend entirely on the circuit. You can use a meter to test a logic circuit, but the results aren't always predictable. Further, many logic circuits change states (pulse) quickly and meters cannot track the voltage switches fast enough.

Logic probes, such as the model in FIG. 23-2, are designed to give a visual and (usually) aural signal of the logic state of a particular circuit line. One LED on the probe

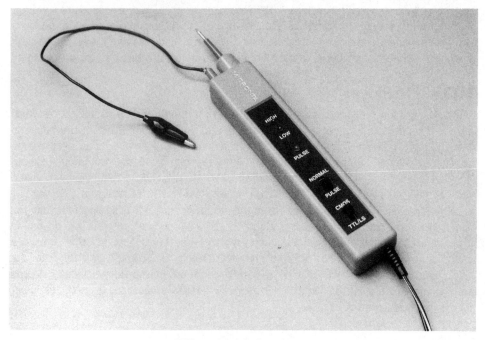

FIG. 23-2. *A logic probe.*

lights up if the logic is 0 (or LOW), another LED lights up if the logic is 1 (or HIGH). Most probes have a built-in buzzer that has a different tone for the two logic levels. That way, you don't need to keep glancing at the probe to see the logic level.

A third LED or tone can indicate a pulsing signal. A good logic probe can detect that a circuit line is pulsing at speeds of up to 10 MHz, which is more than fast enough for laser applications, even when using computer control. The minimum detectable pulse width (the time the pulse remains at one level) is 50 nanoseconds, again more than sufficient.

Although logic probes might sound complex, they are really simple devices and their cost reflects this. You can buy a reasonably good logic probe for under $20. Most probes are not battery operated; rather, they obtain operating voltage from the circuit under test. You can also make a logic probe if you wish. A number of project books provide plans.

Using a Logic Probe

The same safety precautions apply when using a logic probe as they do when using a meter. Be wary when working close to high voltages. Cover them to prevent accidental shock (for obvious reasons, logic probes are not meant for anything but digital circuits, so never apply the leads of the probe to an ac line). Logic probes cannot operate with voltages exceeding about 15 volts dc, so if you are unsure of the voltage level of a particular circuit, test it with a meter first.

Successful use of the logic probe really requires you to have a circuit schematic to refer to. Keep it handy when troubleshooting your projects. It's nearly impossible to blindly use the logic probe on a circuit without knowing what you are testing. And because the probe receives its power from the circuit under test, you need to know where to pick off suitable power. To use the probe, connect the probe's power leads to a voltage source on the board, clip the black ground wire to circuit ground, and touch the tip of the probe against a pin of an integrated circuit or the lead of some other component. For more information on using your probe, consult the manufacturer's instruction sheet.

LOGIC PULSER

A handy troubleshooting accessory when working with digital circuits is the *logic pulser*. This device puts out a timed pulse, letting you see the effect of the pulse on a digital circuit. Normally, you'd use the pulser with a logic probe or an oscilloscope (discussed below). The pulser is switchable between one pulse and continuous pulsing. You can make your own pulser out of a 555 timer IC. FIGURE 23-3 shows a schematic you can use to build your own 555-based pulser; TABLE 23-1 provides a parts list.

Most pulsers obtain their power from the circuit under test. It's important that you remember this. With digital circuits, it's generally a bad idea to present to a device an input signal that is greater than the supply voltage for the device. In other words, if a chip is powered by 5 volts, and you give it a 12-volt pulse, you'll probably ruin the chip. Some circuits work with split (+, −, and ground) power supplies (especially circuits with op amps and digital-to-analog converters). Be sure to connect the leads of the pulser to the correct power points.

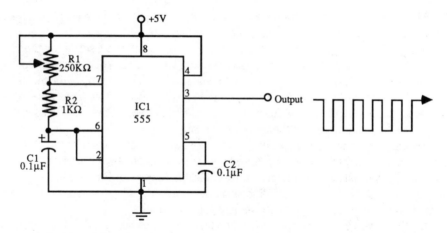

FIG. 23-3. *Schematic diagram for making your own logic pulser. With the components shown, output frequency is approximately 2.5 Hz to 33 Hz.*

Also be sure that you do not pulse a line that has an output but no input. Some integrated circuits are sensitive to unloaded pulses at their output stages, and improper application of the pulse can destroy the chip.

OSCILLOSCOPE

An *oscilloscope* is a pricey tool—good ones start at about $500—and only a small number of electronic and laser hobbyists own one. For really serious work, however, an oscilloscope is an invaluable tool—one that will save you hours of time and frustration.

Things you can do with a scope include some of the things you can do with other test equipment, but oscilloscopes do it all in one box and generally with greater precision. Among the many applications of an oscilloscope, you can:

☆ Test dc or ac voltage levels.
☆ Analyze the waveforms of digital and analog circuits.
☆ Determine the operating frequency of digital, analog, and RF circuits.
☆ Test logic levels.
☆ Visually check the timing of a circuit to see if things are happening in the correct order and at the prescribed time intervals.

Table 23-1. Logic Pulser Parts List

IC1	LM555 timer IC
R1	250 kilohm potentiometer
R2	10 kilohm resistor
C1	2.2 µF electrolytic capacitor
C2	0.1 µF disc capacitor

361

The designs provided in this book don't absolutely require the use of an oscilloscope, but you'll probably want one if you design your own circuits or want to develop your electronic skills. A basic, no-nonsense model is enough, but don't settle for the cheap, single-trace units. A dual-trace (two channel) scope with a 20 to 25 MHz maximum input frequency should do the job nicely. The two channels let you monitor two lines at once, so you can easily compare the input signal and output signal at the same time. You do not need a scope with storage or delayed sweep, although if your model has these features, you're sure to find a use for them sooner or later.

Scopes are not particularly easy to use; they have lots of dials and controls that set operation. Thoroughly familiarize yourself with the operation of your oscilloscope before using it for any construction project or for troubleshooting. Knowing how to set the time-per-division knob is as important as knowing how to turn the scope on. As usual, exercise caution when using the scope with or near high voltages.

Lastly, don't rely on just the instruction manual that came with the set to learn how to use your new oscilloscope. Buy (and read!) a good book on how to effectively use your scope. Appendix B, "Further Reading," lists some books on using oscilloscopes.

FREQUENCY METER

A *frequency meter* (or frequency counter) tests the operating frequency of a circuit. Most models can be used on digital, analog, and RF circuits for a variety of testing chores—from making sure the crystal analog-to-digital circuit is working properly to determining the modulation frequency of your laser beam communication system. You need only a basic frequency meter—a $100 or so investment. Or you can save some money by building a frequency meter kit.

Frequency meters have an upward operating limit, but it's generally well within the region applicable to laser experiments. A frequency meter with a maximum range of up to 50 MHz is enough. A couple of meters are available with an optional *prescaler*, a device that extends the useful operating frequency to well over 100 MHz.

WIRE-WRAPPING TOOL

Making a printed-circuit board for a one-shot application is time consuming, though it can be done with the proper kits and supplies. Conventional point-to-point solder wiring is not an acceptable approach when constructing digital and high-gain analog circuits, which represent the lion's share of electronics you'll be building for your lasers.

The preferred construction method is *wire-wrapping*. Wire-wrapping is a point-to-point wiring system that uses a special tool and extra-fine 28- or 30-gauge wrapping wire. When done properly, wire-wrapped circuits are as sturdy as soldered circuits, and you have the added benefit of making modifications and corrections without the hassle of desoldering and resoldering.

A manual wire-wrapping tool is shown in FIG. 23-4. You insert one end of the stripped wire into a slot in the tool, and place the tool over a square-shaped wrapping post. Give the tool five to ten twirls, and the connection is complete. The edges of the post keep the wire anchored in place. To remove the wire, use the other end of the tool and undo the wrapping.

FIG. 23-4. *A wire-wrapping tool in action.*

Wrapping wire comes in many forms, lengths, and colors, and you need to use special wire-wrapping sockets and posts. See the section below on electronic supplies and components for more details.

BREADBOARD

You should test each of the circuits you want to use with your lasers (including the ones in this book) on a *solderless breadboard* before you commit it to wire-wrap or solder. Breadboards consist of a series of holes with internal contacts spaced 1/10 of an inch apart, which is the most common spacing for ICs. You plug in ICs, resistors, capacitors, transistors, and 20- or 22-gauge wire in the proper contact holes to create your circuit.

Solderless breadboards come in many sizes. For the most flexibility, get a double-width board, one that can accommodate at least 10 ICs. Smaller boards can be used for simple projects; circuits with a high number of components require bigger boards. While you're buying a breadboard, purchase a set of pre-colored wires. The wires come in a variety of lengths and are already stripped and bent for use in breadboards. The set costs $5 to $7, but the price is well worth the time you'll save.

HARDWARE SUPPLIES

A fully functional laser system of just about any description is about 75 percent hardware and 25 percent electronic and electromechanical. Most of your trips to get parts for your laser schemes will be to the local hardware store. Here are some common items you'll want to have around your shop:

Nuts and Bolts

Number 8 and 10 nuts and pan-head stove bolts ($8/32$ and $10/24$, respectively) are good for all-around construction. Get a variety of bolts in ½-, ¾-, 1-, 1¼, and 1½-inch lengths. You may also want to get some 2-inch and 3-inch-long bolts for special applications.

Motor shafts and other heavy-duty applications require ¼-inch 20 or $5/16$-inch hardware. Pan-head stove bolts are the best choice; you don't need hex-head carriage bolts unless you have a specific requirement for them. You can use number 6 ($6/32$) nuts and bolts for small, lightweight applications.

Washers

While you're at the store, stock up on flat washers, fender washers (large washers with small holes), tooth lockwashers and split lockwashers. Get an assortment for the various sizes of nuts and bolts. Split lockwashers are good for heavy-duty applications because they provide more compression locking power. You usually use them with bolt sizes of ¼-inch and larger.

All-Thread Rod

All-thread is 2- to 3-foot lengths of threaded rod stock. It comes in standard thread sizes and pitches. All-thread is good for shafts and linear motion actuators. Get one of each in $8/32$, $10/24$, and ¼-inch 20 threads to start.

Special Nuts

Coupling nuts are just like regular nuts but have been stretched out. They are designed to couple two bolts or pieces of all-thread together, end to end. In lasers, you might use them for a variety of tasks including linear motion actuators and positioning tables.

Locking nuts have a piece of nylon built into them that provides a locking bite when threaded onto a bolt. Locking nuts are preferred over using two nuts tightened together.

EXTRUDED ALUMINUM

For most of your laser designs, you can take advantage of a rather common hardware item: extruded aluminum stock. This aluminum is designed for such things as building bathtub enclosures, picture frames, and other handyman applications and comes in various sizes, thicknesses, and configurations. Length is usually 6, 8, 10, or 12 feet, but if you need less, most hardware stores will cut to order (you save when you buy it in full lengths). The stock is available in plain (dull silver) anodized aluminum and gold anodized aluminum. Get the plain stuff—it's 10 to 25 percent cheaper.

Two particularly handy stocks are $41/64$-by-½-by-$1/16$-inch channel and $57/64$-by-$9/16$-by-$1/16$-inch channel (some stores sell similar stuff with slightly different dimensions). I use these extensively to make parts for optical benches, lens holders, and other laser system parts. Angle stock measuring 1-by-1-by-$1/16$-inches is another frequently used item, usually

364

employed for attaching cross bars and other structural components. No matter what size you eventually settle on for your own designs, keep several feet of the stuff handy at all times. You'll use it often.

If extruded aluminum is not available, another approach is to use shelving standards— the bar-like channel stock used for wall shelving. It's most often available in steel, but some hardware stores carry it in aluminum (silver, gold, and black anodized).

The biggest problem with using shelving standards is that the slots can cause problems when drilling holes for hardware. The drill bits can slip into the slots, causing the hole to be off-center. Some standards have an extra lip on the inside of the channel that can interfere with some of the hardware you use to join the pieces together.

ANGLE BRACKETS

You need a good assortment of ⅜-inch and ½-inch galvanized iron brackets to join the extruded stock or shelving standards together. Use 1½-by-⅜-inch flat corner irons when joining pieces cut at 45-degree angles to make a frame. The 1-by-⅜-inch and 1½-by-⅜-inch corner angle irons are helpful when attaching the stock to baseplates and when securing various components.

ELECTRONIC SUPPLIES AND COMPONENTS

Most of the electronic projects in this book and other books with digital and analog circuits depend on a regular stable of common electronic components. If you do any amount of electronic circuit building, you'll want to stock up on the following standard components. Keeping spares handy prevents you from making repeat trips to the electronics store.

Resistors

Get a good assortment of ¼- or ⅛-watt resistors. Make sure the assortment includes a variety of common values and that there are several of each value. Supplement the assortment with individual purchases of the following resistor values: 270 ohm, 330 ohm, 1 kilohm, 3.3 kilohm, 10 kilohm, and 100 kilohm.

The 270 and 330 ohm values are often used with light-emitting diodes (LEDs) and the remaining values are common to TTL and CMOS digital circuits.

Variable Resistors

Variable resistors, or potentiometers (pots), are relatively cheap and are a boon when designing and troubleshooting circuits. Buy an assortment of the small PC-mount pots (about 80 cents each retail) in the 2.5k, 5k, 10k, 50k, 100k, 25 kilohm and 1 megohm values. You'll find 1 megohm pots often used in op amp circuits, so buy a couple extra of these. I also like to have several extra 10k and 100k pots around because these find heavy use in most all types of circuits.

Capacitors

Like resistors, you'll find yourself returning to the same standard capacitor values project after project. For a well-stocked shop, get a dozen or so each of the following inexpensive ceramic disc capacitors: 0.1, 0.01, and 0.001 μF. The 0.1 and 0.01 μF caps are used extensively as *bypass* components and are absolutely essential when building circuits with TTL ICs. You can never have enough of these.

Many circuits use the in-between values of 0.47, 0.047, and 0.022 μF, so you may want to get a couple of these, too. Power supply, timing, and audio circuits often use larger polarized electrolytic or tantalum capacitors. Buy a few each of 1.0, 2.2, 4.7, 10 and 100 μF values. Some projects call for other values (in the picofarad range and the 1000's of microfarad range). You can buy these as needed unless you find yourself returning to standard values repeatedly.

Transistors

There are thousands of transistors available, and each one has slightly different characteristics than the others. However, most applications need nothing more than ''generic'' transistors for simple switching and amplifying. Common npn signal transistors are the 2N2222 and the 2N3904. Both kinds are available in bulk packages of 10 for about $1. Common pnp signal transistors are the 2N3906 and the 2N2907. Price is the same or a little higher.

I don't discriminate between the plastic and metal can transistors. For example, technically speaking, the plastic 2N2222 are called PN2222 while the metal can version carries the 2N2222 designator. In any case, buy the plastic ones because they're cheaper.

If the circuit you're building specifies another transistor than the generic kind, you might still be able to use one of these if you first look up the specifications of the transistor called for in the schematic. A number of cross-reference guides provide the specifications and replacement-equivalents for popular transistors.

There are common power transistors as well. The npn TIP31 and TIP41 are familiar to most anyone who has dealt with power switching or amplification of up to 1 amp or so. The pnp counterparts are the TIP32 and TIP42. These transistors come in the TO-220 style package.

A common larger capacity npn transistor that can switch 10 amps or more is the 2N3055. It comes in the TO-3 style package and is available everywhere. Price is between 50 cents and $2, depending on the source.

Diodes

Common diodes are the 1N914 for light-duty signal switching applications and the 1N4000 series (1N4001, 1N4002, and so forth). Get several of each and use the proper size to handle the current in the circuit. Refer to a databook on the voltage and power handling capabilities of these diodes.

LEDs

All semiconductors emit light (either visible or infrared), but light-emitting diodes (LEDs) are especially designed for the task. LEDs last longer than regular filament lamps and require less operating current. They are available in a variety of sizes, shapes, and colors. For general applications, the medium-sized red LED is perfect. Buy a few dozen and use them as needed. Some of the projects in this book call for infrared LEDs. These emit no visible light and are used in conjunction with an infrared-sensitive phototransistor or photodiode.

The project in Chapter 4, "Experimenting With Light and Optics," uses a special high-power visible red LED. Refer to that chapter for more information on this LED and where to obtain it.

Integrated Circuits

Integrated circuits let you construct fairly complex circuits from just a couple of components. Although there are literally thousands of different ICs, some with exotic applications, a small handful crops up again and again in hobby projects. You should keep the following ICs in ready stock:

★ **555 timer**. This is by far the most popular integrated circuit for hobby electronics. With just a couple of resistors and capacitors, the NE555 can be made to act as a pulser, a timer, a time delay, a missing pulse detector, and dozens of other useful things. The chip is usually used as a pulse source for digital circuits. It's available in dual versions as the 556 and quadruple versions as the 558. A special CMOS version lets you increase the pulsing rate to 2 MHz.

★ **LM741 op amp**. The LM741 comes second in popularity to the 555. The 741 can be used for signal amplification, differentiation, integration, sample-and-hold, and a host of other useful applications. The 741 is available in a dual version—the 1458. The chip comes in different package configurations. The schematics in this book and those usually found elsewhere, specify the pins for the common 8-pin DIP package. If you are using the 14-pin DIP package or the round can package, check the manufacturer's data sheet for the correct pinouts. Note that there are numerous op amps available, and some have design advantages over the 741.

★ **TTL chips**. TTL ICs are common in computer circuits and other digital applications. There are many types of TTL packages, but you won't use more than 10 or 15 of them unless you're heavily into electronics experimentation. Specifically, the most common and most useful TTL ICs are the 7400, 7401, 7402, 7404, 7407, 7408, 7414, 7430, 7432, 7473, 7474, 74154, 74193, and 74244. All or many of these are available in "TTL chip assortments" through some of the mail-order electronics firms.

★ **CMOS chips**. Because CMOS ICs require less power to operate than the TTL variety, you'll often find them specified for use with low-power laser and remote-control applications. Like TTL, there is a relatively small number of common packages: 4001, 4011, 4013, 4016, 4017, 4027, 4040, 4041, 4049, 4060, 4066, 4069, 4071, and 4081.

Wire

Solid-conductor, insulated, 22-gauge hookup wire can be used in your finished projects as well as connecting wires in breadboards. Buy a few spools in different colors. Solid-conductor wire can be crimped sharply and can break when excessively twisted and flexed. If you expect that wiring in your project might be flexed repeatedly, use stranded wire instead. Heavier 12- to 18-gauge hookup wire is required for connection to heavy-duty batteries, motors, and circuit-board power supply lines.

Wire-wrap wire is available in spool or pre-cut/pre-stripped packages. For ease of use, buy the more expensive pre-cut stuff unless you have a tool that does it for you. Get several of each length. The wire-wrapping tool has its own stripper built in (which you must use instead of a regular wire stripper), so you can always shorten the precut wires as needed. Some special wire-wrapping tools require their own wrapping wire. Check the instruction that came with the tool for details.

CIRCUIT BOARDS

Simple projects can be built onto solder breadboards. These are modeled after the solderless breadboard, so you simply transfer the tested circuit from the solderless breadboard to the solder board. You can cut the board with a hacksaw or razor saw if you don't need all of it.

Larger projects require perforated boards. Get the kind with solder tabs or solder traces on them. You'll be able to secure the components onto the boards with solder. Most perf boards are designed for wire-wrapping.

IC Sockets

You should use sockets for ICs whenever possible. Sockets come in sizes ranging from 8-pin to 40-pin. The sockets with extra-long square leads are for wire-wrapping.

You can also use wire-wrap IC sockets to hold discrete components like resistors, capacitors, diodes, LEDs, and transistors. You can, if you wish, wire-wrap the leads of these components, but because the leads are not square, the small wire doesn't have anything to bite into, so the connection won't be very strong. After assembly and testing and you are sure the circuit works, apply a dab of solder to the leads to hold the wires in place.

SETTING UP SHOP

You'll need a work table to construct the mechanisms and electronic circuits of your lasers and laser systems. Electronic assembly can be indoors or out, but I've found that when working in a carpeted room, it's best to spread another carpet or some protective cover over the floor. When the throw rug fills with solder bits and little pieces of wire and component leads, I can take it outside, beat it with a broom handle, and it's as good as new.

Unlike the manufacturing process, actual use of your laser system should be done indoors, in a controlled environment. You'll be using precision optics, so you should avoid exposing them to dust, dirt, and temperature extremes often found in garages. If you plan on building a sand box holography system as described in Chapters 17 and 18, you must be sure to keep the table indoors and away from moisture, because sand has a tendency to soak up water that can impede the creation of your holographic masterpieces.

Whatever space you choose for your laser lab, make sure all your tools are within easy reach. Keep special tools and supplies in an inexpensive fishing tackle box. Tackle boxes have lots of small compartments for placing screws and other parts.

For best results, your workspace should be an area where the laser system in progress will not be disturbed if you have to leave it for several hours or days (as will usually be the case). The work table should also be one that is off limits or inaccessible to young children or at least an area that can be easily supervised. This is especially true of helium-neon-based lasers and the high-voltage power supplies used with them.

Good lighting is a must. Both mechanical and electronic assembly require detail work, and you need good lighting to see everything properly. Supplement overhead lights with a 60-watt desk lamp. You'll be crouched over the worktable for hours at a time, so a comfortable chair or stool is a must, and be sure the seat is adjusted for the height of the worktable.

24

Buying Laser Parts

Building a laser system from scratch can be difficult or easy—it's up to you. From experience, I've found that the best way to simplify the construction of a laser system is to use standard, off-the-shelf parts like things you can get at the neighborhood hardware store, auto parts store, and electronics store.

Finding parts for your lasers is routine, all things considered, and little thought goes into it. Be forewarned, however, that there are some tricks of the trade, shortcuts, and tips you should consider before you go on a buying spree and stock up for your next project. By shopping carefully and wisely, you can save both time and money in your laser-building endeavors.

PLACES TO BUY

Building laser systems on a budget means you can't take a quick trip to the local industrial laser supply. Even if there is one in your area, the components will cost you more than most people make in a year. Top-grade optical gear carries stiff pricetags, and only well-endowed universities, corporations, and research firms can afford them.

As a laser hobbyist, you must learn how to find and adapt common hardware and other everyday parts to laser components. An old, discarded spyglass, for example, might make a perfect collimating telescope. You might need to refit some of the lenses in the scope, but to help keep costs down, you might find the required lenses at a surplus store or else rob them from another piece of optical equipment. Depending on the quality of the original components as well as the level of your ingenuity, you might end up with a collimating telescope that's every bit as good as one costing several hundred dollars. If you shop wisely, your collimating telescope might only cost you $10.

There are numerous places to buy parts for your laser projects. You'll find the best deals at:

★ hardware stores
★ electronics stores
★ specialty stores
★ surplus stores
★ mail order

Buying at Hardware Stores

The small-town neighborhood hardware store is a great place for nails to fix the back porch or a rake to clean up the yard. But on closer inspection, it seems that while the hardware outlet carries thousands of items, it doesn't have the ones you need at that exact moment. More often than not, you'll find it necessary to make trips to a variety of hardware stores. Some stores cater to a specific segment of do-it-yourselfers and professionals. Some stores are designed expressly to please professional painters, while other are for weekend plumbers and electricians. Realize this specialization and you'll have better luck in finding the parts you need.

Warehouse hardware outlets and builder's supply stores (usually open to the public) are the best source for the wide variety of tools and parts you need for laser experimentation. Items like nuts and bolts are generally available in bulk, so you can save a considerable amount of money.

As you tour the hardware stores in your area, keep a notebook handy and jot down the lines each outlet carries. Then when you find yourself needing a specific item, just refer to your notes. On a regular basis, take an idle stroll through your regular hardware store haunts. Unless the store is very small, you'll always find something new and perhaps laughably useful in laser system design each time you visit.

Buying at Electronics Stores

As recent as ten years ago, electronic parts stores used to be in plentiful supply. Even some automotive outlets carried a full range of tubes, specialty transistors, and other electronic gadgets. Now, Radio Shack remains as the only national electronics store chain. In many towns across the country, its the only thing going.

Radio Shack continues to support electronics experimenters, but they stock only the most common components. If your needs extend beyond resistors, capacitors, and a few integrated circuits, you must turn to other sources. Check the local Yellow Pages under *Electronics-Retail* for a list of electronic parts shops near you.

Radio Shack isn't known for the best prices in electronic parts (although sometimes they have really good bargains), yet the neighborhood independent electronics specialty store might not be much better. It's not unusual for pre-packaged resistors and capacitors to sell for 50 cents to $1 each, and few stores carry cost-saving parts assortments. Unless you need a specific component that isn't available anywhere else near you, stay away from the independent electronics outlet. There are independent stores that don't charge outrageously for parts, or course, but these are the exceptions, not the rule.

All is not lost. Mail order provides a welcome relief to overpriced electronic components. You can find ads for these mail-order firms in the various electronics magazines such as Radio-Electronics and Modern Electronics. Both publications are available at newsstands. Also, several mail-order firms are listed in Appendix A.

Specialty Stores

Specialty stores are those outlets open to the general public that sell items that you won't find in a regular hardware or electronic parts store. Specialty stores don't include surplus outlets (discussed later in this chapter).

What specialty stores are of use to laser hobbyists? Consider these:

★ **Sewing machine repair shops**. Ideal for small gears, cams, levers, and other precision parts. Some shops will sell you broken machines—break the machine down and use the parts for your laser projects.

★ **Used battery depot**. A good source for cheap batteries for your portable laser designs. Most of the batteries are used, but reconditioned. The shop takes in old car and motorcycle batteries and refurbishes them. Selling price is usually between $15 and $25, or 50 to 75 percent less than a new battery.

★ **Junkyards**. Old cars are good sources for powerful dc motors used for windshield wipers, electric windows, and automatic adjustable seats. Bring tools to salvage the parts you want.

★ **Bicycle sales/service shop**. Not the department store that sells bikes, but a *real* professional bicycle shop. Items of interest: control cables, chains, and reflectors.

★ **Industrial parts outlet**. Some places sell gears, bearings, shafts, motors, and other industrial hardware on a one-piece-at-a-time basis. The penalty: high prices.

Shopping the Surplus Store

Surplus is a wonderful thing, but most people shy away from it. Why? If its surplus, as the reasoning goes, it must be worthless junk. That's simply not true. Surplus is exactly what its name implies: extra stock. Because the stock is extra, it's generally priced accordingly—to move it out the door.

Surplus stores that specialize in new and used mechanical and electronic parts (not to be confused with surplus clothing, camping, and government equipment stores) are a pleasure to find. Most areas have at least one such surplus store; some as many as three or four. Get to know each and compare prices. Bear in mind that surplus stores don't have mass-market appeal, so finding them is not always easy. Start by looking in the Yellow Pages under *Electronics* and also under *Surplus*.

Mail-Order Surplus

Some surplus is available through the mail. The number of mail-order surplus outfits that cater to the hobbyist is limited, but you can usually find everything you need if you look carefully enough and are patient. See Appendix A for more information.

While surplus is a great way to stock up on lasers, optical components, dc motors, and other odds and ends, you must shop wisely. Just because the company calls the stuff "surplus" doesn't mean that it's cheap. A *popular* item in a catalog might sell for top dollar.

Always compare prices of similar items offered by various surplus outlets before buying. Consider all the variables, such as the added cost of insurance, postage and handling, and COD fees. Also, be sure that the mail-order firm has a lenient return policy. You should always be able to return the goods if they are not satisfactory to you.

BUYING LASERS AND LASER POWER SUPPLIES

You'll obviously need a laser and a power supply in order to do any laser experiments (you can build a power supply and some types of lasers, but at first it's best to buy until you gain more experience). Lasers are available from a variety of sources including mail order, the local electronic specialty store, and lab supply outlets.

Mail order presents perhaps the least expensive route. Most deal with used and surplus components at a cost savings of 25 to 80 percent of retail list price. Here are some typical prices for components pulled from a recent catalog of laser surplus:

Item	List Price	Sale Price
2 to 3 mW helium-neon tube	$160	$75
12 Vdc He-Ne power supply	$127	$75
Constant-wave laser diode	$37	$10
Cube beam splitter	$42	$10

Prices change constantly because they are driven by the economy and supply and demand. By the time you read this, the sale prices listed above could be considerably higher or lower than the current going price. The list was provided simply to give you an idea of the kinds of savings possible from laser mail-order dealers.

As always, you must exercise care when purchasing through the mail. Avoid sending money to unestablished companies. If you have never heard of the firm before, call them and talk to them about their merchandise, service, and returns policy. Many mail-order laser surplus outfits are small, and the owner himself might answer the phone. If possible, place your first order by COD. Subsequent orders can be placed by COD or check.

By its nature, laser surplus is a variable commodity. Most laser surplus companies publish a catalog or flyer, but the stock can change at any time. It's a good idea to call first to be sure that the merchandise you want is still available. While you are on the phone, you can ask if they have received new products not listed in the catalog. You never know when they will hit the jackpot and you can get in on a great purchase.

It's hard to judge the quality of merchandise pictured or listed in a catalog. Be sure that you can return defective components or those that aren't as they are described in the catalog.

If you are new to lasers, you might feel more comfortable buying the components at a local electronics dealer. Not all dealers stock lasers, of course, and those that usually handle only surplus tubes and supplies. Prices can vary considerably, so it's a good idea to check the going rate against mail-order catalogs. If the average cost in a catalog of a 1 mW He-Ne tube is $75, for example, you'll be forewarned from buying one at $150.

The advantage of buying in person is that you can try out the merchandise before you get it home. If the stock is surplus, you might be able to root through the shelves to find the best tube you can. Recently, I found a perfectly good 8 mW He-Ne tube in the bottom of a junk barrel at a local surplus outlet that cost $10. Luck isn't always this gracious, but a find like this is impossible when dealing with mail order.

A number of mail-order laser surplus dealers are listed in Appendix A. Others advertise in the electronics magazines; refer to recent issues for names and addresses. Finding mail-order companies is fairly easy because they often go to great lengths to promote their name by advertising. A local electronics store won't advertise in the national magazines and might have only a one-line listing in the telephone directory. The Yellow Pages are your best bet, but plan to spend a few hours making some calls. Look up suitable outlets under *Electronics—Retail and Surplus*. Phone them and ask if they deal in laser surplus. The answer will usually be no, but every once in a while you'll hit the jackpot.

Electronic flea marts are becoming more and more popular, particularly among computer and amateur radio enthusiasts. If you aren't aware of any flea marts in your area, ask around at the local electronics stores or see the latest issue of Nuts & Volts and Computer Shopper magazines (see Appendix B).

One flea mart in my area is held at the last Saturday of every month in the parking lot of a large electronics firm. Buyers and sellers come free and most of the stuff is sold by electronics hackers and ham radio operators. Occasionally, one or two dealers come loaded with laser stuff. There's always a crowd around them.

Buying at a flea mart is tricky business because most of the sellers aren't in regular business. Most don't offer any type of guarantee so be sure the gear you buy is in proper working condition before you leave. If you can't test the merchandise at the flea mart, get a receipt that has the telephone number and address of the seller. Announce that you plan to get a replacement if the component you buy proves defective. Few people will argue; they'd do the same if they were on the other side of the table.

APPENDIX A

Sources

When requesting catalogs, indicate field of interest, as many companies publish several interest-specific catalogs. And, don't forget to mention *The Laser Cookbook*.

COMPONENTS, PARTS, AND SYSTEMS

A.C. INTERFACE, INC.
17911 Sampson Lane
Huntington Beach, CA 92647
Representative for Stanley high-output LEDs.

ADVANCED FIBEROPTICS CORP.
7650 East Evans Rd., Ste. B
Scottsdale, AZ 85260
(602) 483-7576
Edu-Kit fiberoptic kit, fiberoptic components, and systems.

ALL ELECTRONICS CORP.
P.O. Box 567
Van Nuys, CA 91408
(800) 826-5432
(818) 904-0524 (in CA)
Retail and surplus components, switches, relays, key-boards, transformers, computer-grade capacitors, cassette player/recorder mechanisms, etc. Mail-order and retail stores in L.A. area. Regular catalog.

ALLEGRO ELECTRONICS SYSTEMS
3E Mine Mountain
Cornwall Bridge, CT 06754
(203) 672-0123
Mail order. Laser equipment. Catalog available.

ALLIED ELECTRONICS
401 E. 8th St.
Ft. Worth, TX 76102
(800) 433-5700
Electronic parts outlet; catalog and regional sales offices. Good source for hard-to-find items.

ALLKIT ELECTRONICS
434 W. 4th St.
West Islip, NY 11795
Electronic components, grab bags, and switches—all at attractive prices.

ALLTRONICS
15460 Union Avenue
San Jose, CA 95124
(408) 371-3053
Mail-order and store in San Jose, CA. New and surplus electronics.

ALPHA PRODUCTS
7904-N Jamaica Ave.
Woodhaven, NY 11421
(800) 221-0916 (orders)
(203) 656-1806 (info)
(718) 296-5916 (NY orders)
Mail-order. Stepper motors and stepper motor controllers (ICs and complete boards), process control devices (relay cards, Touch-Tone decoders, etc.).

AMERICAN DESIGN COMPONENTS
62 Joseph St.
Moonachie, NJ 07074
(201) 939-2710
New and surplus electronics, motors, and computer gear.

AMERICAN SCIENCE CENTER, INC.
601 Linden Place
Evanston, IL 60602
(312) 475-8440
Parent company to surplus outfit Jerryco, American Science Center sells new and surplus lab and optic components. Most of the line is drawn from Edmund Scientific stock at about the same prices. A few interesting buys, such as Ronchi rulings, polarizers, and prisms.

ANALYTIC METHODS
1800 Bloomsbury Ave.
Ocean, NJ 07712
(201) 922-6663
Mail order. Integrated circuits, LEDs, surplus parts, computer cables.

ANCHOR ELECTRONICS
2040 Walsh Ave.
Santa Clara, CA 95050
(408) 727-3693
Electronic components. Catalog available.

BARRETT ELECTRONICS
5312 Buckner Dr.
Lewisville, TX 75028
Mail-order electronic surplus: components, power supplies, computer-grade capacitors.

BCD ELECTRO
P.O. Box 830119
Richardson, TX 75083-0119
Electronic components of all types. Some surplus, including laser systems. Very good prices. Catalog available.

BIGELOW ELECTRONICS
P.O. Box 125
Bluffton, OH 45817-0125
New and surplus electronics, hardware, and HAM gear. Catalog available.

BISHOP GRAPHICS
5388 Sterling Center Dr.
P.O. Box 5007EZ
Westlake Village, CA 91359
(818) 991-2600
EZ Circuit PCB-making supplies.

C & H SALES
2176 E. Colorado Blvd.
Pasadena, CA 91107
(800) 325-9465
Electronic and mechanical surplus including some optics like prisms, mirrors, and lenses. Reasonable prices for most items; regular catalog. Their walk-in store in Pasadena has more items than those listed in the catalog, so if you're in the area, be sure to drop in.

CIRCUIT SPECIALISTS
Box 3047
Scottsdale, AZ 85257
(602) 966-0764
New electronic goodies including hard-to-find Sprague motor control chips (stepper, half-bridge, full-bridge). Also full complement of standard components at good prices.

COMPREHENSIVE GUIDES
7507 Oakdale Ave.
Canoga Park, CA 91306
Laser plans, kits, and supplies.

COMPUTER PARTS MART
3200 Park Blvd.
Palo Alto, CA 94306
(415) 858-1811
Mail-order surplus. Good source for stepper motors, lasers, power supplies, and incremental shaft encoders (pulled from equipment). Regular catalog.

COMPUTER SURPLUS STORE
715 Sycamore Dr.
Milpitas, CA 95035
(408) 434-0168
Surplus electronics. Mostly computers, printers, and power supplies, but also carries components and sometimes even lasers. Walk-in store in Northern California and limited mail order. If you don't see what you want in their flyer, contact them and tell them your needs.

DATAK CORP.
3117 Paterson Plank Rd.
North Bergen, NJ 07047
(201) 863-7667
Supplies for making your own printed-circuit boards including the popular direct-etch dry transfer method.

DIGI-KEY CORP.
701 Brooks Ave. South
P.O. Box 677
Thief River Falls, MN 56701-0677
(800) 344-4539
Mail order. Discount components—everything from crystals to integrated circuits (including 74C926 counter chip) to resistors and capacitors in bulk. Catalog available.

DOKAY COMPUTER PRODUCTS
2100 De la Cruz Blvd.
Santa Clara, CA 95050
(408) 988-0697
(800) 538-8800; (800) 848-8008 (in CA)
Electronic parts (most popular linear, TTL, and CMOS devices) and computer gear (including computer chips). Catalog available.

DOLAN-JENNER INDUSTRIES, INC.
P.O. Box 1020
Woburn, MA 01801
Fiberoptic components.

EDMUND SCIENTIFIC CO.
101 E. Gloucester Pike
Barrington, NJ 08007-1380
(609) 573-6250
Mail order. New and surplus motors, gadgets, and other goodies for laser building. Includes laser tubes, power supplies, complete systems, holography and optics kits, lenses, and much more. Regular catalog (ask for both the Hobby and Industrial versions).

ELECTROVALUE INTERNATIONAL
Box 376
Morris Plains, NJ 07950
(201) 267-1117
Electronics components.

ERAC CO.
8280 Clairemont Mesa Blvd., Suite 117
San Diego, CA 92111
(619) 569-1864
Mail-order and retail store. PC-compatible boards and subsystems, surplus computer boards, computer power supplies and components.

FAIR RADIO SALES
1016 E. Eureka St.
P.O. Box 1105
Lima, OH 45802
(419) 223-9176/(419) 227-6573
An old and established retailer of surplus goods of all types, particularly military communications. A boon to the amateur radio enthusiast and moderately useful to the laser experimenter. Catalog available.

FOBTRON COMPONENTS
17106 S. Broadway
Gardena, CA 90248
Laser components including tubes and power supplies, optics, and mirrors. Flyer available.

FORDHAM RADIO
260 Motor Parkway
Hauppauge, NY 11788
(800) 645-9518
(516) 435-8080 (in NY)
Electronic test equipment, tools, power supplies. Catalog available.

G.B. MICRO
P.O. Box 280298
Dallas, TX 75228
(214) 271-5546
Mail order. Components, 300 kHz crystal, UARTs. Most popular TTL, CMOS, and linear ICs, construction parts.

GENERAL SCIENCE & ENGINEERING
P.O. Box 447
Rochester, NY 14603
(716) 338-7001
Laser components, power supplies, power supply kits, optics, high-output Stanley LED, surplus goodies of all types.

GIANT ELECTRONICS INC.
19 Freeman St.
Newark, NJ 07105
(800) 645-9060
(201) 344-5700
New and surplus electronics, motors, computer peripherals, and more.

GIL ELECTRONICS
P.O. Box 1628
911 Hidden Valley
Soquel, CA 95073
Electronic parts, books, new and surplus computer components.

H&R CORP.
401 E. Erie Ave.
Philadelphia, PA 19134
(215) 426-1708
Surplus mechanical components. Excellent source for heavy-duty dc gear motors. Regular catalog.

HALTED SPECIALTIES CO.
3060 Copper Rd.
Santa Clara, CA 95051
(408) 732-1573
Mail-order and retail stores. New and surplus components, PC-compatible boards and subsystems, ceramic resonators (surplus item), power supplies, most all popular ICs, transistors, etc.

HAL-TRONIX, INC.
12671 Dix-Toledo Highway
P.O. Box 1101
Southgate, MI 48195
(313) 281-7773
Mail order. Surplus computers, computer components, PC-compatible boards and subsystems.

HEATHKIT
P.O. Box 1288
Benton Harbor, MI 49022
(800) 253-0570
Perhaps the best source of professionally produced electronic kits. Heathkit offers a few laser projects including a laser trainer geared to accept a modulated input. The Fischertechnik engineering kits are ideal for use as building blocks in your own laser designs. Also available are testing gear (assembled and kit form) and computers.

HOSFELT ELECTRONICS, INC.
2700 Sunset Blvd.
Steubenville, OH 43952
(800) 524-6464
New and surplus electronics at attractive prices. Catalog available.

INFORMATION UNLIMITED
Box 716
Amherst, NH 03031
(603) 673-4730
Plans and kits for laser and other high-tech gadgets. Good source for reasonably priced new laser tubes. Among the plans are details for building a wide range of Class IV lasers, including CO_2, argon, and tunable dye systems.

JAMECO ELECTRONICS
1355 Shoreway Rd.
Belmont, CA 94002
(415) 592-8097
Mail order. Components, PC-compatible boards and subsystems. Regular catalog.

JDR MICRODEVICES
1224 S. Bascom Ave.
San Jose, CA 95128
(408) 995-5430
Large selection of new components, wire-wrap supplies, PC-compatible boards and subsystems. Mail-order and retail stores in San Jose area.

JERRYCO INC.
607 Linden Place
Evanston, IL 60202
(312) 475-8440
Regular catalog lists hundreds of surplus mechanical and electronic gadgets for lasers. Good source for motors, rechargeable batteries, switches, solenoids, lots more. Don't build a laser system until you get the Jerryco catalog.

J. I. MORRIS CO.
394 Elm St.
Southbridge, MA 01550
(617) 764-4394
Small components and hardware; miniature screws, taps, and nuts/bolts.

JOHN J. MESHNA, JR. CO.
P.O. Box 62
E. Lynn, MA 01904
(617) 595-2272
New and surplus merchandise. Catalog available.

MARTIN P. JONES & ASSOC.
P.O. Box 12685
Lake Park, FL 33403-0685
(407) 848-8236

All sorts of goodies, including electronic components, (transistors, capacitors, resistors, ICs, etc.), optical stuff, and more. Catalog available.

MCM ELECTRONICS
858 E. Congress Park Dr.
Centerville, OH 45459-4072
(513) 434-0031
Test equipment, tools, and supplies. Catalog available.

MEADOWLAKE
25 Blanchard Dr.
Northport, NY 11768
TEC-200 direct-transfer film for making quick integrated circuits from artwork in magazines and this book. Try this stuff out! All you need is a plain paper copier (or access to one).

MEREDITH INSTRUMENTS
P.O. Box 1724
Glendale, AZ 85311
(602) 934-9387
Laser surplus including tubes, power supply, optics, components, more. Stock comes and goes, so call first to make sure they have what you want. Flyer available; very good prices.

MICRO MART, INC.
508 Central Ave.
Westfield, NJ 07090
(201) 654-6008
Electronic components (ICs, transistors, etc.). Their grab bags are better than average, and prices are low.

MOUSER ELECTRONICS/TEXAS DISTRIBUTIN CENTER
2401 Hwy 287 North
North Mansfield, TX 76063
(817) 483-4422

MOUSER ELECTRONICS/CALIFORNIA DISTRIBUTION CENTER
11433 Woodside Ave.
Santee, CA 92071
(619) 449-2222
Mail order. Discount electronic components. Catalog available.

MWK INDUSTRIES
1440 S. State College 3B
Anaheim, CA 92806
(714) 956-8497
Laser tubes, systems, power supplies, and light shows.

OCTE ELECTRONICS
Box 276
Alburg, VT 05440
(514) 739-9328
New and surplus electronic stuff at reasonable prices. OCTE carries a lot of complete components, such as cable converters and power supplies, but individual parts, like batteries and connectors, are also available.

PRECISION ELECTRONICS CORP.
605 Chestnut Street
Union, NJ 07083
(800) 255-8868
(201) 686-4646 (in NJ)
Electronic components such as linear ICs, transistors, and hard-to-find Sanyo and Zenith ICs (mostly for TV).

RADIO SHACK
One Tandy Center
Fort Worth, TX 76102
Nation's largest electronics retailer. Many popular components, though short on ICs. Good source for general electronic needs including some fiberoptic components. Catalog available through store.

R & D ELECTRONICS
1202H Pine Island Rd.
Cape Coral, FL 33909
(813) 772-1441
Mail order. New and surplus electronics (components, switches, ICs).

R & D ELECTRONIC SUPPLY
100 E. Orangethorpe Ave.
Anaheim, CA 92801
(714) 773-0240
Mail order. Power supplies, computer equipment, test equipment.

SHARON INDUSTRIES
1919 Hartog Road
San Jose, CA 95131
(408) 436-0455
Mail-order and retail store. New and surplus electronic components, ICs, computers, PC-compatible boards and subsystems.

SILICON VALLEY SURPLUS
4401 Oakport
Oakland, CA 94601
(415) 261-4506

Surplus electronic, computer, and mechanical goodies. Mail-order and retail store (Oakland, CA).

SMALL PARTS
6901 NE Third Ave.
Miami, FL 33238
(305) 751-0856
A potpourri of small parts, ideally suited for miniature mechanisms. Not cheap but good quality.

SPECTRA LASER SYSTEMS
P.O. Box 6928
Huntington Beach, CA 92615
Laser light-show consultation.

STOCK DRIVE PRODUCTS
55 S. Denton Ave.
New Hyde Park
New York, NY 11040
Gears, sprockets, chains, and more. Available through local Stock Drive distributor. Catalog and engineering guide available.

SYNERGETICS
Box 809
Thatcher, AZ 85552
(602) 428-4073
Columnist Don Lancaster's company, with "hacker's help line." Lancaster offers free technical advice (within limits), a collection of his old and new books (his TTL and CMOS cookbooks are classics), and several "info packs" of useful programming, graphics, and technical stuff.

UNICORN ELECTRONICS
10010 Canoga Ave., Unit B-8
Chatsworth, CA 91311
(800) 824-3432
(818) 341-8833
All popular electronic parts including resistors, capacitors, ICs, and components. Some surplus. All at good prices.

UNITED PRODUCTS, INC.
1123 Valley
Seattle, WA 98109
(206) 682-5025
Mail order. Stepper motors, computer components, test equipment.

WINDSOR DISTRIBUTORS
19 Freeman St.
Newark, NJ 07105
(800) 645-9060
Mail order. Surplus electronics.

SEMICONDUCTOR MANUFACTURERS

Most semiconductor manufacturers maintain a network of regional sales and distribution offices. If you would like more information on a particular product or would like to place an order, contact the manufacturer at the address below and ask for a list of dealers, distributors, and representatives in your area. Most firms will work with individuals, but sales might be subject to minimum orders and/or service charges.

ADVANCED MICRO DEVICES
901 Thompson Place
Sunnyvale, CA 94088
(408) 732-2400

ANALOG DEVICES
One Technology Way
Norwood, MA 02062
(617) 329-4700

BURR-BROWN
P.O. Box 11700
Tucson, AZ 85734
(602) 746-1111

CHERRY SEMICONDUCTOR CORP.
2000 South Country Trail
East Greenwich, RI 02818-0031
(401) 885-3600

CYBERNETIC MICRO SYSTEMS
P.O. Box 3000
San Cregorio, CA 94074
(415) 726-3000

DATA GENERAL CORP.
4400 Computer Dr.
Westborough, MA 01581
(617) 366-1970

EG&G RETICON CORP.
345 Potrero Ave.
Sunnyvale, CA 94086
(408) 738-4266

EXAR INTEGRATED SYSTEMS, INC.
750 Palomar Ave.
Sunnyvale, CA 94088-3575
(408) 732-7970

FAIRCHILD
10400 Ridgeview Ct.
Box 1500
Cupertino, CA 95014
(408) 864-6250

FUJITSU MICROELECTRONICS, INC.
3320 Scott Blvd.
Santa Clara, CA 95054-3197
(408) 727-1700

HARRIS SEMICONDUCTOR
P.O. Box 883
Melbourne, FL 32901
(305) 724-7000

HITACHI AMERICA LTD.
2210 O'Toole Ave.
San Jose, CA 95131
(408) 435-8300

HOBBY SHACK
18480 Bandilier Circle
Fountain Valley, CA 92708

INTEL
3065 Bowers Ave.
Santa Clara, CA 95051
(408) 987-8080

INTERSIL
10600 Ridgeview Ct.
Cupertino, CA 95014
(408) 996-5000

ITT SEMICONDUCTORS
7 Lake St.
Lawrence, MA 01841
(617) 688-1881

MAXIM INTEGRATED PRODUCTS
510 N. Pastoria Ave.
Sunnyvale, CA 94086
(408) 737-7600

MICRO SWITCH
11 West Spring Street
Freeport, IL 61032
(815) 235-6600

MITEL SEMICONDUCTOR
P.O. Box 13320,
Kanata, Ontario, K2K 1X5
Canada

MITSUBISHI ELECTRONICS AMERICA
1050 E. Arques Ave.
Sunnyvale, CA 94086
(408) 730-5900

MONOLITHIC MEMORIES
2151 Mission College Blvd.
Santa Clara, CA 95054
(408) 970-9700

MOSTEK
1310 Electronics Ave.
Carrollton, TX 75006
(214) 466-6000

MOTOROLA
5005 E. McDowell Rd.
Phoenix, AZ 85008
(602) 244-7100

NATIONAL SEMICONDUCTOR
2900 Semiconductor Dr.
Santa Clara, CA 95051
(408) 721-5000

NCR
8181 Byers Rd.
Miamisburg, OH 45342
(513) 866-7217

NEC ELECTRONICS, INC.
401 Ellis St.
Mountain View, CA 94039-7241
(415) 960-6000

OKI SEMICONDUCTOR
650 E. Mary Ave.
Sunnyvale, CA 94086
(408) 720-1900

PENWALT PIEZO
Box C
Prussia, PA 19406
(215) 337-6710

PLESSEY SOLID STATE
9 Parker St.
Irvine, CA 92718
(714) 472-0303

PRECISION MONOLITHICS, INC.
1500 Space Park Dr.
Santa Clara, CA 95052-8020
(408) 727-9222

RAYTHEON SEMICONDUCTOR
350 Ellis St.
Mountain View, CA 94039-7016
(415) 968-9211

RETICON
245 Potrero Ave.
Sunnyvale, CA 94086
(408) 738-4266

ROCKWELL INTL.
4311 Jamboree Rd.
P.O. Box C
Newport Beach, 92658-8902
(714) 833-4700

SGS SEMICONDUCTOR
1000 E. Bell Rd.
Phoenix, AZ 85022
(602) 867-6100

SHARP ELECTRONICS CORP.
10 Sharp Plaza
Paramus, NJ 07652

SIGNETICS
811 E. Arques Ave.
Sunnyvale, CA 94088-3409
(408) 991-2000

SILICONIX
2201 Laurelwood Rd.
Santa Clara, CA 95054
(408) 988-8000

SPRAGUE ELECTRIC
115 NE Cutoff
Worcester, MA 01613-2036
(617) 853-5000

SPRAGUE SOLID STATE
3900 Welsh Rd.
Willow Grove, PA 19090
(215) 657-8400

TEXAS INSTRUMENTS
Literature Response Center
P.O. Box 809066
Dallas, TX 75228
(214) 232-3200

3M ELECTRONIC PRODUCTS DIVISION
P.O. Box 2963
Austin, TX 78769
(512) 834-6708

TRW SEMICONDUCTORS
P.O. Box 2472
La Jolla, CA 92038
(619) 457-1000

TOSHIBA AMERICA INC.
2692 Dow Ave.
Tustin, CA 92680
(714) 832-0102

XICOR
851 Buckeye Ct.
Milpitas, CA 95035
(408) 432-8888

ZILOG, INC.
210 Hacienda Ave.
Campbell, CA 95008-6609
(408) 370-8000

LASER AND LASER COMPONENTS MANUFACTURERS AND RETAILERS

ADVANCED CONTROL SYSTEMS CORP.
205 Oak St.
Pembrooke, MA 02359
(617) 826-4477

ADVANCED KINETICS
1231 Victoria St.
Costa Mesa, CA 92627
(714) 646-7165

AG ELECTRO-OPTICS LTD.
Tarporley
Cheshire CW6 0HX
United Kingdom
(08293) 3305/3678

COHERENT INC.
Laser Products Division
3210 Porter Dr.
P.O. Box 10321
Palo Alto, CA 94303
(415) 493-2111

CONTROL TECHNICS CORP.
22600-C Lambert St.
Unit 909
El Toro, CA 92630
(714) 770-9911

COOPER LASERSONICS
Laser Products Division
48503 Millmont Dr.
Fremont, CA 94538
(415) 770-0800

CVI LASER CORP.
P.O. Box 11308
200 Dorado Pl., S.E.
Albuquerque, NM 87192
(505) 296-9541

EASTMAN KODAK COMPANY
Photographic Products Group
Rochester, NY 14650

FAIRLIGHT B.V.
P.O. Box 81037
NL-3009 GA Rotterdam
The Netherlands
(010) 4206444

GLENDALE OPTICAL CO.
130 Crossways Park Dr.
Woodbury, NY 11797

HAMMAMATSU CORP.
360 Foothill Rd.
P.O. Box 6910
Bridgewater, NJ 08807
(201) 231-0960

HARDIN OPTICAL CO.
P.O. Box 219
Bandon, OR 97411
(503) 347-3307

HUGHES AIRCRAFT CO.
P.O. Box 45066
7200 Hughes Terrace
Los Angeles, CA 90045-0066
(213) 568-6838

ISOTECH INC.
3858 Benner Rd.
Miamisburg, OH 45342
(513) 859-1808

LASER APPLICATIONS INC.
Division of Lasermetrics, Inc.
3500 Aloma Ave., Ste. D-9
Winter Park, FL 32792
(305) 678-8995

LASER DEVICES INC.
#5 Hangar Way
Watsonville, CA 95076
(408) 722-8300

LASER DRIVE INC.
5465 William Flynn Hwy.
Gibsonia, PA 15044
(412) 443-7688

LASER LINES LTD.
Beaumont Close
Banbury, Oxon, OX16 7TQ
United Kingdom
(0295) 67755

LASERMETRICS INC.
Electro-Optics Division
196 Coolidge Ave.
Englewood, NJ 07631
(201) 894-0550

LASER POWER OPTICS
12777 High Bluff Dr.
San Diego, CA 92130-2016
(619) 755-0700

LASER SCIENCE INC.
80 Prospect St.
Cambridge, CA 02139
(617) 868-4350

MELLES GRIOT/OPTICS
1770 Kettering St.
Irvine, CA 92714
(714) 261-5600

MELLES GRIOT/GAS LASERS
2251 Rutherford Rd.
Carlsbad, CA 92008
(619) 438-2131

METROLOGIC INSTRUMENTS INC.
Laser Products Division
143 Harding Ave.
Bellmawr, NJ 08031
(609) 933-0100

NEC ELECTRONICS
401 Ellis St.
Mountain View, CA 94039
(415) 960-6000

NEWPORT CORP.
P.O. Box 8020
18235 Mt. Baldy Circle
Fountain Valley, CA 92728-9811
(714) 963-9811

J.A. NOLL CO.
Box 312
Monroeville, PA 15146
(412) 856-7566

NON-LINEAR DEVICES
126 Andrew Ave.
Oakland, NJ 07436
(201) 337-0666

OPTIKON CORP. LTD.
410 Conestogo Rd.
Waterloo, Ontario Canada N2L 4E2
(519) 885-2551

POLYTEK OPTRONICS INC.
3001 Redhill Ave.
Bldg. R, Suite 102
Costa Mesa, CA 92626
(714) 850-1831

PRECISION OPTICAL
869 W. 17th St.
Costa Mesa, CA 92627
(714) 631-6800

QEI
5A2 Damonmill Sq.
Concord, MA 01742
(617) 369-8081

SPECTRA DIODE LABS
3333 N. First St.
San Jose, CA 95134-1995
(408) 432-0203

SPECTRA-PHYSICS INC.
Laser Analytics Division
25 Wiggins Ave.
Bedford, MA 01730
(617) 275-2650

SPEIRS ROBERTSON & CO. LTD.
Laser Division
Moliver House, Oakley Rd.
Bromham, Bedford
United Kingdom
(02302) 3410

OTHER IMPORTANT ADDRESSES

AMERICAN NATIONAL STANDARDS INSTITUTE (ANSI)
1430 Broadway
New York, NY 10018
Sets and maintains standards; useful data on laser practices and specifications.

LASER INSTITUTE OF AMERICA
5151 Monroe St.
Toledo, OH 43623
(419) 882-8706
Provides publications of interest to users, manufacturers, and sellers of lasers. Several good publications include Laser Safety Guide, Laser Safety Reference Book, and Fundamentals of Lasers. Ask for a current price list and availability of titles.

CENTER FOR DEVICES AND RADIOLOGICAL HEALTH
(CDRH)   (formerly Bureau of Radiological Health)
Optical Radiation Products Section
HF2-312
8757 Georgia Ave.
Silver Springs, MD 20910
(301) 427-8228
A department of the Food and Drug Administration (FDA) that regulates the commercial manufacture and use of lasers. *If you manufacture, sell, or demonstrate laser systems, you must comply to minimum standards.*

U.S. DEPARTMENT OF COMMERCE
Patent and Trademark Office
Washington, DC 20231
Provides general information on patents and copies thereof.

APPENDIX B

Further Reading

Here is a selected list of magazines and books that can enrich your understanding and enjoyment of all facets of lasers and laser applications.

MAGAZINES

Computer Shopper
5211 S. Washington Ave.
P.O. Box F
Titusville, FL 32781
Monthly magazine (some call it a bible) for computer enthusiasts. Probably containing more ads than articles, Computer Shopper carries updated listings of swap meets and bulletin boards as well as classified advertising from small surplus dealers.

Electronics Today
1300 Don Mills Rd.
Toronto, Ontario M3B 3M8
Canada
General-interest electronics magazine with emphasis on hobby how-to articles.

Hands-on Electronics
500 Bi-County Blvd.
Farmingdale, NY 11735
Monthly magazine put out by the editors of Radio-Electronics. The articles and construction projects are aimed at beginning electronics enthusiasts. Few articles are specifically on lasers, but some of the circuits can be adapted for laser projects.

Laser Focus
1001 Watertown St.
Newton, MA 02165
A monthly trade magazine for the laser industry. Available by paid or audited (free to qualified readers) subscription. Check a good public library for back issues.

Lightwave
The Journal of Fiber Optics
235 Bear Hill Rd.
Waltham, MA 02154
A monthly trade journal on fiberoptics and its applications.

Modern Electronics
76 North Broadway
Hicksville, NY 11801

Monthly magazine for electronics hobbyists. Don't miss the regular columns by hobby electronic guru Forrest Mims III. Many of the editors used to be involved with Popular Electronics before that magazine changed over to computer-only coverage (and then ceased publication). Check back issues of Modern Electronics for Don Lancaster's fascinating columns (he moved his Hardware Hacker column to Radio-Electronics in late 1987).

Nuts & Volts
P.O. Box 1111
Placeutia, CA 92670

Monthly "magazine" that contains only advertising—both display ads and classifieds. Often carries ads for new and used laser equipment and lists upcoming swap meets (both computer and electronic). Subscription price is fairly low.

Physics Today
335 East 45 St.
New York, NY 10017

Monthly magazine with technical slant; no how-to's but plenty of theory on all subjects of a physical nature. The annual buyer's guide is useful for names and addresses of companies specializing in lasers, optics, and related products.

QST
225 Main St.
Newington, CT 06111

Monthly magazine aimed at the amateur radio enthusiast, but often carries articles of interest to the laser experimenter. Look for stories on electromagnetic spectrum, fiberoptics, project construction, remote control, etc.

Spectra
P.O. Box 1146
Pittsfield, MA 01202

A monthly trade journal that focuses on lasers and other related topics. Available by paid or audited (free to qualified readers) subscription. Check a good public library for back issues.

Radio-Electronics
500 Bi-County Blvd.
Farmingdale, NY 11735

Monthly magazine for electronic hobbyists. Occasional article on lasers. Be sure to read the column by Don Lancaster of Synergetics. Although he seldom addresses lasers in specific, his tips and hands-on help are invaluable for general electronics experimentation. He also provides sources for hard-to-find parts and information.

BOOKS

How to Read Electronic Circuit Diagrams—2nd Edition; Brown, Lawrence, and Whitson
TAB BOOKS, Catalog #2880
How to read and interpret schematic diagrams.

Build Your Own Laser, Phaser, Ion Ray Gun, & Other Working Space-Age Projects; Robert E. Iannini
TAB BOOKS, Catalog #1604
A guide to making high-tech gadgets, with some insight and designs for laser-based projects. Iannini, who runs a mail-order firm, presents six laser projects plus a few others (such as infrared light detection) that can be used with laser systems. A companion book, Build Your Own Working Fiberoptic and Laser Space-Age Projects *(Catalog #2724) offers more laser-based designs.*

Circuit Scrapbook; Forrest Mims III
McGraw-Hill
More of Mims' Popular Electronics columns—these from 1979 to 1981. Several good circuits and designs that can be adapted for laser work. Also check out the sequel, Circuit Scrapbook II *(Howard W. Sams Co.). This volume includes several chapters on experimenting with solid state laser diodes. Be sure to read the sections on laser diode handling precautions.*

CMOS Cookbook; Don Lancaster
Howard W. Sams Co.
A classic in its own time, the CMOS Cookbook *presents useful design theory and practical circuits for many popular CMOS chips. The companion book,* TTL Cookbook, *is equally as helpful.*

Computer Peripherals That You Can Build; Gordon W. Wolfe
TAB BOOKS, Catalog #2749
Step-by-step instructions for building a variety of functional computer peripherals.

Elementary Plane Surveying; Davis & Kelly
McGraw-Hill
Overview of plane surveying, including tools of the trade, taking measurements, and doing the calculations. No specific information on using laser-based surveying equipment, but the information presented can be used as a surveying primer.

Engineer's Mini-Notebook; Forrest Mims III
Radio Shack book series
The Engineer's Mini-Notebooks *is a series of small books written by Forrest Mims that cover a wide variety of hobby electronics: using the NE555 timer to optoelectronics circuits to op-amp circuits, and more. The entire set is a must-have, and besides, they're cheap.*

44 Power Supplies for Your Electronic Projects; Traister and Mayo
TAB BOOKS, Catalog #2922
Forty-four complete power supplies designed for general electronics projects.

Fundamentals of Optics; Jenkins and White
McGraw-Hill
An in-depth and scholarly look at optics. Broken down into logical subjects of geometrical, wave, and quantum optics.

Guide to Practical Holography; Outwater and Hamersueld
Pentangle Press
A good, easy-to-understand manual on hologram-making.

Hologram Book, The; Joseph E. Kasper and Steven A. Feller
Prentice-Hall
Introduction to lasers and laser light and detailed instructions for making holograms.

Holography Handbook; Unterseher, Hansen, and Schlesinger
Ross Books

Perhaps the best book on amateur holography. It even comes with a white-light reflective rainbow hologram. Although the artwork is too "folksy" for my taste, the designs are technically sound. If you are interested in holography, you need this book.

How to Troubleshoot & Repair Electronic Circuits; Robert L. Goodman
TAB BOOKS, Catalog #1218

General troubleshooters guide, both analog and digital.

IBM PC Connection; James W. Coffron
Sybex Books

A good beginner's guide on connecting the IBM PC (or compatible) to the outside world. Information on circuit-building, programming, and troubleshooting.

Introduction to Lasers and Masers; A. E. Siegman
McGraw-Hill

Although a little old (copyright 1971), this book provides good details of the inner works of all major types of lasers, including crystal, glass, gas, and semiconductor.

Lasers: The Incredible Light Machines; Forrest M. Mims III

Introduction to lasers; includes a short history.

Lasers: The Light Fantastic, Second Edition; Hallmark and Horn
TAB BOOKS, Catalog #2905

A good introduction to the mechanics and applications of lasers. Interesting chapters on laser gyroscopes, quantum mechanics, and lasers in space.

Laser Experimenter's Handbook; Delton T. Horn
TAB BOOKS, Catalog #3115

The hows and whys of lasers. Includes six projects using laser diodes.

Laser and Light; W. H. Freeman
W. H. Freeman

Material drawn from back issues of Scientific American on laser principles.

Magic of Holography; Philip Heckman
Atheneum

A concise and easy-to-understand explanation of holography and its applications. Although the book lacks specific information on how to make your own holograms, there are plenty of setup illustrations you can use as blueprints.

Masers and Lasers; H. Arthur Klein
J.B. Lippincott Company

Although dated (copyright 1963), this book provides some interesting aspects about lasers not found in other books and gives an easy-to-understand overview of lasers, electromagnetic spectrum, and early pioneering work in light physics.

Optical Holography; Robert L. Collier
Academic Press

A technical and engaging text on optical (as opposed to acoustic) holography. A little old (the original copyright is 1971), but most of the information is still valid.

Optics; N. V. Klein
John Wiley & Sons

A highly technical text about optics with lots of formulas and math on optics design.

Practical Interfacing Projects with the Commodore Computers; Robert H. Luetzow
   TAB BOOKS, Catalog #1983
   How to use a Commodore 64 computer to control appliances, robotic devices, and more.

Principles & Practice of Laser Technology; Hrand M. Mucheryan
   TAB BOOKS, Catalog #1529
   An introduction to lasers and laser applications. Heavy on the industrial side of laser use.

Programmer's Problem Solver, for the IBM PC, XT, & AT; Robert Jourdain
   Brady
   A technical book on the inner-working of the IBM PC, with special emphasis on programming in BA-SIC, assembly, and machine code. Extensive section on parallel ports.

Principles of Holography; Howard M. Smith
   John Wiley & Sons
   A technical overview of holography, covering its history, application, and setup, written by a research associate at Kodak. Good source for application formulas and information on holographic emulsions.

Robot Builder's Bonanza; Gordon McComb
   TAB BOOKS, Catalog #2800
   My earlier book on a compendium of projects useful to the robot experimenter. Some of the information, such as dc motor operation and remote control, can be suitably adapted to laser work.

Silliconnection: Coming of Age in the Electronic Era; Forrest M. Mims III
   McGraw-Hill
   A history and personal autobiography of the electronics revolution. Mims, a writer and inventor, contributed to the development of the first personal computer and was employed at the Air Force Weapons Laboratory in New Mexico where he worked with early solid-state laser diodes. This book includes the interesting story of how the National Enquirer offered to pay Mims to build a lightbeam listening device using a laser and phototransistor receiver to tap into the private conversations of the late Howard Hughes.

Understanding Digital Electronics; R. H. Warring
   TAB BOOKS, Catalog #1593
   Introduction to the principles of digital theory.

Glossary

aberration—A defect in an optical component that can degrade the purity of the light passing through it.

absolute zero—The lowest "possible" temperature, equal to minus 273.16 degrees C (−459.69 degrees F).

ac—Alternating current. Current that fluctuates to positive and negative values about a zero point. The current available at wall outlets.

access time—The time required to get a piece of information from a memory chip or disk drive.

address—A number that indicates the location of a piece of information in computer memory.

amplitude—The relative strength (usually voltage) of an analog signal. Amplitude can be expressed as either a negative or positive number referenced to a particular standard.

analog—A continuous electrical signal representing a condition (such as temperature or the position of game control paddles). Unlike a digital signal that is discrete and has only two levels, an analog signal can have an infinite number of levels.

analog-to-digital converter (ADC)—An electrical circuit (usually one integrated circuit) that transforms analog signals to their digital equivalents. A digital-to-analog converter performs the reverse function.

angstrom—A unit of measurement used to describe wavelength or size; equals 100 billionths of a meter (10^{-10} meter).

anode—The positive terminal of a laser, light-emitting diode, laser diode, or other electronic component. See also *cathode*.

antihalation backing—A semi-opaque material or coating applied to the back side of photographic film that prevents the spreading (halo) of light during exposure.

ASCII—Acronym for American Standard Code for Information Interchange. Used by all personal computers.

assembler—A program that converts the computer's memory into binary code for execution. Acts as a compiler for assembly language.

assembly language—Machine language codes translated into mnemonic codes that are easier for programmers to remember.

attenuation—The restriction or loss of electrical or optical power through a medium or circuit. Optical attenuation occurs when light passes through a length of optical fiber.

axial ray—A ray of light that travels along the optical axis.

bandwidth—A continuous range of frequencies or wavelengths, defined by an upper and lower limit.

BASIC—Acronym for Beginner's All-purpose Symbolic Instruction Code. Programming language for computers developed as a simplified form of FORTRAN. BASIC is not standard and can vary from computer to computer. The BASIC dialect for IBM PC and compatible computers is Microsoft's GW-BASIC; the dialect for the Commodore 64 is a specialized Commodore BASIC.

baud—A measure of the rate at which digital data is transmitted, in bits per second.

band-pass filter—A filter that blocks both high and low frequencies but passes a middle band. Can pertain to both electrical signals or light. See also *low-pass filter* and *high-pass filter*.

beam—A collection of rays (light waves) moving in one direction.

beam splitter—An optical component that divides a beam two or more different ways. Beam splitters are rated by the ratio of light they reflect and transmit. For example a 50:50 beam splitter splits a beam equally in two directions (usually by reflection and transmission).

beamwidth—The linear width of the beam, usually specified as the region over which the beam intensity falls within predetermined limits.

binary—A numbering system with just two digits: 1 and 0. Computers group these 1s and 0s together to create more complex forms of data.

birefringence—The property of splitting a beam of light in two directions due to double refraction.

bit—Acronym for *bi*nary dig*it*. Represents either of two binary states—1 or 0. Bits are usually grouped in sets of eight (called a *byte*) for easier manipulation.

blackbody—A substance or medium that absorbs light and thermal energy completely. The blackbody is a technical impossibility but is often used as an ideal model for light and temperature studies.

bleeder resistor—A resistor placed across the output of a power supply to drain the current from the filtering capacitors when power is removed.

boot—A program on disk, tape, or in a permanent portion of the computer's memory (ROM), used to start the computer and get it ready for applications or programming software.

Brewster's Angle—The angle at which a transparent material, such as glass or quartz, is placed with respect to the normal of incident light so that both refracted and reflected rays are perpendicular. Brewster's Angle windows placed in gas lasers cause the light output to be polarized in one plane.

bus—A set of electrical contacts that carry a variety of computer or analog signals.

byte—A group of bits, usually eight, universally used to represent a character.

candle—A unit of luminous intensity. Specifically, one candle is equal to $\frac{1}{60}$th of one square centimeter of projected area of a blackbody radiator operating at the temperate of solidification of platinum.

carrier—A radio frequency signal superimposed over another signal.

carrier frequency—The frequency of the carrier signal.

392

cathode—The negative terminal of a laser, light-emitting diode, laser diode, or other electronic component.

CCD—Short for *charge-c*oupled *de*vice, a type of integrated circuit, that is sensitive to light and used as an imaging device.

chip—An integrated circuit, such as that used for computer memory or the microprocessor.

chopper—A device or electronic circuit that provides a pulsating or on/off characteristic to a beam of light or electric current. A mechanical chopper is an incremental shaft encoder, where the light from an LED or laser to a photodiode is intermittently blocked by slits in a wheel.

chromatic aberration—A type of lens distortion where colors are focused at different points due to refraction.

clad—A refractive coating over the interior core of an optical fiber.

clock—An electronic metronome used for the purpose of timing signals in a circuit. The clock times out signals so that they occur at a steady, predictable rate and maintain the proper sequence of events.

coherence—The property of identical phase and time relationships.

coherence length—The greatest distance that the beam from a laser will remain temporally coherent. Coherency is often related as a percentage of deviation, not an absolute; therefore, the coherence length is subject to the acceptable range of coherency for a given application.

coherent radiation—Electromagnetic radiation where the waves are in phase in both time and space (temporal coherence and spatial coherence); as opposed to non-coherent radiation whose waves are out of phase in space and/or time.

collimate—To make parallel.

collimator—An optical component that forces a diverging or converging light beam, as from a laser, to travel in parallel lines. The collimator lens (sometimes a mirror) is placed at its focal distance from the light source.

common—The ground point of a circuit, or a common path for an electrical signal.

compiler—A program that converts high-level language (like BASIC) into the binary code required by the computer. Compiler instructions are known as "object code." The original high-level language program is known as the "source code."

concave lens—A type with at least one surface that curves inward. Concave lenses include plano-concave and double-concave. A concave lens diverges parallel light rays to a virtual (not real) focal point.

conductivity—The measure of the ability of a material to carry an electric current, expressed as ohms or ohms per meter.

constructive waves—Waves reaching an area in the same phase. Since each wave is in phase with the others, the waves act to constructively add to one another, thereby increasing amplitude. See also *destructive waves*.

continuous wave—A type of laser whose output is continuous as opposed to pulsed; an uninterrupted beam of laser light.

convergence—The bending of light by a lens or mirror to a common point.

converging lens—Generally, a lens that is thicker in the middle than the edges.

convex—A type of lens with at least one surface that curves outward. Convex lenses include plano-convex and double-convex. A convex lens converges parallel light rays to a common real focal point.

corona—The blue glow surrounding a conductor when high voltages are present. The corona is caused by the ionization of the surrounding air and is usually accompanied by a crackling or "spitting" noise.

coupling loss—The loss of light output at the mechanical connections of fiberoptics.

CPU—Acronym for *central processing unit*. The part of the computer responsible for processing, storing, and retrieving data. The *microprocessor* is often referred to as the CPU, but the term

CPU can also encompass the computer proper minus the keyboard, disk drives, monitor, and so forth.

critical angle—The maximum angle of incidence for which a beam of light will be transmitted from one medium (such as air) to another medium (such as an optical fiber or prism).

cylindrical lens—A lens that expands or reduces an image in one axis only.

dc—Direct current. Current such as that from a battery where the voltage level remains the same (either positive or negative) with respect to ground.

decibel (dB)—A unit of electrical measurement used extensively in audio applications. An increase of 3 dB is a doubling of electrical (or signal) strength; an increase of 10 dB is a doubling of perceived loudness.

default—In computer applications, the normal.

demodulate—The separation of the message signal from a carrier wave.

destructive waves—Waves reaching an area that are not in the same phase. Because the waves are out of phase, they act to partially or completely cancel each other out, thereby diminishing amplitude. See also *constructive waves*.

dichroic filter—An optical filter using organic or chemical dyes that transmits light at selected wavelengths but blocks others.

diffraction—A change in the direction of a beam of light when encountering an edge or opening of an object. For example, diffraction takes place when passing through a small aperture.

diffraction grating—An optical component that splits a light beam into many discrete parts.

digital—Information expressed in binary—ON and OFF—form. The OFF state is usually indicated as a numeral 0; the ONC state is usually indicated as a numeral 1. Many 0s and 1s can be grouped together to represent any other number or value.

digital-to-analog converter (DAC)—An electrical circuit (usually one integrated circuit) that transforms digital signals to their analog equivalents. An analog-to-digital converter performs the reverse function.

diode laser—A solid-state injection laser, similar to a light-emitting diode (LED).

dispersion—Usually applies to the separation of polychromatic light into individual monochromatic frequencies. A prism is often used to disperse white light into the visible spectrum.

divergence—The bending of light rays away from each other; the spreading of light.

diverger—An optical component that spreads light rays.

diverging lens—Generally, a lens that is thinner in the middle than at the edges.

double-concave—A type of negative lens where both faces curve inward. Also called a *bi-concave* lens.

duty cycle—In electronics, the ratio of the ON and OFF states of a circuit. The higher the duty cycle (approaching 100 percent), the longer the circuit remains in the ON state.

electroluminescence—In a semiconductor, the direct conversion of current into light.

electromagnetic waves—Transverse (up and down) waves having both an electric and magnetic component. Each component is perpendicular to one another, and both are perpendicular to the direction of propagation.

electro-optic effect—The general change in the refractive index of a material when that material is subjected to an electric field. Also referred to as Pockels and Kerr effects. Similar to acousto-optic and magneto-optic effects, where an acoustic or magnetic pulse changes a material's refractive index.

emulsion—In photography, a chemical coating, usually on clear acetate or glass, that is sensitive to light.

etalon—An extremely flat optical component with parallel surfaces engineered to increase the coherence length of a laser by eliminating modes that are slightly out of phase.

excited state—The state of an atom that occurs when outside energy, such as from an electric field, causes the atom to contain more energy than normal.

f/number—A number that expresses the relative light-gathering power of a lens. The f/number is calculated by dividing the diameter of the lens by the focal length of the lens.

fiberoptic—A solid glass or plastic "tube" that conducts light along its length.

fiber bundle—Many individual strands of optical fibers grouped together. The bundle can be coherent or non-coherent, depending on its ability to transmit an undistorted optical image through the fibers.

filter—1. An electrical circuit designed to prevent the passage of certain frequencies. 2. An optical component designed to block the passage of light at certain frequencies.

focus coil—An electromechanical coil, often used in compact disc and video disc players, that moves an objective lens up and down for proper focus. Surplus coils can often be used in laser experiments.

focal length—The distance from the center of a lens or curved mirror to the point where light converges to a point. Lenses with a positive focal point cause the light rays to form at a specific point; lenses with a negative focal point only appear to focus at an imaginary point.

focal point—That point in space where rays converge to a common point (positive lens system) or from which they appear to diverge (negative lens system).

foot-candle—A unit of illumination. One foot-candle is defined as the amount of light that falls on an area of one square foot on which there is uniform distribution of one lumen.

foot-lambert—A unit of brightness or luminance. One foot-lambert is defined as the uniform distribution of a surface emitting or reflecting light at the rate of one lumen per square foot.

Foucault focusing—A system of focusing whereby the separation of two beams of light changes according to variances in the distance of the focal point.

frequency—The repetition rate of an electromagnetic wave, whether it be light or an electric signal.

fringe—An individual interference band, created by one cycle of waves out of phase (constructive and destructive waves).

front-surface mirror—Also called a *first-surface mirror*. A mirror where the reflective coating is on the front side of the glass. Light rays do not pass through the glass medium before striking the reflective layer, thereby eliminating the ghost image that otherwise appears.

Gabor zone plate—Usually a photographic film made from the pattern formed by a plane wave interfering with a spherical wave. One of many types of holograms.

gain—Amplification, usually expressed in dB.

getter—A supplemental component placed in tubes (including gas laser tubes) that vaporizes any remaining air and gas or that scrubs out impurities after the tube is sealed.

graded index—The design of an optical fiber where the density (and therefore refractive index) of the cladding is graded through its width. Graded indexing is the opposite of step indexing and is said to improve the light-passing capacity of the fiber.

ground—Refers to the point of (usually) zero voltage, and can apply to a power circuit or a signal circuit.

ground state (or level)—The normal unexcited state of an atom.

half wavelength—One half of a wavelength of light; the distance between the crest and trough of a single wave of light.

hertz—Abbreviated Hz. A unit of measurement used for expressing frequency or cycles per second, named after German physicist H.R. Hertz. One hertz is equal to one cycle per second. Also commonly used with the letters "k," "M," and "G" to indicate thousands, millions, and billions, respectively.

He-Ne—Shorthand for helium-neon laser.

hexadecimal—A counting system based on 16 numerals. In computers, hexadecimal counting (or "hex" for short) uses the numerals 0 through 9, plus A through F.

high-pass filter—A filter that passes only high frequencies. Can pertain to both electrical signals or light. See also *low-pass filter* and *band-pass filter*.

holography—A form of photography that uses sound or light waves to record a three-dimensional image of an object. Holography records the instantaneous intensity and phase of reflected and/or transmitted waves onto photographic film.

IC—*I*ntegrated *c*ircuit. Also called a chip. A complete electrical circuit housed in a self-contained package. See also *LSI*.

illumination—Light falling onto a defined area; the photometric counterpart to irradiance. Illumination is commonly expressed in foot-candles.

impedance—The degree of resistance that an alternating current will encounter when passing through a circuit, device, or wire. Impedance is expressed in ohms.

incandescence—Typically, the creation of light by passing an electric current through a wire filament.

incoherent—Lacking the property of coherency.

index of refraction—A relative or absolute ratio of the velocity of a specific wavelength of light within a particular medium and that of air or vacuum.

infrared (IR)—The portion of the electromagnetic spectrum between about 7,800 angstroms (780 nanometers) to about 1,000,000 angstroms (100,000 nanometers or 1 mm). The infrared band is between the visible light spectrum and microwave spectrum. Infrared (or IR) illumination beyond about 880 nm is largely invisible to the human eye.

injection laser—Synonymous with diode laser.

I/O—Short for *i*nput/*o*utput. Refers to the paths by which information enters a system (input) and leaves the system (output).

interference—1. An unwanted signal combining with a desired signal. 2. Light waves out of phase mixing and causing banding or fringing. The superimposing of one light wave on the other, causing the loss of light energy at some points and the reinforcement of light energy at others.

interferometer—An apparatus designed to mix two beams of coherent light (usually from a laser) to study the interference patterns that result. Interferometry is the study of the effects of interference fringes.

irradiance—The amount of light radiating on a defined area; the radiometric counterpart of illumination. Irradiance is usually expressed in watts per square centimeter (watts/cm^2).

kilobyte—Term used to denote "1,000" bytes, or 1K (precisely, 1024 bytes).

kilohertz—A unit of electrical frequency equal to one thousand cycles per second, abbreviated kHz.

laser—A mechanical device that produces intense, coherent radiation (mostly in the visible light and infrared regions). Laser is an acronym for *l*ight *a*mplification by *s*timulated *e*mission of *r*adiation.

lasing—The process of producing laser light.

LED—Short for *l*ight *e*mitting *d*iode, a unique type of semiconductor that is made to emit a bright beam of light. Often used as a panel indicator, but also employed in remote-sensing systems and infrared remote-control devices.

lens—An optical device used to refract (bend) light waves in a specific direction and amount.

lightguide—An optical fiber.

logic—Primarily used to indicate digital circuits or components that accept one or more signals and act on those signals in a predefined, orderly fashion.

longitudinal wave—A wave that oscillates and travels in the same direction, as opposed to *transverse wave*.

low-pass filter—A filter that passes only low frequencies. Can pertain to electrical signals or light. See also *high-pass filter* and *band-pass filter*.

LSI—*L*arge *s*cale *i*ntegration. A complex integrated circuit (IC) that is a combination of many ICs and other electronic components that are normally packaged separately. LSI (and VLSI, for *v*ery *l*arge *s*cale *i*ntegration) chips are used in the latest video equipment and held proprietary by the manufacturer.

megabyte—A term used to indicate millions of bytes; one megabyte (1 Mb) is equal to 1,048,567 bytes, or 1,024 kilobytes.

megahertz—A unit of electrical frequency equal to one million cycles per second; abbreviated MHz.

meridional ray—A light ray that crosses through the optical axis.

micrometer—A unit of metric measurement used to define one millionth of a meter, or 10^{-6} meter. Usually abbreviated as μ.

micron—A unit of metric measurement used to define one millionth of a meter, or 10^{-6} meter. Usually abbreviated as μ. Since about 1965, ''micron'' has fallen into disfavor as a metric term and has been replaced by *micrometer*.

microprocessor—The brain of a computer or computerized system. Performs all of the mathematical and logical operations necessary for the functioning of the system. Sometimes called the *CPU*.

mode—The degree to which the beam of a laser is spatially coherent. Two common modes are TEM_{00}, characterized as spatial coherence across the diameter of a beam and an even spread of light, and *multimode*, where the light appears dark in the center and has rings on the outside edges.

modulation—Altering the characteristic of a wave by imposing another wave on top of it. Often used to carry intelligent communication signals over electromagnetic (radio and light) waves.

monitor photodiode—A special photodiode sandwiched behind the laser diode for the purpose of monitoring the light output of the laser.

monochromatic—Possessing only one color, or more specifically, a specific wavelength (or a very restricted range) in the electromagnetic spectrum.

motherboard—The main printed board in a computer that contains the microprocessor, memory, and support electronics.

negative lens—A lens that has a negative (virtual) focal point. Negative lenses cause light to diverge. See also *positive lens*.

Newton's Rings—A series of rings that appear due to interference caused by two closely spaced parallel surfaces. Newton's Rings occur mainly when sandwiching glass or plastic pieces together.

noise—An undesired signal. In audio circuits, noise is usually hiss and hum; in high-gain amplification circuits, noise is often the result of thermal effects on the junctions of semiconductors. This noise impairs and limits amplification.

normal—In optics, an imaginary line drawn perpendicular to the surface of a lens, mirror, or other optical component. Often referred to as *line normal*.

objective lens—The lens used to focus a beam of light on a subject.

ohm—The unit of measure of impedance or resistance.

optical axis—Basically, the path taken by light rays as it passes through or reflects off the components of an optical system.

optical fiber—The same as *fiberoptic*.

optical filter—Any type of a number of mediums that restrict the passage of light at specific wavelengths.

parallax—The shift in the perspective of an object when the viewing position is slightly changed.

parallel—A type of input/output scheme where data is transferred eight (or more) bits at any time.

paraxial ray—A ray that's close to, and nearly parallel with, the optical axis.

peak wavelength—The wavelength at which the radiant power of a source is at its highest. Peak wavelength is generally expressed in nanometers or angstroms.

perpendicular—At right angles.

phase—The position in time of a sound, electrical, or light wave in relation to another wave. Expressed in degrees. Waves are sinusoidal—with peaks and valleys. Zero degree phase is when the peaks and valleys of both waves are even. A phase of 180 degrees is when the peaks of one wave coincide with the valleys of another.

photodetector—Any of a number of different types of electrical devices that are sensitive to visible light or infrared radiation. The photodiode, phototransistor, silicon solar cell, and cadmium-sulfide cell are common types of photodetectors.

photon—A "packet," or quanta, of light energy, used to describe the particle-like characteristics of light.

piezoelectric effect—An effect inherent in certain substances (mainly crystals and ceramics) where an application of current causes vibrational motion. Conversely, vibrational motion (including stress, bending, and friction) causes an output voltage.

point source—Radiation (usually light) whose maximum width or dispersion is less than $\frac{1}{10}$th the distance between the source and the receiver. Generally any source of light that can be considered originating as an infinitely small point.

polarizer—A material that blocks light waves traveling in a particular plane. The material may also alter the polarization characteristics, changing it between plane, elliptical, and circular.

polarized beam splitter (PBS)—An optical component used in some compact disc players that consists of two prisms with a common 45-degree face. Polarizing elements in the beam splitter are oriented so that only properly polarized light passes through.

population inversion—A condition where more atoms are in the abnormal high-energy excited state than those atoms that are in the normal low-energy ground state. Population inversion is a must in order for lasing to occur.

positive lens—A lens that has a positive (real) focal point. Positive lenses cause light to converge. See also *negative lens*.

pps—Short for *p*ulses *p*er *s*econd, expressed as hertz if the pulses are electrical.

prism—An optical component usually in the form of a wedge that redirects a light beam by diffraction and reflection.

Q-switch—A mechanism that inhibits oscillation within a laser until a certain amount of energy is stored. When the desired energy level is reached, the laser is permitted to output its light as a short, high-output pulse.

quantization—The conversion of the instantaneous amplitude of an instantaneous moment of sound (sample) into its approximate digital equivalent.

quantum—A bundle of energy; photons.

quarter wave plate (QWP)—An optical component that shifts the polarity of light 90 degrees.

RAM—Acronym for *r*andom *a*ccess *m*emory. A type of temporary memory used for storing data, either entered by the user or loaded from the software. RAM is volatile (its contents are lost when the power is removed).

real image—Light focused in space.

rectilinear propagation—Radiation traveling in a straight line.

reflection—The "bouncing" of light off the surface of a medium.

refraction—The bending of light as it passes from one medium to another. Refraction occurs because the velocity of light changes depending on the density of the medium it is passing through.

resistance—Opposition to direct electrical current (dc), expressed in ohms.

ROM—Acronym for *read only memory*. A type of permanent memory where program instruction can be stored and accessed any time. Unlike RAM, ROM is not volatile; the contents of ROM are not lost when the power is removed.

sample-and-hold—An electronic circuit used to sample incoming data and hold it momentarily until the next sampling interval. Often used to maintain a voltage level for a period of time for analog-to-digital conversion.

sampling—The measurement of the instantaneous amplitude of a signal, at regular intervals.

serial—A type of input/output scheme where data is transferred one bit at a time.

servo—An electronic circuit that modifies its output in accordance to a constantly varying input signal.

signal—The desired portion of electrical information.

signal-to-noise (S/N)—The relationship, expressed as a ratio in dB, between signal and noise.

skew ray—A light ray that does not cross or come into contact with the optical axis.

spatial coherence—1. The uniform arrangement of electromagnetic waves as they travel through space. 2. Coherence across the diameter of the beam.

spatial filter—A type of optical component used to pass only a small portion of a light beam. The filter, consisting of a pinhole and focusing lens, cleans up the beam by eliminating the effects of dirt, grease, and scratches on lenses and spatial incoherence across the diameter of a laser beam.

speckle—The grainy appearance of reflected laser light, caused by light reflecting off a small area of the object and interfering with itself (also caused by local interference within the eye).

spontaneous emission—Radiation, usually visible light or infrared, which is emitted when atoms at an excited state drop back down to a ground state.

step index—The design of an optical fiber where the density (and therefore refractive index) of the cladding changes abruptly through its width. Step indexing is the opposite of *graded indexing*.

stimulated emission—Radiation emitted when atoms at an excited state drop to a ground state while in the presence of radiant energy at the same frequency.

temporal coherence—The uniformity of electromagnetic waves over time.

transmittance—The ratio, expressed in percentage, of the radiant power emitted from a source to the total radiant power received.

transverse wave—A wave that oscillates at right angles to the path along which the wave travels. Most light waves are transverse, as opposed to *longitudinal*.

ultraviolet—The portion of the electromagnetic spectrum immediately above visible light and below X-rays.

vacuum—The absence of matter, including air.

virtual image—An image formed by diverging light rays that appear to emerge from a point.

waist—The diameter of a laser beam.

wave—In lasers, an oscillation, in the form of a sine wave, of electromagnetic radiation through a medium (or through a vacuum).

wave train—A series of identical waves.

wave front—A shape (either 2D or 3D) formed by wave points of identical phase meeting at some point in space. Also see *wavelet*.

wavelength—The distance between valleys and peaks of an electromagnetic or acoustic wave.

wavelet—The leading edge of a wave that combines with the leading edges of neighboring waves to form a wavefront.

Index

Edited by Lisa A. Doyle

Other Bestsellers or Related Interest

GORDON McCOMB'S TIPS & TECHNIQUES FOR THE ELECTRONICS HOBBYIST
—Gordon McComb

This volume covers every facet of your electronics hobby from setting up a shop to making essential equipment. This single, concise handbook will answer almost all of your questions. You'll find general information on electronics practice, important formulas, tips on how to identify components, and more. You can use it as a source for ideas, as a textbook on electronics techniques and procedures, and a databook on electronics formulas, functions, and components. 288 pages, 307 illustrations. Book No. 3485, $17.95 paperback, $27.95 hardcover

SUPERCONDUCTIVITY: The Threshold of a New Technology—Jonathan L. Mayo

Superconductivity is generating an excitement not seen in the scientific world for decades! Experts are predicting advances in state-of-the-art technology that will make most existing electrical and electronic technologies obsolete! This book is one of the most complete and thorough introductions to a multifaceted phenomenon that covers the full spectrum of superconductivity and superconductive technology. 160 pages, 58 illustrations. Book No. 3022, $12.95 paperback only

FIBEROPTICS AND LASER HANDBOOK
—2nd Edition
—Edward L. Safford, Jr., and John A. McCann

Explore the dramatic impact that lasers and fiberoptics have on our daily lives—PLUS, exciting ideas for your own experiments! Now, with the help of experts Safford and McCann, you'll discover the most current concepts, practices, and applications of fiberoptics, lasers, and electromagnetic radiation technology. Included are terms and definitions, discussions of the types and operations of current systems, and amazingly simple experiments you can conduct! 240 pages, 108 illustrations. Book No. 2981, $18.95 paperback only

ROBOT BUILDER'S BONANZA: 99 Inexpensive Robotics Projects—Gordon McComb

Where others might only see useless surplus parts you can imagine a new "life form." Now, there's a book that will help you make your ideas and dreams a reality. With the help of *Robot Builder's Bonanza* you can truly express your creativity. This fascinating guide offers you a complete, unique collection of tested and proven product modules that you can mix and match to create an almost endless variety of highly intelligent and workable robot creatures. 336 pages, 283 illustrations. Book No. 2800, $16.95 paperback only

101 SOLDERLESS BREADBOARDING PROJECTS—Delton T. Horn

Would you like to build your own electronic circuits but can't find projects that allow for creative experimentation? Want to do more than just duplicate someone else's ideas? In anticipation of your needs, Delton T. Horn has put together the ideal project *ideas* book! It gives you the option of customizing each project. With over 100 circuits and circuit variations, you can design and build practical, useful devices from scratch! 220 pages, 273 illustrations. Book No. 2985, $15.95 paperback, $24.95 hardcover

20 INNOVATIVE ELECTRONICS PROJECTS FOR YOUR HOME—Joseph O'Connell

More than just a collection of 20 projects, this book provides helpful hints and sound advice for the experimenter and home hobbyist. Particular emphasis is placed on unique yet truly useful devices that are justifiably time- and cost-efficient. Projects include a protected outlet box (for your computer system), a variable AC power controller, a remote volume control, a fluorescent bike light, and a pair of active minispeakers with built-in amplifiers. 256 pages, 130 illustrations. Book No. 2947, $13.95 paperback only

BEYOND THE TRANSISTOR: 133 Electronic Projects—Rufus P. Turner and Brinton L. Rutherford

Strongly emphasized in this second edition are the essential basics of electronic theory and practice. This is a guide that will give its reader the unique advantage of being able to keep up to date with the many rapid advances continuously taking place in the electronics field. It is an excellent reference for the beginner, student, or hobbyist. 240 pages, 173 illustrations. Book No. 2887, $9.95 paperback, $16.95 hardcover

BUILD YOUR OWN WORKING FIBEROPTIC, INFRARED AND LASER SPACE-AGE PROJECTS—Robert E. Iannini

Here are plans for a variety of useful electronic and scientific devices, including a high sensitivity laser light detector and a high voltage laboratory generator (useful in all sorts of laser, plasma ion, and particle applications as well as for lighting displays and special effects). And that's just the beginning of the exciting space-age technology that you'll put to work! 288 pages, 198 illustrations. Book No. 2724, $16.95 paperback only

Look for These and Other TAB Books at Your Local Bookstore

To Order Call Toll Free 1-800-822-8158
(in PA, AK, and Canada call 717-794-2191)

or write to TAB Books, Blue Ridge Summit, PA 17294-0840.

Title	Product No.	Quantity	Price

☐ Check or money order made payable to TAB Books

Charge my ☐ VISA ☐ MasterCard ☐ American Express

Acct. No. _____ Exp. _____

Signature: _____

Name: _____

Address: _____

City: _____

State: _____ Zip: _____

Subtotal $ _____

Postage and Handling
($3.00 in U.S., $5.00 outside U.S.) $ _____

Add applicable state and local
sales tax $ _____

TOTAL $ _____

TAB Books catalog free with purchase; otherwise send $1.00 in check or money order and receive $1.00 credit on your next purchase.

Orders outside U.S. must pay with international money order in U.S. dollars.

TAB Guarantee: If for any reason you are not satisfied with the book(s) you order, simply return it (them) within 15 days and receive a full refund. **BC**